ELEVENTH EDITION

Benson's
Microbiological
Applications

Laboratory Manual in General Microbiology

Alfred E. Brown

Auburn University

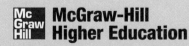

McGraw-Hill Higher Education

Boston   Burr Ridge, IL   Dubuque, IA   New York   San Francisco   St. Louis
Bangkok   Bogotá   Caracas   Kuala Lumpur   Lisbon   London   Madrid   Mexico City
Milan   Montreal   New Delhi   Santiago   Seoul   Singapore   Sydney   Taipei   Toronto

McGraw-Hill
Higher Education

BENSON'S MICROBIOLOGICAL APPLICATIONS: LABORATORY MANUAL IN GENERAL MICROBIOLOGY, COMPLETE VERSION, ELEVENTH EDITION

Published by McGraw-Hill, a business unit of The McGraw-Hill Companies, Inc., 1221 Avenue of the Americas, New York, NY 10020. Copyright © 2009 by The McGraw-Hill Companies, Inc. All rights reserved. Previous editions © 2007, 2005, 2002, 1998, 1994, 1990, 1985, . No part of this publication may be reproduced or distributed in any form or by any means, or stored in a database or retrieval system, without the prior written consent of The McGraw-Hill Companies, Inc., including, but not limited to, in any network or other electronic storage or transmission, or broadcast for distance learning.

Some ancillaries, including electronic and print components, may not be available to customers outside the United States.

♲ This book is printed on acid-free paper.

1 2 3 4 5 6 7 8 9 0 VNH/VNH 0 9 8

ISBN 978–0–07–352255–5
MHID 0–07–352255–4

Publisher: *Michelle Watnick*
Senior Sponsoring Editor: *James F. Connely*
Director of Development: *Kristine Tibbetts*
Developmental Editor: *Darlene M. Schueller*
Senior Marketing Manager: *Tami Petsche*
Project Manager: *Joyce Watters*
Senior Production Supervisor: *Sherry L. Kane*
Associate Design Coordinator: *Brenda A. Rolwes*
Cover/Interior Designer: *Studio Montage*
(USE) Cover Image: *Lawrence Lawry/Getty Images RF*
Senior Photo Research Coordinator: *Lori Hancock*
Compositor: *Aptara*
Typeface: *10/12 Times Roman*
Printer: *Von Hoffmann Press*

Some of the laboratory experiments included in this text may be hazardous if materials are handled improperly or if procedures are conducted incorrectly. Safety precautions are necessary when you are working with chemicals, glass test tubes, hot water baths, sharp instruments, and the like, or for any procedures that generally require caution. Your school may have set regulations regarding safety procedures that your instructor will explain to you. Should you have any problems with materials or procedures, please ask your instructor for help.

www.mhhe.com

Alfred E. Brown

Professor, Auburn University

Ph.D., UCLA

Research focus: Biology and physiology of photosynthetic bacteria; mode of action of herbicides that interfere with photosynthetic electron transport in the purple, nonsulfur bacteria; effects of herbicides that inhibit branched-chain amino acid synthesis in soil bacteria; physiology and ecology of cyanobacteria.

"The primary focus of my research has been the physiology of the photosynthetic, purple, nonsulfur bacteria. Photosynthetic electron transport in these bacteria has been studied using atrazine, an **s**-triazine herbicide, which acts as a quinone antagonist in both plants and photosynthetic bacteria. In the bacteria, atrazine replaces ubiquinone in the binding site associated with the L-subunit, one of the reaction center proteins, thus inhibiting electron transport. Resistance to atrazine develops as a result of a mutation in which a single amino acid change occurs in the L-subunit. In these mutants, ubiquinone can still bind to the L-subunit, but the herbicide has a diminished binding capacity, thus allowing electron transport to occur.

My laboratory has also characterized populations of soil bacteria that are naturally resistant to another class of herbicides, the sulfonylureas. These compounds inhibit acetolactate synthase (ALS), the first enzyme in the pathway for the synthesis of the branched-chain amino acids, valine and isoleucine. The ALS in these bacteria appears to be different from the well-characterized ALS described in *Escherichia coli* that is inhibited by these compounds.

More recently, my laboratory has worked with a company that manufactures roofing shingles. The shingles become heavily contaminated with certain species of cyanobacteria that are selected for by the limestone incorporated into the shingles. These cyanobacteria can live in the extremes of temperature and dessication found on the roof environment. Growth on the shingles causes significant discoloration, leading to financial losses for the industry. We have characterized some of the species that occur on shingles and investigated compounds that will inhibit the growth of these bacteria.

Teaching experience: Dr. Brown's teaching experience includes general microbiology and medical microbiology with laboratory sections, methods in microbiology, undergraduate research, applied and environmental microbiology with laboratory sections, photosynthesis, microbial physiology, and a graduate course in biomembranes.

CONTENTS

CONTENTS ■

PREFACE

Benson's Microbial Applications has been the "gold standard" of microbiology laboratory manuals for over 30 years. This manual has a number of attractive features that resulted in its adoption in universities, colleges, and community colleges for a wide variety of microbiology courses. These features include "user friendly" diagrams that students can easily follow, clear instructions, and an excellent array of reliable exercises suitable for beginning or advanced microbiology courses.

In revising the tenth edition, I have tried to maintain these important strengths and further enhance the manual by updating exercises and making the bacterial nomenclature more consistent with the first edition of *Bergey's Manual of Systematic Bacteriology*. The second edition of *Bergey's Manual* is in publication but is incomplete at this time.

Organizational Changes

Changes to the 11th edition include:

- The safety rules have been rewritten to be more consistent with rules and regulations that apply to student laboratories.
- A new table of the organisms used in the manual has been introduced. The table includes the current accepted name of the organism, its ATCC number, Gram stain and morphology, its habitat, its biosafety level, and the exercise in which the organism is used.
- The exercises on the protozoa and algae (Exercise 6) and the fungi (Exercise 8) in the Survey of Microorganisms have been revised. The new exercises have been updated with the current taxonomy of these two groups of microorganisms.
- The existing photographs of the simple, acid-fast, Gram, capsule, and spore stains in Part 2, Survey of Microorganisms, have been replaced with new photos. Also, a new photo of a negative stain has been added.
- The introductory material in Exercise 21, Cultivation of Anaerobes, has been expanded to include a revised listing of organisms and a discussion of how oxygen affects different groups of bacteria. The discussion also includes comments on the nature of sensitivity of anaerobes to oxygen.
- Exercise 34 on The Effects of Lysozyme on Bacterial Cells has been revised. The new exercise includes an expanded discussion on wall structure and the role of teichoic acids in the gram-positive bacterial cell wall. The exercise also includes testing human saliva as a source of lysozyme. The effects of lysozyme on gram-positive and negative cells are discussed and how the presence of teichoic acids in the cell wall of *Staphylococcus aureus* inhibits the action of lysozyme.
- The table on antimicrobic disks in Exercise 36, Antimicrobic Sensitivity Testing, has been expanded to include more antimicrobics. Also included are the potency values for each antimicrobic disk and specific values for sensitive, intermediate, and resistant categories for various organisms. The introductory material now includes definitions for antibiotics, antimicrobics, and semi-synthetic and synthetic compounds.
- Figures and flow charts in Exercises 41 and 42 for the identification of an unknown have been revised and modified to make them more easy to follow.
- The information on the nitrogen cycle has been revised to include details on nitrogenase, the enzyme responsible for the fixation of nitrogen. Figure 50.1 describing the nitrogen cycle has been modified to include more details regarding the various organisms involved in the nitrogen cycle.
- The introductory material in Exercise 60, Bacterial Counts of Food, has been expanded. This section now includes a discussion on the association of bacteria with foods. It also addresses how foods are important in foodborne diseases and how counts may or not be important in transmission of disease.
- A map of the plasmid pUC19 has been added to Exercise 67, Bacterial Transformation.
- The introductory section of Exercise 73, Slide Agglutination Test, has been expanded to include a more detailed discussion of antigens and the antibody response.

Laboratory Reports

The laboratory reports are designed to guide and reinforce student learning and provide a convenient place for recording data. The reports consist of observation sections, results sections, short answer questions, multiple-choice questions, and fill-in-the-blank questions. Instructors may choose to either post the answer key from the instructor's manual so students can check their work or collect and grade the laboratory reports.

Instructor's Manual

The accompanying instructor's manual provides: (1) a listing of equipment and supplies needed for performing all of the experiments, (2) procedures for setting up the experiments, and (3) answers to all the questions for the laboratory reports. The instructor's manual can be found at www.mhhe.com/labcentral. Please contact your sales representative for additional information.

Acknowledgements

I want to express my deep gratitude to Dr. Kathy Lawrence and Dr. Christine Sundermann of Auburn University for revising the exercises on fungi and protozoa and algae respectively.

I would also like to thank Dr. Mike Miller, Director of the Auburn University Research Instrumentation Facility for photographing the various stains in Part 4, Staining and Observation of Micro-organisms.

I would also like to thank Georgeann Ellis, Coordinator of the Microbiology laboratories at Auburn University who made corrections to the manual and offered many helpful suggestions.

Sang-Jin Suh from Auburn University provided me with invaluable input on the bacterial genetics and biotechnology exercises.

I also want to extend thanks to Barry Chess from Pasadena City College for updating the instructor's manual.

I wish to express my utmost gratitude to Dr. Jolie Stepaniak for reviewing each exercise in the manual and making many helpful recommendations that helped improve the 11th edition.

The updates and improvements in this edition were guided by the helpful reviews of the following instructors. Their input was critical to the decisions that shaped this edition of *Benson's Microbiological Applications*.

Mohammed K. Abbas *Schoolcraft College*
Raj Boopathy *Nicholls State University*
Marc A. Brodkin *Widener University*
Paula J. Burns *Mendocino Community College*
Barry Chess *Pasadena City College*
Neena Din *Loyola College*
David W. Essar *Winona State University*
Allen Lee Farrand *Bellevue Community College*
K. Michael Foos *Indiana University East*
Denise Y. Friedman *Hudson Valley Community College*
Phillip E. Funk *DePaul University*
Robert F. Gessner *Valencia Community College*
Darryl V. Grennell *Alcorn State University*
Janice L. Horton *Missouri State University*
Chike A. Igboechi *Medgar Evers College, CUNY*
Stanley Kikkert *Mesa Community College*
Kevin B. Kiser *Cape Fear Community College*
Emily L. Lilly *University of Massachusetts Dartmouth*
Donald G. Lindmark *Cleveland State University*
Sue A. Looney *University of Alaska Anchorage*
Mary V. Mawn *Hudson Valley Community College*
Yilei Qian *Indiana University South Bend*
Laura B. Regassa *Georgia Southern University*
Timberley Roane *University of Colorado at Denver and Health Sciences Center*
Pushpa Samkutty *Southern University*
Lisa M. Schechter *University of Missouri-St. Louis*
Lois V. Sealy *Valencia Community College*
Susan Skelly *Rutgers University*
John G. Steiert *Missouri State University*
Jolie A. Stepaniak *Henry Ford Community College*
Gregory Weigel *University of Central Florida*
Roseann S. White *University of Central Florida*
John M. Zamora *Middle Tennessee State University*

I would like to thank all the people at McGraw-Hill for their efforts and support. It is always a pleasure to work with such a professional and competent group of people. Special thanks go to Michelle. Watnick, Publisher; Jim Connely Sr., Sponsoring Editor; Darlene M. Schueller, Developmental Editor; Tami Petsche, Marketing Manager; Joyce Watters, Project Manager; Lori Hancock, Photo Research Coordinator; and Brenda Rolwes, Designer; and many others who worked "behind the scenes."

BASIC MICROBIOLOGY LABORATORY SAFETY

Every student and instructor must focus on the need for safety in the microbiology laboratory. While the lab is a fascinating and exciting learning environment, there are hazards that must be acknowledged and rules that must be followed to prevent accidents and contamination with microbes. The following outline will provide every member of the laboratory section the information required to assure a safe learning environment.

Microbiological laboratories are special, often unique environments that may pose identifiable infectious disease risks to persons who work in or near them. Infections have been contacted in the laboratory throughout the history of microbiology. Early reports described laboratory-associated cases of typhoid, cholera, glanders, brucellosis, and tetanus to name a few. Recent reports have documented laboratory-acquired cases in laboratory workers and health-care personnel involving *Bacillus anthrasis, Bordetella pertusis, Brucella, Burkholderia pseudomallei, Campylobacter, Chlamydia,* and toxins from *Clostridium tetani, Clostridium botulinum,* and *Corynebacterium diphtheriae.* While we have a greater knowledge of these agents and antibiotics with which to treat them, safety and handling still remain primary issues.

The "Biohazard" symbol must be affixed to any container or equipment used to store or transport potentially infectious materials.

Courtesy of the Centers for Disease Control.

The term "containment" is used to describe the safe methods and procedures for handling and managing microorganisms in the laboratory. An important laboratory procedure practiced by all microbiologists that will guarantee containment is **aseptic technique,** which prevents workers from contaminating themselves with microorganisms, ensures that others and the work area do not become contaminated, and also ensures that microbial cultures do not become unnecessarily contaminated with unwanted organisms. Primary containment involves personnel and the immediate laboratory and is provided by good microbiological technique and use of appropriate safety equipment. Secondary containment is also important because it guarantees that infectious agents do not escape from the laboratory and contaminate the environment external to the lab. Containment, therefore, relies on good microbiological technique and laboratory protocol as well as elements of laboratory design.

Biosafety Levels (BSL)

The recommended biosafety level(s) for handling microorganisms represent the potential of the agent to cause disease and the conditions under which the agent should be safely handled. The Centers for Disease Control recommends biosafety levels for microorganisms and viruses. These levels take into account many factors such as virulence, pathogenicity, antibiotic resistance patterns, vaccine and treatment availability, and other factors. The recommended biosafety levels are as follows:

1. **BSL 1**—agents not known to cause disease in healthy adults; standard microbiological practices (SMP) apply; no safety equipment required; sinks required. Examples: *Bacillus subtilis, Micrococcus lutetus*

2. **BSL 2**—agents associated with human disease; standard microbiological practices apply plus limited access, biohazard signs, sharps precautions, and a biosafety manual required. Biosafety cabinet (BSC) used for aerosol/splash generating operations; lab coats, gloves, face protection required; contaminated waste is autoclaved. **All microorganisms used in the exercises in this manual are classified as BSL 1 or BSL 2.** Examples: *Staphylococcus aureus, Streptococcus pyogenes*

Note: Although some of the organisms that students will culture and work with are classified as BSL 2, these organisms are laboratory strains that do not pose the same threat of infection as primary isolates of the same organism taken from patients in clinical samples. Hence, these laboratory strains can, in most cases, be handled using normal procedures and equipment found in the vast majority of student teaching laboratories. However, it should be emphasized that many bacteria are opportunistic pathogens and therefore all microorganisms should be handled by observing proper techniques and precautions.

3. **BSL 3**—indigenous/exotic agents that may have serious or lethal consequences and with a potential for aerosol transmission. BSL 2 practices plus controlled access; decontamination of all waste and lab clothing before laundering; determination of baseline antibody titers to agents; biosafety cabinets used for all specimen manipulations; respiratory protection used as needed; physical separation from access corridors; double door access; negative airflow into the lab; exhaust air not recirculated. Examples: *Mycobacterium tuberculosis* and vesicular stomatitis virus (VSV)

4. **BSL 4**—dangerous/exotic agents of a life-threatening nature or unknown risk of transmission; BSL 3 practices plus clothing change before

entering the laboratory; shower required before leaving the lab; all materials decontaminated on exit; positive pressure personnel suit required for entry; separated/isolated building; dedicated air supply/exhaust and decontamination systems. Examples: Ebola and Lassa viruses

Each of the biosafety levels consist of combinations of laboratory practices and techniques, safety equipment and laboratory facilities. Each combination is specifically appropriate for the operations performed and the documented or suspected routes of transmission of the infectious agents. Common to all biosafety levels are standard practices, especially aseptic technique.

Standard Laboratory Rules and Practices

1. Students should store all books and materials not used in the laboratory in areas or receptacles designated for that purpose. Only a lab notebook, the laboratory manual, and pen/pencil should be brought to the student work area.
2. Eating, drinking, chewing gum, and smoking are not allowed in the laboratory. Students must also avoid handling contact lenses or applying make-up while in the laboratory.
3. Safety equipment:
   a. Some labs will require that lab coats be worn in the laboratory at all times. Others may make this optional or not required. Lab coats can protect a student from contamination by microorganisms that he/she is working with and prevent contamination from stains and chemicals. At the end of the laboratory session, lab coats are usually stored in the lab in a manner prescribed by the instructor. Lab coats, gloves, and safety equipment should not be worn outside of the laboratory unless properly decontaminated first.
   b. You may be required to wear gloves while performing the lab exercises. They protect the hands against contamination by microorganisms and prevent the hands from coming in direct contact with stains and other reagents.
   c. Face protection/safety glasses may be required by some instructors while you are performing experiments. Safety glasses can prevent materials from coming in contact with the eyes. They must be especially worn when working with ultraviolet light to prevent eye damage because they block out UV rays. If procedures involve the potential for splash/aerosols, face protection should be worn.
   d. Know the location of eye wash and shower stations in the event of an accident that requires the use of this equipment. Also know the location of first aid kits.

4. Sandals or open-toe shoes are not be worn in the laboratory. Accidental dropping of objects could result in serious injury.
5. Students with long hair should tie the hair back to avoid accidents when working with Bunsen burners/open flames. Long hair can also be a source of contamination when working with cultures.
6. Before beginning the activities for the day, work areas should be wiped down with the disinfectant that is provided for that purpose. Likewise, when work is finished for the day, the work area should be treated with disinfectant to ensure that any contamination from the exercise performed is destroyed. Aviod contamination of the work surface by not placing contaminated pipettes, loops/needles, or swabs on the work surface. Dispose of contaminated paper towels used for swabbing in the biohazard container.
7. Use extreme caution when working with open flames. The flame on a Bunsen burner is often difficult to see when not in use. Caution is imperative when working with alcohol and open flames. Alcohol is highly flammable and fires can easily result when using glass rods that have been dipped in alcohol. **Always make sure the gas is turned off before leaving the laboratory.**
8. Any cuts or injuries on the hands must be covered with band-aids to prevent contamination. If you injure or cut yourself during the laboratory, notify the laboratory instructor immediately.
9. Pipetting by mouth is prohibited in the lab. All pipetting must be performed with pipette-aids. Be especially careful when inserting glass pipettes into pipette aids as the pipette can break and cause a serious injury.
10. Know the location of exits and fire extinguishers in the laboratory.
11. Most importantly, read the exercise and understand the laboratory protocol before coming to laboratory. In this way you will be familiar with potential hazards in the exercise.
12. When working with microfuges, be familiar with their safe operation and make sure that all microfuge tubes are securely capped before centrifuging.
13. When working with electrophoresis equipment, follow the directions carefully to aviod electric shock.
14. If you have any allergies or medical conditions that might be complicated by participating in the laboratory, inform the instructor. Women who are pregnant should discuss the matter of enrolling in the lab with their family physician and the laboratory instructor.

Disposal of Biological Wastes

Dispose of all contaminated materials properly and in the appropriate containers:

1. Biohazard containers—biohazard containers are to be lined with clear autoclave bags; disposable petri plates, used gloves, and any materials such as contaminated paper towels, etc., should be discarded in these containers; no glassware, test tubes, or sharp items are to disposed of in biohazard containers.
2. Sharpkeepers—sharps, slides, coverslips, broken glass, disposable pipettes, and Pasteur pipettes should be discarded in these containers. If instructed to do so, you can discard contaminated swabs, wooden sticks, and microfuge tubes in the sharpkeepers.
3. Discard shelves, carts, etc.—contaminated culture tubes and glassware used to store media and other glassware should be placed in these areas for decontamination and washing.
4. Trash cans—any noncontaminated materials, paper, or trash should be discarded in these containers. Under no circumstances should laboratory waste should be disposed of in trash cans.

Discard other materials as directed by your instructor. This may involve placing materials such as slides contaminated with blood in disinfectant baths before these materials can be discarded.

Emergencies

Surface Contamination

1. Report all spills immediately to the laboratory instructor.
2. Cover the spill with paper towels and saturate the paper towels with disinfectant.
3. Allow the disinfectant to act for at least 20 minutes.
4. Remove any glass or solid material with forceps or scoop and discard the waste in an appropriate manner.

Personnel Contamination

1. Notify lab instructor.
2. Clean exposed area with soap/water, eye wash (eyes) or saline (mouth).
3. Apply first aid and treat as an emergency.

Biosafety Levels for Selected Infectious Agents

Biosafety Level (BSL)	Typical Risk	Organism
BSL 1	Not likely to pose a disease risk to healthy adults.	*Achromobacter denitrificans* *Alcaligenes faecalis* *Bacillus cereus* *Bacillus subtilis* *Corynebacterium pseudodiphtheriticum* *Enterobacter aerogenes* *Enterococcus faecalis* *Micrococcus luteus* *Neisseria sicca* *Proteus vulgaris* *Pseudomonas aeruginosa* *Staphylococcus epidermidis* *Staphylococcus saprophyticus*
BSL 2	Poses a moderate risk to healthy adults; unlikely to spread throughout community; effective treatment readily available.	*Escherichia coli* *Klebsiella pneumoniae* *Mycobacterium phlei* *Salmonella enterica var. Typhimurium* *Shigella flexneri* *Staphylococcus aureus* *Streptococcus pneumoniae* *Streptococcus pyogenes*
BSL 3	Can cause disease in healthy adults; may spread to community; effective treatment readily available.	*Blastomyces dermatitidis* *Chlamydia trachomatis* *Coccidioides immitis* *Coxiella burnetii* *Francisella tularensis* *Histoplasma capsulatum* *Mycobacterium bovis* *Mycobacterium tuberculosis* *Pseudomonas mallei* *Rickettsia canadensis* *Rickettsia prowazekii* *Yersinia pestis*
BSL 4	Can cause disease in healthy adults; poses a lethal risk and does not respond to vaccines or antimicrobial therapy.	Filovirus *Herpesvirus simiae* Lassa virus Marburg virus

Microorganisms Used or Isolated in the Lab Exercises in this Manual

Organism	Gram Stain and Morphology	Habitat	BSL	Lab Exercise
Alcaligenes faecalis ATCC 19018	Negative rod	Decomposing organic material, feces	1	31
Azotobacter insignis ATCC 12523	Negative rod	Soil, water	1	52
Azotobacter nigricans ATCC 35009	Negative rod	Soil, water	1	52
Azotobacter vinelandii ATCC 12518	Negative rod	Soil, water	1	52
Bacillus cereus var. mycoides ATCC 21929	Positive rod in chains	Soil	1	59
Bacillus coagulans ATCC 10778	Positive rod	Spoiled food, silage	1	64
Bacillus megaterium ATCC 35985	Positive rod	Soil, water	1	13, 15, 16, 19, 30, 33
Bacillus subtilis ATCC 31578	Positive rod	Soil, decomposing organic matter	1	16, 21, 42
Candida glabrata ATCC 200918	Yeast	Human oral cavity	1	31
Chromobacterium violaceum ATCC 12572	Negative rod	Soil and water; opportunistic pathogen in humans	1	10
Citrobacter freundii ATCC 8090	Negative rod	Humans, animals, soil water; sewage opportunistic pathogen	1	72
Clostridium beijerinckii ATCC 14949	Positive rod	Soil	1	21
Clostridium sporogenes ATCC 10000	Positive rod	Soil, animal feces	1	21, 57, 64
Corynebacterium xerosis ATCC 373	Positive rods, club-shaped	Conjunctiva, skin	1	12
Desulfovibrio desulfuricans ATCC 13541	Negative, curved rods	Soil, sewage, water	1	56
Enterobacter aerogenes ATCC 29007	Positive cocci, occur in chains	Feces of man and animals	2	21
Enterococcus faecium ATCC 19433	Positive cocci in pairs, short chains	Feces of humans and animals	2	71
Enterococcus faecalis ATCC 10741	Negative rods	Water, sewage, soil, dairy products	2	41, 43, 61
Escherichia coli ATCC 31446	Negative rods	Sewage, intestinal tract of warm-blooded animals	2	9, 19, 21, 22, 26, 27, 29, 30, 31, 34, 36, 41, 42, 58, 59, 61, 64, 66, 67, 68

Microorganisms Used or Isolated in the Lab Exercises in this Manual (continued)

Organism	Gram Stain and Morphology	Habitat	BSL	Lab Exercise
Geobacillus stearothermophilus ATCC 12976	Variable rods	Soil, spoiled food	1	64
Halobacterium salinarium ATCC 19700	Gives gram-negative reaction; rods	Salted fish, hides, meats	1	32
Klebsiella pneumoniae ATCC 10273	Negative rods	Intestinal tract of humans; respiratory and intestinal pathogen in humans	2	14
Micrococcus luteus ATCC 12698	Positive cocci that occur in pairs	Mammalian skin	1	18, 78
Moraxella (Branhamella) catarrhalis ATCC 232446	Negative cocci that often occur in pairs with flattened sides	Pharynx of humans	1	15
Mycobacterium smegmatis ATCC 14468	Positive rods; may be Y-shaped or branched	Smegma of humans	1	15, 17
Paracoccus denitrificans ATCC 15543	Negative spherical cells or short rods	Soil	1	53
Penicillium chrysogenum ATCC 9478	Filamentous fungus	Soil	1	59
Physarum polycephalum ATCC 204388	Slime mold	Decaying leaves	1	23
Proteus vulgaris ATCC 12454	Negative rods	Intestines of humans, and animals; soil and polluted waters	1	18, 36, 42, 43, 58
Pseudomonas aeruginosa ATCC 47053	Negative rods	Soil and water; opportunistic pathogen in humans	1	15, 36, 37, 41
Pseudomonas fluorescens ATCC 11150	Negative rods	Soil, water, spoiled food; clinical specimens	1	59
Saccharomyces cerevisiae ATCC 76455	Yeast		1	31
Salmonella enterica subsp. *enterica* serovar *Typhimurium* ATCC 35988	Negative rods	Most frequent agent of *Salmonella* gastroenteritis in humans	2	73
Serratia marcescens ATCC 13880	Negative rods	Opportunistic pathogen in humans	1	
Shigella flexneri ATCC 29903	Negative rods	Pathogen of humans	2	72
Staphylococcus aureus ATCC 35556	Positive cocci, irregular clusters	Skin, nose, GI tract of humans, pathogen	2	10, 13, 15, 17, 21, 28, 30, 31, 33 34, 36, 37, 41, 42 43, 57, 58, 59, 70 71, 73
Staphylococcus epidermidis ATCC 155	Positive cocci that occur in pairs and tetrads	Human skin, animals; opportunistic pathogen	1	49, 70

Microorganisms Used or Isolated in the Lab Exercises in this Manual

Organism	Gram Stain and Morphology	Habitat	BSL	Lab Exercise
Staphylococcus saprophyticus ATCC 15305	Positive cocci that occur singly and in pairs	Human skin; opportunistic pathogen in the urinary tract	1	70
Streptococcus agalactiae ATCC 14364	Positive cocci; occurs in long chains	Upper respiratory and vaginal tract of humans, cattle; pathogen	2	71
Streptococcus bovis ATCC 35034	Positive cocci; pairs and chains	Cattle, sheep, pigs; occasional pathogen in humans	2	71
Streptococcus dysagalactiae subspecies *equisimilis* ATCC 35666	Positive cocci; short to long chains	mastitis in cattle	2	71
Streptococcus equi ATCC 39506	Positive cocci; cocco-bacilli; occur in pairs and chains	Pathogen of horses	2	71
Streptococcus mitis ATCC 49456	Positive cocci in pairs and chains	Oral cavity of humans	2	71
Streptococcus mutans ATCC 33402	Positive cocci in pairs and chains	Tooth surface of humans, causes dental caries	2	71
Streptococcus pneumoniae ATCC 10015	Positive cocci in pairs	Human pathogen	2	71
Streptococcus pyogenes ATCC 12202	Positive cocci in chains	Human respiratory tract; pathogen	2	71
Streptococcus salivarius ATCC 25975	Positive cocci in short and long chains	Tongue and saliva	2	71
Thermoanaerobacterium thermosaccharolyticum ATCC 7965	Negative rods; single cells or pairs	Soil, spoiled canned foods	1	64

Microscopy

Although there are many kinds of microscopes available to the micro-biologist today, only four types will be described here for our use: the brightfield, darkfield, phase-contrast, and fluorescence microscopes. If you have had extensive exposure to microscopy in previous courses, this unit may not be of great value to you; however, if the study of microorganisms is a new field of study for you, there is a great deal of information that you need to acquire about the proper use of these instruments.

Microscopes in a college laboratory represent a considerable investment and require special care to prevent damage to the lenses and mechanical parts. A microscope may be used by several people during the day and moved from the work area to storage; which results in a much greater chance for damage to the instrument than if the microscope were used by only a single person.

The complexity of some of the more expensive microscopes also requires that certain adjustments be made periodically. Knowing how to make these adjustments to get the equipment to perform properly is very important. An attempt is made in the five exercises of this unit to provide the necessary assistance for getting the most out of the equipment.

Microscopy should be as fascinating to the beginner as it is to the professional of long standing; however, only with intelligent under-standing can the beginner approach the achievement that occurs with years of experience.

Brightfield Microscopy

A microscope that allows light rays to pass directly to the eye without being deflected by an intervening opaque plate in the condenser is called a *brightfield microscope.* This is the conventional type of instrument encountered by students in beginning courses in biology; it is also the first type to be used in this laboratory.

All brightfield microscopes have certain things in common, yet they differ somewhat in mechanical operation. Similarities and differences of various makes are discussed in this exercise so that you will know how to use the instrument that is available to you. Before attending the first laboratory session in which the microscope is used, read over this exercise and answer all the questions on the Laboratory Report. Your instructor may require that the Laboratory Report be handed in prior to doing any laboratory work.

Care of the Instrument

Microscopes represent considerable investment and can be damaged easily if certain precautions are not observed. The following suggestions cover most hazards.

Transport   When carrying your microscope from one part of the room to another, use both hands to hold the instrument, as illustrated in figure 1.1. If it is carried with only one hand and allowed to dangle at your side, there is always the danger of collision with furniture or some other object. And, *under no circumstances should one attempt to carry two microscopes at one time.*

Clutter   Keep your workstation uncluttered while doing microscopy. Keep unnecessary books and other materials away from your work area. A clear work area promotes efficiency and results in fewer accidents.

Electric Cord   Microscopes have been known to tumble off of tabletops when students have entangled a foot in a dangling electric cord. Don't let the electric cord on your microscope dangle in such a way as to risk foot entanglement.

Lens Care   At the beginning of each laboratory period, check the lenses to make sure they are clean. At the end of each lab session, be sure to wipe any immersion

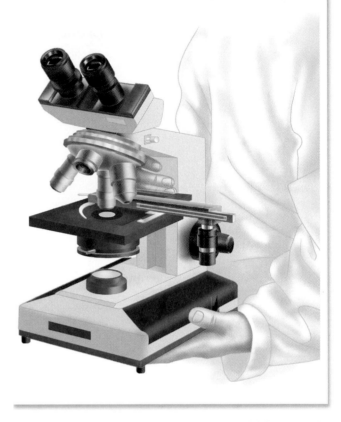

Figure 1.1   The microscope should be held firmly with both hands while being carried.

oil off the immersion lens if it has been used. More specifics about lens care are provided on page 6.

Dust Protection   In most laboratories dustcovers are used to protect the instruments during storage. If one is available, place it over the microscope at the end of the period.

Components

Before we discuss the procedures for using a microscope, let's identify the principal parts of the instrument as illustrated in figure 1.2.

Framework   All microscopes have a basic frame structure, which includes the **arm** and **base.** To this framework all other parts are attached. On many of

Figure 1.2 **The compound microscope.**
(Courtesy of the Olympus Corporation, Lake Success, NY)

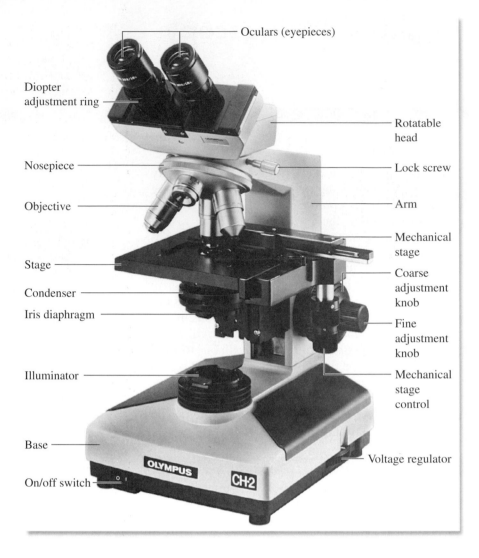

the older microscopes the base is not rigidly attached to the arm as is the case in figure 1.2; instead, a pivot point is present that enables one to tilt the arm backward to adjust the eyepoint height.

Stage The horizontal platform that supports the microscope slide is called the **stage.** Note that it has a clamping device, the **mechanical stage,** which is used for holding and moving the slide around on the stage. Note, also, the location of the **mechanical stage control** in figure 1.2.

Light Source In the base of most microscopes is positioned some kind of light source. Ideally, the lamp should have a **voltage control** to vary the intensity of light. The microscope in figure 1.2 has a knurled wheel on the right side of its base to regulate the voltage supplied to the light bulb.

Most microscopes have some provision for reducing light intensity with a **neutral density filter.** Such

a filter is often needed to reduce the intensity of light below the lower limit allowed by the voltage control. On microscopes such as the Olympus CH-2, one can simply place a neutral density filter over the light source in the base. On some microscopes a filter is built into the base.

Lens Systems All compound microscopes have three lens systems: the oculars, the objectives, and the condenser. Figure 1.3 illustrates the light path through these three systems.

The **ocular,** or eyepiece, is a complex piece, located at the top of the instrument, that consists of two or more internal lenses and usually has a magnification of 10×. Most modern microscopes (figure 1.2) have two ocular (binocular) lenses.

Three or more **objectives** are usually present. Note that they are attached to a rotatable **nosepiece,** which makes it possible to move them into position over a slide. Objectives on most laboratory micro-

trolling light intensity. On the Olympus microscope in figure 1.2, the diaphragm is controlled by turning a knurled ring. On some microscopes, a diaphragm lever is present. Figure 1.3 illustrates the location of the condenser and diaphragm.

Focusing Knobs The concentrically arranged **coarse adjustment** and **fine adjustment knobs** on the side of the microscope are used for bringing objects into focus when studying an object on a slide. On some microscopes, these knobs are not positioned concentrically as shown here.

Ocular Adjustments On binocular microscopes, one must be able to change the distance between the oculars and to make diopter changes for eye differences. On most microscopes, the interocular distance is changed by simply pulling apart or pushing together the oculars.

To make diopter adjustments, one focuses first with the right eye only. Without touching the focusing knobs, diopter adjustments are then made on the left eye by turning the knurled **diopter adjustment ring** (figure 1.2) on the left ocular until a sharp image is seen. One should now be able to see sharp images with both eyes.

Resolution

It would appear that the magnification of a microscope is only limited by the magnifying power of a lens system. However, in reality the limit for most light microscopes is 1000× which is set by an intrinsic property of lenses called **resolving power.** The resolving power of a lens is its ability to completely separate two objects in a microscopic field. The resolving power is given by the formula $d = 0.5\ \lambda/NA$. The limit of resolution, d, or the distance between the two objects, is a function of two properties: the wavelength of the light used to observe a specimen, λ, and a property of lenses called the **numerical aperture** or NA. Numerical aperture is a mathematical expression that describes how the condenser lens concentrates and focuses the light rays from the light source. Its value is maximized when the light rays are focused into a cone of light that then passes through the specimen into the objective lens. However, because some light is refracted or bent as it passes from glass into air, the refracted light rays are lost, and as a result the numerical aperture is diminished (figure 1.4). The greater the loss of refracted light, the lower the numerical aperture. The final result is that the resolving power is greatly reduced.

For any light microscope, the limit of resolution is about 0.2 μm. This means that two objects closer than 0.2 μm would not be seen as two distinct objects.

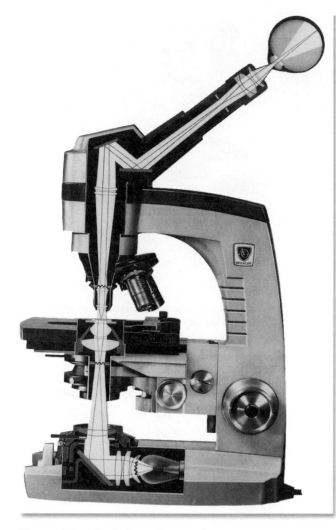

Figure 1.3 The light pathway of a microscope.

scopes have magnifications of 10×, 40×, and 100×, designated as **low power, high-dry,** and **oil immersion,** respectively. Some microscopes will have a fourth objective for rapid scanning of microscopic fields that is only 4×.

The total magnification of a compound microscope is determined by multiplying the power of the ocular lens times the power of the objective lens used. Thus, the magnification of a microscope in which the oil immersion lens is being used is:

$$10 \times 100 = 1000$$

The third lens system is the **condenser,** which is located under the stage. It collects and directs the light from the lamp to the slide being studied. Unlike the ocular and objective lenses, the condenser lens does not affect the magnifying power of the compound microscope. The condenser can be moved up and down by a knob under the stage. A **diaphragm** within the condenser regulates the amount of light that reaches the slide. Microscopes that lack a voltage control on the light source rely entirely on the diaphragm for con-

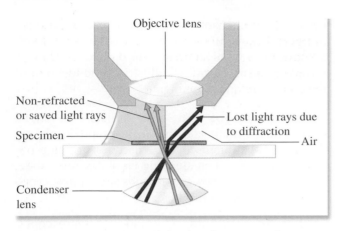

Figure 1.4 Immersion oil, having the same refractive index as glass, prevents light loss due to diffraction.

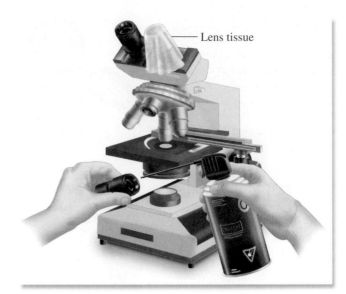

Figure 1.5 When oculars are removed for cleaning, cover the ocular opening with lens tissue. A blast from an air syringe or gas cannister removes dust and lint.

Because bacterial cells are about 1 μm, the cells can be resolved by the light microscope but that is not the case for internal structures in bacterial cells that are smaller than 0.2 μm.

In order to maximize the resolving power from a lens system, the following should be considered:

- A **blue filter** should be placed over the light source because the shorter wave length of the resulting light will provide maximum resolution.
- The condenser should be kept at the highest position that allows the maximum amount of light to enter the objective lens and therefore limit the amount of light lost due to refraction.
- The diaphragm should not be stopped down too much. While closing the diaphragm improves the contrast, it also reduces the numerical aperture.
- **Immersion oil** should be used between the slide and the 100× objective lens. This is a special oil that has the same refractive index as glass. When placed between the specimen and objective lens, the oil forms a continuous lens system that limits the loss of light due to refraction.

The bottom line is that for magnification to increase, resolution must also increase. Thus, a greater magnification cannot simply be achieved by adding a 20× ocular lens.

Lens Care

Keeping the lenses of your microscope clean is a constant concern. Unless all lenses are kept free of dust, oil, and other contaminants, they are unable to achieve the degree of resolution that is intended. Consider the following suggestions for cleaning the various lens components:

Cleaning Tissues   Only lint-free, optically safe tissues should be used to clean lenses. Tissues free of abrasive grit fall in this category. Booklets of lens tissue are most widely used for this purpose. Although several types of boxed tissues are also safe, *use only the type of tissue that is recommended by your instructor* (figure 1.5).

Solvents   Various liquids can be used for cleaning microscope lenses. Green soap with warm water works very well. Xylene is universally acceptable. Alcohol and acetone are also recommended, but often with some reservations. Acetone is a powerful solvent that could possibly dissolve the lens mounting cement in some objective lenses if it were used too liberally. When it is used it should be used sparingly. Your instructor will inform you as to what solvents can be used on the lenses of your microscope.

Oculars   The best way to determine if your eyepiece is clean is to rotate it between the thumb and forefinger as you look through the microscope. A rotating pattern will be evidence of dirt.

If cleaning the top lens of the ocular with lens tissue fails to remove the debris, one should try cleaning the lower lens with lens tissue and blowing off any excess lint with an air syringe or gas cannister. *Whenever the ocular is removed from the microscope, it is imperative that a piece of lens tissue be placed over the open end of the microscope as illustrated in figure 1.5.*

Objectives Objective lenses often become soiled by materials from slides or fingers. A piece of lens tissue moistened with green soap and water, or one of the acceptable solvents mentioned on page 6, will usually remove whatever is on the lens. Sometimes a cotton swab with a solvent will work better than lens tissue. At any time that the image on the slide is unclear or cloudy, assume at once that the objective you are using is soiled.

Condenser Dust often accumulates on the top surface of the condenser; thus, wiping it off occasionally with lens tissue is desirable.

Procedures

If your microscope has three objectives, you have three magnification options: (1) low-power, or 100× total magnification, (2) high-dry magnification, which is 400× total with a 40× objective, and (3) 1000× total magnification with a 100× oil immersion objective.

Whether you use the low-power objective or the oil immersion objective will depend on how much magnification is necessary. Generally speaking, however, it is best to start with the low-power objective and progress to the higher magnifications as your study progresses. Consider the following suggestions for setting up your microscope and making microscopic observations.

Low-Power Examination The main reason for starting with the low-power objective is to enable you to explore the slide to look for the object you are planning to study. Once you have found what you are looking for, you can proceed to higher magnifications. Use the following steps when exploring a slide with the low-power objective:

1. Position the slide on the stage with the material to be studied on the *upper* surface of the slide. Figure 1.6 illustrates how the slide must be held in place by the mechanical stage retainer lever.
2. Turn on the light source, using a *minimum* amount of voltage. If necessary, reposition the slide so that the stained material on the slide is in the *exact center* of the light source.
3. Check the condenser to see that it has been raised to its highest point.
4. If the low-power objective is not directly over the center of the stage, rotate it into position. Be sure that as you rotate the objective into position it clicks into its locked position.
5. Turn the coarse adjustment knob to lower the objective until it stops. A built-in stop will prevent the objective from touching the slide.

6. While looking down through the ocular (or oculars), bring the object into focus by turning the fine adjustment focusing knob. Don't readjust the coarse adjustment knob. If you are using a binocular microscope, it will also be necessary to adjust the interocular distance and diopter adjustment to match your eyes.
7. Manipulate the diaphragm lever to reduce or increase the light intensity to produce the clearest, sharpest image. Note that as you close down the diaphragm to reduce the light intensity, the contrast improves and the depth of field increases. Stopping down the diaphragm when using the low-power objective does not decrease resolution.
8. Once an image is visible, move the slide about to search out what you are looking for. The slide is moved by turning the knobs that move the mechanical stage.
9. Check the cleanliness of the ocular, using the procedure outlined earlier.
10. Once you have identified the structures to be studied and wish to increase the magnification, you may proceed to either high-dry or oil immersion magnification. However, before changing objectives, *be sure to center the object you wish to observe.*

High-Dry Examination To proceed from low-power to high-dry magnification, all that is necessary is to rotate the high-dry objective into position and open up the diaphragm somewhat. It may be necessary to make a minor adjustment with the fine adjustment knob to sharpen up the image, but *the coarse ad-justment knob should not be touched.*

Good quality modern microscopes are **parfocal**. This means that the image will remain in focus when changing from a lower-power objective lens to a higher-power lens. Only minimal focusing should be necessary with the fine focus adjustment.

When increasing the lighting, be sure to open up the diaphragm first instead of increasing the voltage on your lamp; the reason is that *lamp life is greatly extended when used at low voltage.* If the field is not

Figure 1.6 The slide must be properly positioned as the retainer lever is moved to the right.

Table 1.1 Relationship of Working Distance to Magnification

LENS	MAGNIFICATION	FOCAL LENGTH (mm)	WORKING DISTANCE (mm)
Low-power	10×	16.0	7.7
High-dry	40×	4.0	0.3
Oil immersion	100×	1.8	0.12

bright enough after opening the diaphragm, feel free to increase the voltage. A final point: Keep the condenser at its highest point.

Oil Immersion Techniques The oil immersion lens derives its name from the fact that a special mineral oil is interposed between the specimen and the 100× objective lens. As stated previously, this reduces light diffraction and maximizes the numerical aperture to improve the resolution. The use of oil in this way enhances the resolving power of the microscope. Figure 1.4 reveals this phenomenon.

With parfocal objectives one can go directly to oil immersion from either low-power or high-dry. On some microscopes, however, going from low-power to high-power and then to oil immersion is better. Once the microscope has been brought into focus at one magnification, the oil immersion lens can be rotated into position without fear of striking the slide.

Before rotating the oil immersion lens into position, however, a drop of immersion oil must be placed on the slide. An oil immersion lens should never be used without oil. Incidentally, if the oil appears cloudy, it should be discarded.

When using the oil immersion lens, it is best to open the diaphragm as much as possible. Stopping down the diaphragm tends to limit the resolving power of the optics. In addition, the condenser must be kept at its highest point. If different colored filters are available for the lamp housing, it is best to use blue or greenish filters to enhance the resolving power.

Since the oil immersion lens will be used extensively in all bacteriological studies, it is of paramount importance that you learn how to use this lens properly. Using this lens takes a little practice due to the difficulties usually encountered in manipulating the lighting. It is important for all beginning students

to appreciate that the working distance of a lens, the distance between the lens and microscope slide, decreases significantly as the magnification of the lens increases (table 1.1). Hence, the potential for damage to the oil immersion lens because of a collision with the microscope slide is very great. A final comment of importance: At the end of the laboratory period remove all immersion oil from the lens tip with lens tissue.

Putting It Away

When you take a microscope from the cabinet at the beginning of the period, you expect it to be clean and in proper working condition. The next person to use the instrument after you have used it will expect the same consideration. A few moments of care at the end of the period will ensure these conditions. Check over the following list of items at the end of each period before you return the microscope to the cabinet.

1. Remove the slide from the stage.
2. If immersion oil has been used, wipe it off the lens and stage with lens tissue. (Do not wipe oil off slides you wish to keep. Simply put them into a slide box and let the oil drain off.)
3. Rotate the low-power objective into position.
4. If the microscope has been inclined, return it to an erect position.
5. If the microscope has a built-in movable lamp, raise the lamp to its highest position.
6. If the microscope has a long attached electric cord, wrap it around the base.
7. Adjust the mechanical stage so that it does not project too far on either side.
8. Replace the dustcover.
9. If the microscope has a separate transformer, return it to its designated place.
10. Return the microscope to its correct place in the cabinet.

Laboratory Report

Before the microscope is to be used in the laboratory, answer all the questions on Laboratory Report 1 that pertain to brightfield microscopy. Preparation on your part prior to going to the laboratory will greatly facilitate your understanding. Your instructor may wish to collect this report at the *beginning of the period* on the first day that the microscope is to be used in class.

Laboratory Report

Student: _____

Date: _____ Section: _____

1 Brightfield Microscopy

A. Short Answer Questions

1. Describe the position of your hands when carrying the microscope to and from your laboratory bench.

   _____

   _____

2. Differentiate between the limit of resolution of the typical light microscope and that of the unaided human eye.

   _____

   _____

3. (a) What two adjustments can be made to the condenser? (b) What effect do these adjustments have on the image?

   _____

   _____

4. Why are condenser adjustments generally preferred over the use of the voltage control?

   _____

   _____

5. When using the oil immersion lens, what four procedures can be implemented to achieve the maximum resolution?

   _____

   _____

   _____

6. Why is it advisable to start first with the low-power lens when viewing a slide?

   _____

   _____

7. Why is it necessary to use oil in conjunction with the oil immersion lens and not with the other objectives?

   _____

   _____

8. What is the relationship between the working distance of an objective lens and its magnification power?

   _____

   _____

B. Matching Questions

Match the lens (condenser, high-dry, low-power, ocular, or oil immersion) to its description. Choices may be used more than once.

1. This objective lens provides the highest magnification.
2. This objective lens provides the second highest magnification.
3. This objective lens provides the lowest magnification.
4. This objective lens has the shortest working distance.
5. The coarse focus knob should be adjusted only when using this objective lens.
6. This lens collects and focuses light from the lamp onto the specimen on the slide.
7. This lens, also known as the eyepiece, often comes in pairs.
8. Diopter adjustments can be made to this lens.
9. A diaphragm is used to regulate light passing through this lens.

C. True-False

1. Acetone is the safest solvent for cleaning an objective lens.
2. Only lint-free, optically safe tissue should be used to wipe off microscope lenses.
3. The total magnification capability of a light microscope is only limited by the magnifying power of the lens system.
4. The coarse focus knob can be used to adjust the focus when using any of the objective lenses.
5. Once focus is achieved at one magnification, a higher-power objective lens can be rotated into position without fear of striking the slide.

D. Multiple Choice

Select the answer that best completes the following statements.

1. The resolving power of a microscope is a function of
   a. the magnifying power of the lenses.
   b. the numerical aperture of the lenses.
   c. the wavelength of light.
   d. Both (a) and (b) are correct.
   e. Both (b) and (c) are correct.

2. The coarse and fine focus knobs adjust the distance between
   a. the objective and ocular lenses.
   b. the ocular lenses.
   c. the ocular lenses and your eyes.
   d. the stage and the condenser lens.
   e. the stage and the objective lens.

3. A microscope that maintains focus when the objective magnification is increased is called
   a. binocular.
   b. myopic.
   c. parfocal.
   d. refractive.
   e. resolute.

ANSWERS

Matching

1. _____
2. _____
3. _____
4. _____
5. _____
6. _____
7. _____
8. _____
9. _____

True-False

1. _____
2. _____
3. _____
4. _____
5. _____

Multiple Choice

1. _____
2. _____
3. _____

4. The total magnification achieved when using a 100× oil immersion lens with 10× binocular eyepieces is
   a. 10×.
   b. 100×.
   c. 200×.
   d. 1000×.
   e. 2000×.

5. The most useful adjustment for increasing image contrast in low-power magnification is
   a. closing down the diaphragm.
   b. closing one eye.
   c. opening up the diaphragm.
   d. placing a drop of oil on the slide.
   e. using a blue filter.

6. Before the oil immersion lens is rotated into place, you should
   a. center the object of interest in the preceding lens.
   b. lower the stage with use of the coarse focus adjustment knob.
   c. place a drop of oil on the slide.
   d. Both (a) and (c) are correct.
   e. All are correct.

4. _____

5. _____

6. _____

Darkfield Microscopy

Delicate transparent living organisms can be more easily observed with darkfield microscopy than with conventional brightfield microscopy. This method is particularly useful when one is attempting to identify spirochaetes in an exudate from a syphilitic lesion. Figure 2.1 illustrates the appearance of these organisms under such illumination. This effect may be produced by placing a darkfield stop below the regular condenser or by replacing the condenser with a specially constructed one.

Another application of darkfield microscopy is in the fluorescence microscope (Exercise 4). Although fluorescence may be seen without a dark field, it is greatly enhanced with this application.

To achieve the darkfield effect it is necessary to alter the light rays that approach the objective in such a way that only oblique rays strike the objects being viewed. The obliquity of the rays must be so extreme that if no objects are in the field, the background is completely light-free. Objects in the field become brightly illuminated by the rays that are reflected up through the lens system of the microscope.

Although there are several different methods for producing a dark field, only two devices will be described here: the star diaphragm and the cardioid condenser. The availability of equipment will determine the method to be used in this laboratory.

The Star Diaphragm

One of the simplest ways to produce the darkfield effect is to insert a star diaphragm into the filter slot of the condenser housing as shown in figure 2.2. This device has an opaque disk in the center that blocks the central rays of light. Figure 2.3 reveals the effect of this stop on the light rays passing through the condenser. If such a device is not available, one can be made by cutting round disks of opaque paper of different sizes that are cemented to transparent celluloid disks that will fit into the slot. If the microscope normally has a diffusion disk in this slot, it is best to replace it with rigid clear celluloid or glass.

An interesting modification of this technique is to use colored celluloid stops instead of opaque paper. Backgrounds of blue, red, or any color can be produced in this way.

In setting up this type of darkfield illumination, it is necessary to keep these points in mind:

1. Limit this technique to the study of large organisms that can be seen easily with low-power magnification. *Good resolution with higher-powered objectives is difficult with this method.*

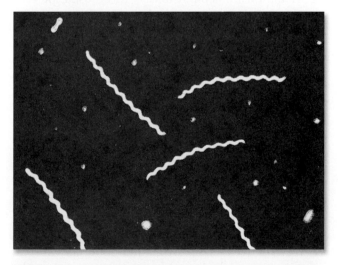

Figure 2.1 Transparent living microorganisms, such as the syphilis spirochaete, can be seen much more easily when observed in a dark field.

Figure 2.2 The insertion of a star diaphragm into the filter slot of the condenser will produce a dark field suitable for low magnifications. © The McGraw-Hill Companies/Auburn University Photographic Service

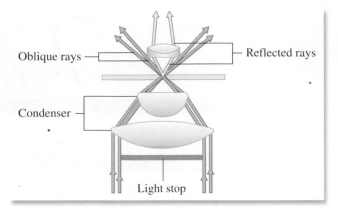

Figure 2.3 The star diaphragm allows only peripheral light rays to pass up through the condenser. This method requires maximum illumination.

2. Keep the diaphragm wide open and use as much light as possible. If the microscope has a voltage regulator, you will find that the higher voltages will produce better results.
3. Be sure to center the stop as precisely as possible.
4. Move the condenser up and down to produce the best effects.

The Cardioid Condenser

The difficulty that results from using the star diaphragm or opaque paper disks with high-dry and oil immersion objectives is that the oblique rays are not as carefully metered as is necessary for the higher magnifications. Special condensers such as the cardioid or paraboloid types must be used. Since the cardioid type is the most frequently used type, its use will be described here.

Figure 2.4 illustrates the light path through such a condenser. Note that the light rays entering the lower element of the condenser are reflected first off a convex mirrored surface and then off a second

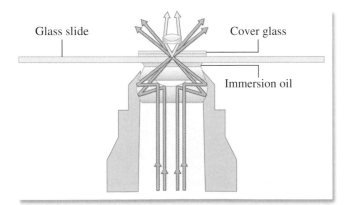

Figure 2.4 A cardioid condenser provides greater light concentration for oblique illumination than the star diaphragm.

concave surface to produce the desired oblique rays of light. Once the condenser has been installed in the microscope, the following steps should be followed to produce ideal illumination.

Materials

- slides and cover glasses of excellent quality (slides of 1.15–1.25 mm thickness and No. 1 cover glasses)

1. Adjust the upper surface of the condenser to a height just below stage level.
2. Place a clear glass slide in position over the condenser.
3. Focus the 10× objective on the top of the condenser until a bright ring comes into focus.
4. Center the bright ring so that it is concentric with the field edge by adjusting the centering screws on the darkfield condenser. If the condenser has a light source built into it, it will also be necessary to center it as well to achieve even illumination.
5. Remove the clear glass slide.
6. If a funnel stop is available for the oil immersion objective lens, remove the oil immersion objective and insert the funnel stop. (This stop serves to reduce the numerical aperture of the oil immersion objective to a value that is less than that of the condenser.)
7. Place a drop of immersion oil on the upper surface of the condenser and place the slide on top of the oil. The following preconditions in slide usage must be adhered to:
   - Slides and cover glasses should be optically perfect. Scratches and imperfections will cause annoying diffractions of light rays.
   - Slides and cover glasses must be free of dirt or grease of any kind.
   - A cover glass should always be used.
8. If the oil immersion lens is to be used, place a drop of oil on the cover glass.
9. If the field does not appear dark and lacks contrast, return to the 10× objective and check the ring concentricity and light source centration. If contrast is still lacking after these adjustments, the specimen is probably too thick.
10. If sharp focus is difficult to achieve under oil immersion, try using a thinner cover glass and adding more oil to the top of the cover glass and bottom of the slide.

Laboratory Report

This exercise may be used in conjunction with Part 2 when studying the various types of organisms. After reading over this exercise and doing any special assignments made by your instructor, answer the questions in Laboratory Report 1 about darkfield microscopy.

Laboratory Report

Student: _____

Date: _____ Section: _____

2   Darkfield Microscopy

A. Short Answer Questions

1. For which types of specimens is darkfield microscopy preferred over brightfield microscopy?

    _____

    _____

    _____

2. If a darkfield condenser causes all light rays to bypass the objective, where does the light come from that makes an object visible in a dark field?

    _____

    _____

    _____

3. What advantage does a cardioid condenser have over a star diaphragm?

    _____

    _____

    _____

4. If accessories for darkfield microscopy were not available, how would you construct a simple star diaphragm?

    _____

    _____

    _____

Phase-Contrast Microscopy

If one tries to observe cells without the benefit of staining, very little contrast or detail can be seen because the cells appear transparent against the aqueous medium in which they are usually suspended. Staining increases the contrast between the cell and its surrounding medium, allowing the observer to see more cellular detail, including some inclusions and various organelles. However, staining usually results in cell death, which means we are unable to observe living cells or their activities, and staining can also lead to undesirable artifacts.

A microscope that is able to differentiate the transparent protoplasmic structures and enhance the contrast between a cell and its surroundings, without the necessity of staining, is the **phase-contrast microscope.** This microscope was developed by the Dutch physicist, Frits Zernike in the 1930s. For his discovery of phase-contrast microscopy, he was awarded the Nobel Prize in 1953. Today it is the microscope of choice for viewing living cells and their activities such as motility. Figure 3.1 illustrates the contrast differences between brightfield and phase-contrast images. In this exercise, you will learn to use the phase-contrast microscope and observe the activities of living cells.

To understand how a phase-contrast microscope works, it is necessary to review some of the physical properties of light and how it interacts with matter such as biological material. Light energy can be represented as a waveform that has both an amplitude and a characteristic wavelength (illustration 1, figure 3.2). Some objects can reduce the amplitude of a light wave, and they would appear as dark objects in a microscope. In contrast, light can pass through matter without affecting the amplitude, and these objects would appear transparent in a microscope. However, as light passes through some of the transparent objects, it can be slowed down by $\frac{1}{4}$ wavelength, resulting in a **phase shift** of the light's wavelength (illustration 2, figure 3.2). For a cell, the phase shift without a reduction in amplitude results in the cell having a different refractive index than its surroundings. However the phase shifts caused by biological material are usually too small to be seen as contrast differences in a brightfield microscope. Therefore, in a brightfield microscope, cells appear transparent rather than opaque against their surroundings. Since biological material lacks any appreciable contrast, it becomes necessary to stain

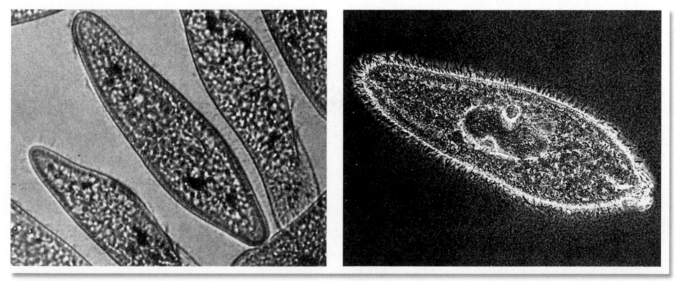

(a) Brightfield (b) Phase-contrast

Figure 3.1 Comparison of brightfield and phase-contrast images.

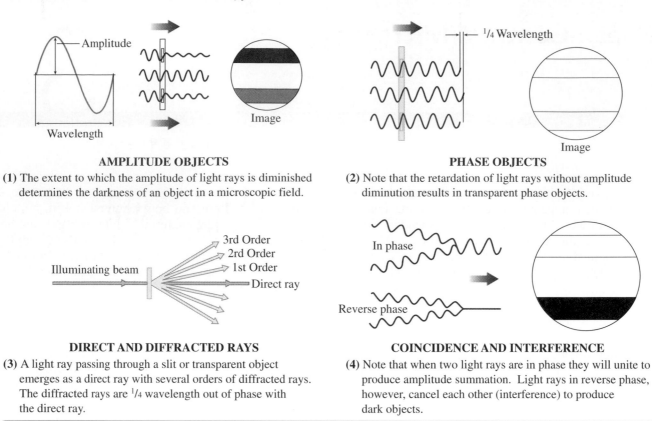

AMPLITUDE OBJECTS

(1) The extent to which the amplitude of light rays is diminished determines the darkness of an object in a microscopic field.

PHASE OBJECTS

(2) Note that the retardation of light rays without amplitude diminution results in transparent phase objects.

DIRECT AND DIFFRACTED RAYS

(3) A light ray passing through a slit or transparent object emerges as a direct ray with several orders of diffracted rays. The diffracted rays are $\frac{1}{4}$ wavelength out of phase with the direct ray.

COINCIDENCE AND INTERFERENCE

(4) Note that when two light rays are in phase they will unite to produce amplitude summation. Light rays in reverse phase, however, cancel each other (interference) to produce dark objects.

Figure 3.2 The utilization of light rays in phase-contrast microscopy.

cells with various dyes in order to study them. However, Zernike took advantage of the $\frac{1}{4}$ wavelength phase shift to enhance the small contrast differences in the various components that comprise a cell, making them visible in his microscope. This involved manipulating the light rays that were shifted and those that were unchanged as they emerged from biological material.

Two Types of Light Rays

Light rays passing through a transparent object emerge as either direct or diffracted rays. Those rays that pass straight through unaffected by the medium are called **direct rays.** They are unaltered in amplitude and phase. The rays that are bent because they are retarded by the medium (due to density differences) emerge from the object as **diffracted rays.** It is these specific light rays that are retarded $\frac{1}{4}$ wavelength. Illustration 3, figure 3.2, illustrates these two types of light rays.

If the direct and diffracted light waves are brought into exact phase with each other, the result is **coincidence** with the resultant amplitude of the converged waves being the sum of the two waves. This increase in amplitude will produce increased brightness of the object in the field. In contrast, if two light waves of

equal amplitude are in reverse phase ($\frac{1}{2}$ wavelength off), their amplitudes will cancel each other to produce a dark object. This is called **interference.** Illustration 4, figure 3.2 shows these two conditions.

The Zernike Microscope

In constructing his first phase-contrast microscope, Zernike experimented with various configurations of diaphragms and various materials that could be used to retard or advance the direct light rays. Figure 3.3 illustrates the optical system of a typical modern phase-contrast microscope. It differs from a conventional brightfield microscope by having (1) a different type of diaphragm and (2) a phase plate.

The diaphragm consists of an **annular stop** that allows only a hollow cone of light rays to pass through the condenser to the specimen on the slide. The phase plate is a special optical disk located on the objective lens near its rear focal plane. It has a phase ring that advances or retards the direct light rays $\frac{1}{4}$ wavelength.

Note in figure 3.3 that the direct rays converge on the phase ring to be advanced or retarded $\frac{1}{4}$ wavelength. These rays emerge as solid lines from the object on the slide. This ring on the phase plate is

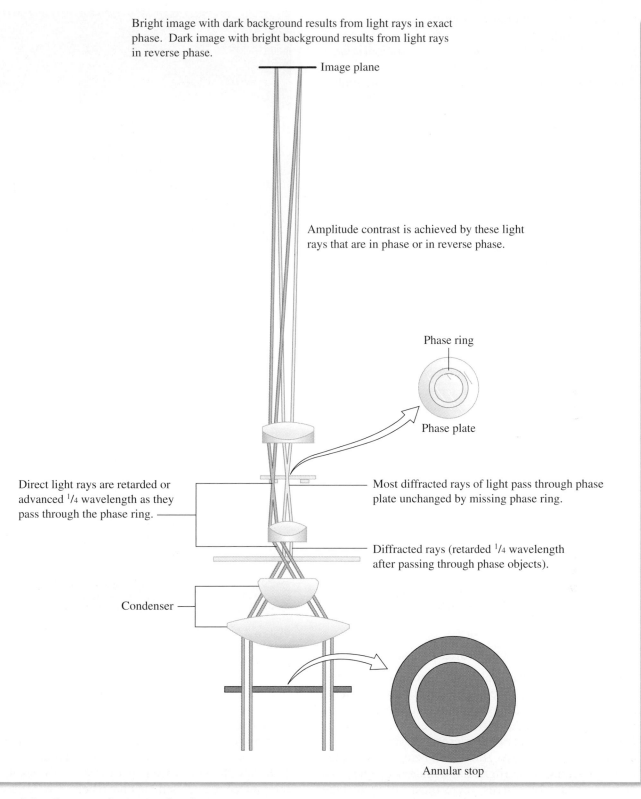

Bright image with dark background results from light rays in exact phase. Dark image with bright background results from light rays in reverse phase.

— Image plane

Amplitude contrast is achieved by these light rays that are in phase or in reverse phase.

Phase ring

Phase plate

Direct light rays are retarded or advanced $^1/_4$ wavelength as they pass through the phase ring. —

Most diffracted rays of light pass through phase plate unchanged by missing phase ring.

Diffracted rays (retarded $^1/_4$ wavelength after passing through phase objects).

Condenser —

Annular stop

Figure 3.3 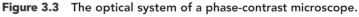 The optical system of a phase-contrast microscope.

coated with a material that will produce the desired phase shift. The diffracted rays, on the other hand, which have already been retarded $\frac{1}{4}$ wavelength by the phase object on the slide, completely miss the phase ring and are not affected by the phase plate. It should be clear, then, that depending on the type of phase-contrast microscope, the convergence of diffracted and direct rays on the image plane will result in either a brighter image (*amplitude summation*) or a darker image (*amplitude interference* or *reverse phase*). The former is referred to as **bright-phase** microscopy; the latter as **dark-phase** microscopy. The apparent brightness or darkness, incidentally, is proportional to the square of the amplitude; thus, the image will be four times as bright or dark as seen through a brightfield microscope.

It should be added here that the phase plates of some microscopes have coatings to change the phase of the diffracted rays. In any event, the end result will be the same: to achieve coincidence or interference of direct and diffracted rays.

Microscope Adjustments

If the annular stop under the condenser of a phase-contrast microscope can be moved out of position, this instrument can also be used for brightfield studies. Although a phase-contrast objective has a phase ring attached to the top surface of one of its lenses, the presence of that ring does not impair the resolution of the objective when it is used in the brightfield mode. It is for this reason that manufacturers have designed phase-contrast microscopes in such a way that they can be quickly converted to brightfield operation.

To make a microscope function efficiently in both phase-contrast and brightfield situations, one must master the following procedures:

- lining up the annular ring and phase rings so that they are perfectly concentric,
- adjusting the light source so that maximum illumination is achieved for both phase-contrast and brightfield usage, and
- being able to shift back and forth easily from phase-contrast to brightfield modes. The following suggestions should be helpful in coping with these problems.

Alignment of Annulus and Phase Ring

Unless the annular ring below the condenser is aligned perfectly with the phase ring in each objective lens, good phase-contrast imagery cannot be achieved. Figure 3.4 illustrates the difference between nonalignment and alignment. If a microscope has only one phase-contrast objective, there will be only one annular stop that has to be aligned. If a microscope has two or more phase objectives, there must be a substage unit with separate annular stops for each phase objective, and alignment procedure must be performed separately for each objective and its annular stop.

Since the objective cannot be moved once it is locked in position, all adjustments are made to the annular stop. On some microscopes the adjustment may be made with tools, as illustrated in figure 3.5.

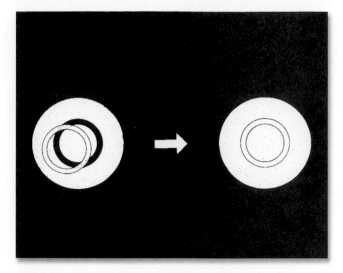

Figure 3.4 The image on the right illustrates the appearance of the rings when perfect alignment of phase ring and annulus diaphragm has been achieved.

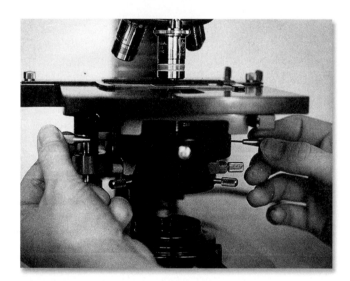

Figure 3.5 Alignment of the annulus diaphragm and phase ring is accomplished with a pair of Allen-type screwdrivers on this American Optical microscope.

On other microscopes, in figure 3.6, the annular rings are moved into position with special knobs on the substage unit. Since the method of adjustment varies from one brand of microscope to another, one has to follow the instructions provided by the manufacturer. Once the adjustments have been made, they are rigidly set and needn't be changed unless someone inadvertently disturbs them.

To observe ring alignment, one can replace the eyepiece with a **centering telescope** as shown in figure 3.7. With this unit in place, the two rings can be brought into sharp focus by rotating the focusing ring on the telescope. Refocusing is necessary for

Figure 3.6 Alignment of the annulus and phase ring is achieved by adjusting the two knobs as shown. ©The McGraw-Hill Companies/Auburn University Photographic Service

Figure 3.8 Some microscopes have an aperture viewing unit that can be used instead of a centering telescope for observing the orientation of the phase ring and annular ring.

Figure 3.7 If the ocular of a phase-contrast microscope is replaced with a centering telescope, the orientation of the phase ring and annular ring can be viewed. ©The McGraw-Hill Companies/Auburn University Photographic Service

each objective and its matching annular stop. Some manufacturers, such as American Optical, provide an aperture viewing unit (figure 3.8), which enables one to observe the rings without using a centering telescope. Zeiss microscopes have a unit called the ***Optovar,*** which is located in a position similar to the American Optical unit that serves the same purpose.

Light Source Adjustment

For both brightfield and phase-contrast modes, it is essential that optimum lighting be achieved. This is no great problem for a simple setup such as the American Optical instrument shown in figure 3.9. For multiple phase objective microscopes, however, there are many more adjustments that need to be made. A few suggestions that highlight some of the problems and solutions follow:

1. Since blue light provides better images for both phase-contrast and brightfield modes, make certain that a blue filter is placed in the filter holder that is positioned in the light path. If the microscope has no filter holder, placing the filter over the light source on the base will help.

2. Brightness of field under phase-contrast is controlled by adjusting the voltage or the iris diaphragm on the base. Considerably more light is required for phase-contrast than for brightfield since so much light is blocked out by the annular stop.

3. The evenness of illumination on some microscopes, seen on these pages, can be adjusted by removing the lamp housing from the microscope and focusing the light spot on a piece of translucent white paper. For the detailed steps in this

Figure 3.9 The annular stop on this American Optical microscope is located on a slideway. When pushed in, the annular stop is in position.

procedure, one should consult the instruction manual that comes with the microscope. Light source adjustments of this nature are not necessary for the simpler types of microscopes.

Working Procedures

Once the light source is correct and the phase elements are centered you are finally ready to examine slide preparations. Keep in mind that from now on most of the adjustments described earlier should not be altered; however, if misalignment has occurred due to mishandling, it will be necessary to refer back to alignment procedures. The following guidelines should be adhered to in all phase-contrast studies:

• Use only optically perfect slides and cover glasses (no bubbles or striae in the glass).
• Be sure that slides and cover glasses are completely free of grease or chemicals.

• Use wet mount slides instead of hanging drop preparations. The latter leave much to be desired. Culture broths containing bacteria or protozoan suspensions are ideal for wet mounts.
• In general, limit observations to living cells. In most instances, stained slides are not satisfactory.

The first time you use phase-contrast optics to examine a wet mount, follow these suggestions:

1. Place the wet mount slide on the stage and bring the material into focus, *using brightfield optics* at low-power magnification.
2. Once the image is in focus, switch to phase optics at the same magnification. Remember, it is necessary to place in position the matching annular stop.
3. Adjust the light intensity, first with the base diaphragm and then with the voltage regulator. In most instances, you will need to increase the amount of light for phase-contrast.
4. Switch to higher magnifications, much in the same way you do for brightfield optics, except that you have to rotate a matching annular stop into position.
5. If an oil immersion phase objective is used, add immersion oil to the top of the condenser as well as to the top of the cover glass.
6. Don't be disturbed by the "halo effect" that you observe with phase optics. Halos are normal.

Laboratory Report

This exercise may be used in conjunction with Part 2 in studying various types of organisms. Organelles in protozoans and algae will show up more distinctly than with brightfield optics. After reading this exercise and doing any special assignments made by your instructor, answer the questions in Laboratory Report 3.

Laboratory Report

Student: _____

Date: _____   Section: _____

3   Phase-Contrast Microscopy

A. Short Answer Questions

1. Staining of cells is often performed to enhance images acquired by brightfield microscopy. Phase-contrast microscopy does not require cell staining. Why is this advantageous?

    _____

    _____

2. As light passes through a transparent object, how are direct and diffracted light rays produced? How much phase shift occurs?

    _____

    _____

3. How do coincidence and interference of light rays differ? What type of image does each produce? How does that contribute to a sharper image?

    _____

    _____

    _____

4. Differentiate between bright-phase and dark-phase microscopy in terms of phase shift.

    _____

    _____

    _____

5. Which two items can be used to check the alignment of the annulus and phase ring?

    _____

    _____

B. Multiple Choice

Select the answer that best completes the following statements.

1. A phase-contrast microscope differs from a brightfield microscope by having a
    a. blue filter in the ocular lens.
    b. diaphragm with an annular stop.
    c. phase plate in the objective lens.
    d. Both (b) and (c) are correct.
    e. All are correct.

ANSWERS

Multiple Choice

1. _____

2. If direct rays passing through an object are advanced $\frac{1}{4}$ wavelength by the phase ring, the diffracted rays are
   a. in phase with the direct rays.
   b. $\frac{1}{2}$ wavelength out of phase with the direct rays.
   c. $\frac{3}{4}$ wavelength out of phase with direct rays.
   d. in reverse phase with the direct rays.
   e. Both (b) and (d) are correct.

3. Amplitude summation occurs in phase-contrast optics when both direct and diffracted rays are
   a. in phase.
   b. in reverse phase.
   c. off $\frac{1}{4}$ wavelength.
   d. None of these are correct.

4. The phase-contrast microscope is best suited for observing
   a. living organisms in an uncovered drop on a slide.
   b. stained slides with cover glasses.
   c. living organisms in hanging drop slide preparations.
   d. living organisms on a slide with a cover glass.

2. _____

3. _____

4. _____

Fluorescence Microscopy

The fluorescence microscope is a unique instrument that is indispensible in research and some diagnostic procedures. For example, immunofluorescence techniques have made laboratory diagnosis of many diseases such as AIDS, Legionnaire's disease, and anthrax much simpler with this type of microscope than with the conventional microscopes described in previous exercises in this manual. The use of fluorescent-tagged antibodies has also provided a means to detect the location of specific proteins inside of cells or on their surface. However, fluorescence microscopy is not without pitfalls. Before one can use the fluorescent microscope and manipulate the fluorescent stains necessary to visualize images, one needs a good basic understanding of the method, its limitations, and its hazards. One of the greatest dangers is the potential for serious eye injury, which can occur from the UV light source if the instrument is not used properly.

A fluorescent microscope differs from an ordinary brightfield microscope in several aspects. (A modern fluorescent microscope is shown in figure 4.1.) First, it uses a powerful mercury vapor light source that generates ultraviolet light. Second, a darkfield condenser is usually used in place of the Abbe brightfield condenser. The third difference is that it employs a combination of filters that selects for the ultraviolet light that excites fluorescent compounds in the specimen but allows only the resulting flourescent light to reach the observer. In the following exercise, you will learn the general principles of fluorescence microscopy and how to use the microscope.

The Principle of Fluorescence

When a molecule absorbs energy, its electrons become excited and are promoted to a high-energy (singlet) state. When the electrons return to the low-energy ground state, the absorbed energy is not lost, as dictated by the First Law of Thermodynamics, but can appear as another form of energy (figure 4.2). This new form of energy can be heat, light, or chemical energy. If light is emitted as the molecule subsequently returns to its ground state, it is called **photoluminescence.** The photoluminescence can be in the form of either **fluorescence** or **phosphorescence.** If the excited molecule returns to the ground state in less

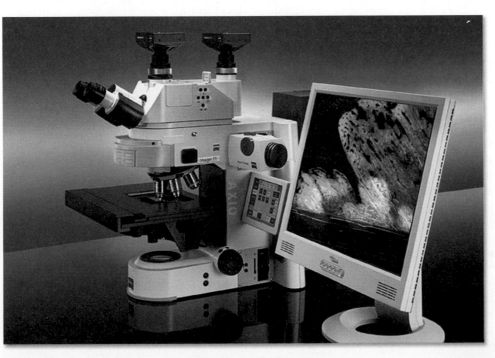

Figure 4.1 An advanced fluorescent stereo microscope. Picture used with the permission of Carl Zeiss AG

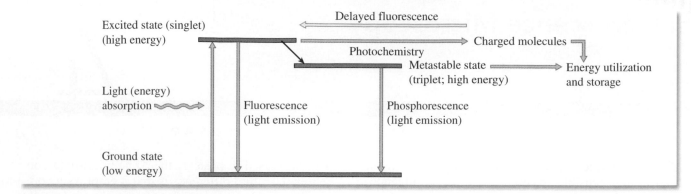

Figure 4.2 Spectra of light absorption and emission.

than 10^{-4} seconds, the light emitted is called fluorescence and the wavelength of the emitted fluorescence is always a *longer* wavelength than the exciting light. For phosphorescence, however, the excited molecule does not immediately return to the ground state but drops down to a metastable (triplet), high-energy state and then decays to the ground state. The time interval between the absorption of exciting radiation and emission of phosphorescence light is greater than 10^{-4} seconds and, hence, the half-life for phosphorescence is much longer than fluorescence. Phosphorescence is the basis for wrist watches that glow in the dark.

Microbiological material that is to be studied with fluorescence microscopy must be intrinsically fluorescent or stained with special dyes called **fluorochromes.** Oftentimes, fluorochromes such as auromine O, acridine orange, and fluoroscein are chemically linked to biological molecules such as antibodies and these are used to tag specific proteins on or in the cell. Molecules such as bacteriochlorophyll and some cofactors in methane bacteria naturally fluoresce when excited. As a result, photosynthetic bacteria and some of the methane bacteria are visible in a fluorescent microscope owing to their intrinsic fluorescence. The molecular structure of a molecule as well as some physical parameters have a great deal to do with whether the molecule will fluoresce. One of the more significant problems with fluorescence is **quenching.** With continued exposure of fluorescent molecules to ultraviolet light, the fluorescence will eventually diminish and disappear due to quenching of the fluorescence. As a result, objects become more difficult to make out. Chemical compounds such as acids can contribute to quenching.

Microscope Components

Figure 4.3 illustrates, diagrammatically, the light pathway of a fluorescence microscope. The essential components are the light source, heat filter, exciter filter, condenser, and barrier filter. The characteristics and functions of each item follow.

Light Source An essential component of a fluorescent microscope is a mercury vapor arc lamp. Such a bulb is necessary because it produces shorter wavelengths of light (ultraviolet and blue) that are needed for good fluorescence. To produce the arc in a mercury vapor lamp, voltages of 18,000 volts are required and hence a power supply transformer is always used.

The wavelengths produced by these lamps include the ultraviolet range, 200–400 nm, the visible range, 400–780 nm, and long infrared radiation above 780 nm. The ultraviolet rays are the most dangerous because they can cause mutations in DNA and they can cause serious eye damage.

Mercury vapor arc lamps are expensive and can be dangerous. Because these lamps contain pressurized gas, they have the potential to explode. Also, direct exposure of the eyes to these lamps can cause serious damage. Knowledge of these hazards is essential for safe operation of the microscope. One should never attempt to use a fluorescent microscope without a complete understanding of its operation. However, if one understands and uses the necessary precautions, the fluorescent microscope can be a valuable tool to the researcher and diagnostician.

Heat Filter The infrared rays generated by the mercury vapor arc lamp produce a considerable amount of heat. These rays serve no useful purpose in fluorescence and place considerable stress on the filters within the system. To remove these rays, a heat-absorbing filter is the first element in front of the condensers. Ultraviolet rays, as well as most of the visible spectrum, pass through this filter unimpeded.

Exciter Filter After the light has been cooled down by the heat filter, it passes through the exciter filter, which absorbs all the wavelengths except the short ones needed to excite the fluorochrome on the slide. These filters are very dark and are designed to let through only the green, blue, violet, or ultraviolet rays. If the exciter filter is intended for visible light

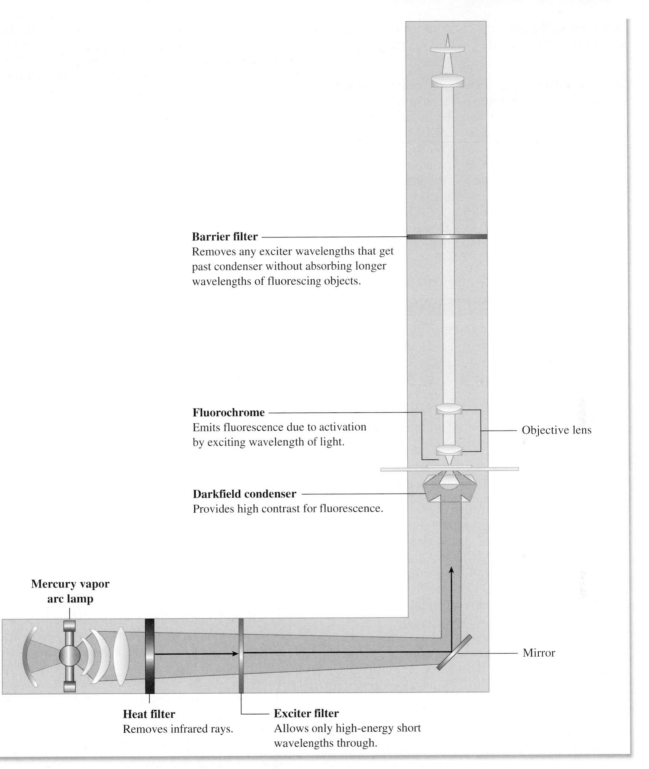

Barrier filter
Removes any exciter wavelengths that get past condenser without absorbing longer wavelengths of fluorescing objects.

Fluorochrome
Emits fluorescence due to activation by exciting wavelength of light.

— Objective lens

Darkfield condenser
Provides high contrast for fluorescence.

Mercury vapor arc lamp

— Mirror

Heat filter
Removes infrared rays.

Exciter filter
Allows only high-energy short wavelengths through.

Figure 4.3 The light pathway of a fluorescence microscope.

(blue, green, or violet) transmission, it will also allow ultraviolet transmittance.

Condenser   To achieve the best contrast of a fluorescent object in the microscopic field, a darkfield condenser is used. It must be kept in mind that weak fluorescence of an object in a bright field would be difficult to see. The dark background produced by the darkfield condenser, thus, provides the desired contrast. Another bonus of this type of condenser is that the majority of the ultraviolet light rays is deflected by the condenser, protecting the observer's eyes. To achieve this, the numerical aperture of the objective is always 0.05 less than that of the condenser.

Barrier Filter   This filter is situated between the objective and the eyepiece to remove all remnants of the exiting light so that only the fluorescence is seen. When ultraviolet excitation is employed with its very dark, almost black-appearing exciter filters, the corresponding barrier filters appear almost colorless. On the other hand, when blue exciter filters are used, the matching barrier filters have a yellow to deep orange color. In both instances, the significant fact is that the barrier filter should cut off precisely the shorter exciter wavelengths without affecting the longer fluorescence wavelengths.

Use of the Microscope

As in the case of most sophisticated equipment of this type, it is best to consult the manufacturer's instruction manual before using it. Although different makes of fluorescence microscopes are essentially alike in principle, they may differ considerably in the fine points of operation. Since it is not possible to be explicit about the operation of all makes, all that will be attempted here is to generalize.

Some Precautions   To protect yourself and others, keep the following points in mind:

1. Remember that the pressurized mercury arc lamp is literally a potential bomb. Design of the equipment is such, however, that with good judgment, no injury should result. When these lamps are cold, they are relatively safe, but when hot, the inside pressure increases to eight atmospheres, or 112 pounds per square inch.

   The point to keep in mind is this—*never attempt to inspect the lamp while it is hot.* Let it cool completely before opening up the lamp housing. Usually, 15 to 20 minutes cooling time is sufficient.
2. *Never expose your eyes to the direct rays of the mercury arc lamp.* Equipment design is such that the bulb is always shielded against the scattering of its rays. Remember that the unfiltered light from one of these lamps is rich in both ultraviolet and infrared rays—both of which are damaging to the eyes. *Severe retinal burns can result from exposure to the mercury arc rays.*
3. Be sure that the barrier filter is always in place when looking down through the microscope. *Removal of the barrier filter or exciter filter or both filters while looking through the microscope could cause eye injury.* It is possible to make mistakes

of this nature if one is not completely familiar with the instrument. Remember, the function of the barrier filter is to prevent traces of ultraviolet light from reaching the eyes without blocking wavelengths of fluorescence.

Warm-up Period   The lamps in fluorescence microscopes require a warm-up period. When they are first turned on, the illumination is very low, but it increases to maximum in about 2 minutes. *Optimum illumination occurs when the equipment has been operating for 30 minutes or more.* Most manufacturers recommend leaving the instruments turned on for an hour or more when using them. It is not considered good economy to turn the instrument on and off several times within a 2- or 3-hour period.

Keeping a Log   The life expectancy of a mercury arc lamp is around 400 hours. A log should be kept of the number of hours that the instrument is used so that inspection can be made of the bulb at approximately 200 hours. A card or piece of paper should be kept conveniently near the instrument so that the individual using the instrument is reminded to record the time that the instrument is turned on and off.

Filter Selection   The most frequently used filter combination is the bluish Schott BG12 (AO #702) exciter and the yellowish Schott OG1 barrier filters. Figure 4.4 shows the wavelength transmission of each of these filters. Note that the exciter filter gives peak emission of light in the 400-nm area of the spectrum. These rays are violet. It allows practically no green or yellow wavelengths through. The shortest wavelengths that this barrier filter lets through are green to greenish-yellow.

If a darker background is desired than is being achieved with the above filters, one may add a pale blue Schott BG38 to the system. It may be placed on either

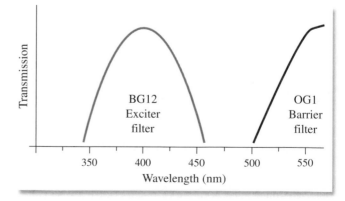

Figure 4.4   Spectral transmissions of BG12 and OG1 filters.

side of the heat filter, depending on the type of equipment being used. If it is placed between the lamp and heat filter, it will also function as another heat filter.

Examination When looking for material on the slide, it is best to use low- or high-power objectives. It may be desirable to move the illuminator out of position and use incandescent lighting for this phase of the work. Once the desirable field has been located, the mercury vapor arc illuminator can be moved into position.

Keep in mind that there is no diaphragm control on darkfield condensers. Some instruments are supplied with neutral density filters to reduce light intensity. The best system of illumination control, however, is achieved with objectives that have a built-in iris control. These objectives have a knurled ring that can be rotated to control the contrast.

For optimum results, it is essential that oil be used between the condenser and the slide. And, of course, if the oil immersion lens is used, the oil must also be interposed between the slide and the objective. It is also important that special low-fluorescing immersion oil be used. *Ordinary immersion oil should be avoided.*

Although the ocular of a fluorescence microscope is usually 10×, one should not hesitate to try other size oculars if they are available. With brightfield microscopes, it is generally accepted that nothing is gained by going beyond 1000× magnification. In a fluorescence microscope, however, the image is formed in a manner quite different from its brightfield counterpart obviating the need for following the 1000× rule. The only loss by using the higher magnification is some brightness.

Laboratory Report

Complete Laboratory Report 4.

Laboratory Report

Student: _____

Date: _____ Section: _____

4 Fluorescence Microscopy

A. Short Answer Questions

1. Differentiate between phosphorescence and fluorescence.

   _____

   _____

2. What relationship exists between the wavelength of light and its energy?

   _____

   _____

3. What is quenching?

   _____

   _____

4. Describe two hazards associated with use of mercury vapor arc lamps.

   _____

   _____

5. Describe the length and importance of the "warm-up period" for the fluorescence microscope lamp.

   _____

   _____

6. What are two advantages of using a darkfield condenser for fluorescence microscopy?

   _____

   _____

7. What special considerations should be made when using the oil immersion lens for fluorescence microscopy?

   _____

   _____

Fluorescence Microscopy (continued)

B. Multiple Choice

Select the answer that best completes the following statements.

1. Specific proteins on or in a cell can be tagged with fluorochromes that have been chemically linked to
   a. antibodies.
   b. chlorophyll.
   c. methane.
   d. Both (b) and (c) are correct.
   e. All are correct.

2. The heat filter is used to eliminate nonuseful, heat-generating
   a. infrared rays.
   b. ultraviolet rays.
   c. visible spectrum rays.
   d. Both (b) and (c) are correct.
   e. All are correct.

3. Before reaching the ocular, the light generated by the mercury vapor arc lamp travels through the
   a. barrier filter, heat filter, condenser, slide, objective lens, and then the exciter filter.
   b. exciter filter, barrier filter, slide, condenser, objective lens, and then the heat filter.
   c. exciter filter, condenser, heat filter, slide, objective lens, and then the barrier filter.
   d. heat filter, exciter filter, condenser, barrier filter, slide, and then the objective lens.
   e. heat filter, exciter filter, condenser, slide, objective lens, and then the barrier filter.

4. The wavelength of fluorescent light rays is
   a. always longer than the exciting wavelength.
   b. always shorter than the exciting wavelength.
   c. about the same length as the exciting wavelength.
   d. sometimes shorter and at other times longer in wavelength.

5. The barrier filter in the fluorescence microscope is kept in position to
   a. block all light rays from getting through.
   b. allow fluorescence light rays to pass through.
   c. screen out exciting light rays.
   d. Both (b) and (c) are correct.

Answers

Multiple Choice

1. _____

2. _____

3. _____

4. _____

5. _____

Microscopic Measurements

Figure 5.1 The ocular micrometer with retaining ring is inserted into the base of the eyepiece.

Figure 5.2 Stage micrometer is positioned by centering the small glass disk over the light source.

With an ocular micrometer properly installed in the eyepiece of your microscope, it is a simple matter to measure the size of microorganisms that are seen in the microscopic field. An **ocular micrometer** consists of a circular disk of glass that has graduations engraved on its upper surface. These graduations appear as shown in illustration B, figure 5.4. On some microscopes one has to disassemble the ocular so that the disk can be placed on a shelf in the ocular tube between the two lenses. On most microscopes, however, the ocular micrometer is simply inserted into the bottom of the ocular, as shown in figure 5.1. Before one can use the micrometer, it is necessary to calibrate it for each of the objectives by using a stage micrometer.

The principal purpose of this exercise is to show you how to calibrate an ocular micrometer for the various objectives on your microscope.

Calibration Procedure

The distance between the lines of an ocular micrometer is an arbitrary value that has meaning only if the ocular micrometer is calibrated for the objective that is being used. A **stage micrometer** (figure 5.2), also known as an *objective micrometer,* has lines inscribed on it that are exactly 0.01 mm (10 µm) apart. Illustration C, figure 5.4, shows these graduations.

To calibrate the ocular micrometer for a given objective, it is necessary to superimpose the two scales

and determine how many of the ocular graduations coincide with one graduation on the scale of the stage micrometer. Illustration A in figure 5.4 shows how the two scales appear when they are properly aligned in the microscopic field. In this case, seven ocular divisions match up with one stage micrometer division of 0.01 mm to give an ocular value of 0.01/7, or 0.00143 mm. Since there are 1000 micrometers in 1 millimeter, these divisions are 1.43 µm apart.

With this information known, the stage micrometer is replaced with a slide of organisms (figure 5.3)

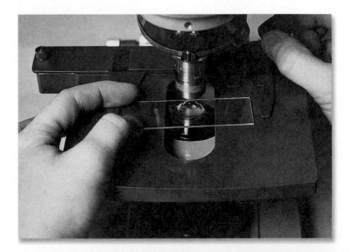

Figure 5.3 After calibration is completed, the stage micrometer is replaced with a slide for measurements.

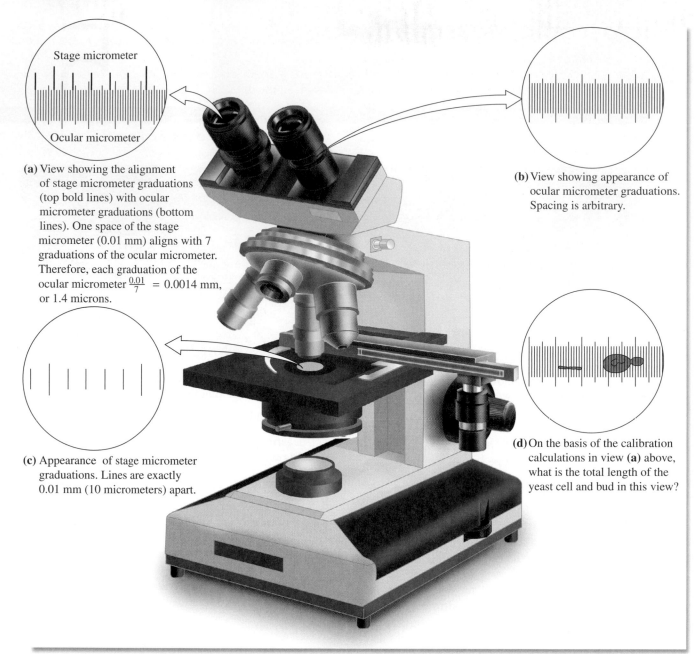

(a) View showing the alignment of stage micrometer graduations (top bold lines) with ocular micrometer graduations (bottom lines). One space of the stage micrometer (0.01 mm) aligns with 7 graduations of the ocular micrometer. Therefore, each graduation of the ocular micrometer $\frac{0.01}{7}$ = 0.0014 mm, or 1.4 microns.

(b) View showing appearance of ocular micrometer graduations. Spacing is arbitrary.

(c) Appearance of stage micrometer graduations. Lines are exactly 0.01 mm (10 micrometers) apart.

(d) On the basis of the calibration calculations in view **(a)** above, what is the total length of the yeast cell and bud in this view?

Figure 5.4 Calibration of ocular micrometer.

to be measured. Illustration D, figure 5.4, shows how a field of microorganisms might appear with the ocular micrometer in the eyepiece. To determine the size of an organism, then, it is a simple matter to count the graduations and multiply this number by the known distance between the graduations. When calibrating the objectives of a microscope, proceed as follows.

Materials

- ocular micrometer or eyepiece that contains a micrometer disk
- stage micrometer

1. If eyepieces are available that contain ocular micrometers, replace the eyepiece in your microscope with one of them. If it is necessary to insert an

ocular micrometer in your eyepiece, find out from your instructor whether it is to be inserted below the bottom lens or placed between the two lenses within the eyepiece. In either case, great care must be taken to avoid dropping the eyepiece or reassembling the lenses incorrectly. *Only with your instructor's prior approval shall eyepieces be disassembled.* Be sure that the graduations are on the upper surface of the glass disk.

2. Place the stage micrometer on the stage and center it exactly over the light source.

3. With the low-power (10×) objective in position, bring the graduations of the stage micrometer into focus, *using the coarse adjustment knob. Reduce the lighting.* Note: If the microscope has an automatic stop, do not use it as you normally would for regular microscope slides. The stage micrometer slide is too thick to allow it to function properly.

4. Rotate the eyepiece until the graduations of the ocular micrometer lie parallel to the lines of the stage micrometer.

5. If the **low-power objective** is the objective to be calibrated, proceed to step 8.

6. If the **high-dry objective** is to be calibrated, swing it into position and proceed to step 8.

7. If the **oil immersion lens** is to be calibrated, place a drop of immersion oil on the stage micrometer, swing the oil immersion lens into position, and bring the lines into focus; then, proceed to the next step.

8. Move the stage micrometer laterally until the lines at one end coincide. Then look for another line on the ocular micrometer that coincides *exactly* with one on the stage micrometer. Occasionally, one stage micrometer division will include an even number of ocular divisions, as shown in illustration A of figure 5.4. In most instances, however, several stage graduations will be involved. In this case, divide the number of stage micrometer divisions by the number of ocular divisions that coincide. The figure you get will be that part of a stage micrometer division that is seen in an ocular division. This value must then be multiplied by 0.01 mm to get the amount of each ocular division.

Example: 3 divisions of the stage micrometer line up with 20 divisions of the ocular micrometer.

$$\text{Each ocular division} = \tfrac{3}{20} \times 0.01$$
$$= 0.0015 \text{ mm}$$
$$= 1.5 \text{ } \mu m$$

Replace the stage micrometer with slides of organisms to be measured.

Measuring Assignments

Organisms such as protozoans, algae, fungi, and bacteria in the next few exercises may need to be measured. If your instructor requires that measurements be made, you will be referred to this exercise.

Later on you will be working with unknowns. In some cases measurements of the unknown organisms will be pertinent to identification.

If trial measurements are to be made at this time, your instructor will make appropriate assignments. **Important:** Remove the ocular micrometer from your microscope at the end of the laboratory period.

Laboratory Report

Answer the questions in Laboratory Report 5.

Laboratory Report

Student: _____

Date: _____ Section: _____

5 Microscopic Measurements

A. Short Answer Questions

1. How do the graduations differ between ocular and stage micrometers? _____
   _____
   _____

2. If thirteen ocular divisions line up with two divisions of the stage micrometer, what is the diameter (μm) of a cell that spans sixteen ocular divisions? _____
   _____
   _____

3. Why must the entire calibration procedure be performed for each objective? _____
   _____
   _____

Survey of Microorganisms

Microorganisms abound in the environment. Eukaryotic microbes such as protozoa, algae, diatoms, and amoebas are plentiful in ponds and lakes. Bacteria are found associated with animals, occur abundantly in the soil and in water systems, and have even been isolated from core samples taken from deep within the earth's crust. Bacteria are also present in the air where they are distributed by convection currents that transport them from other environments. The Archaea, modern day relatives of early microorganisms, occupy some of the most extreme environments such as acidic-volcanic hot springs, anaerobic environments devoid of any oxygen, and lakes and salt marshes excessively high in sodium chloride. Cyanobacteria are photosynthetic prokaryotes that can be found growing in ponds and lakes, on limestone rocks, and even on the shingles that protect the roofs of our homes. Fungi are a very diverse group of microorganisms that are found in most common environments. For example, they degrade complex molecules in the soil, thus contributing to its fertility. Sometimes, however, they can be nuisance organisms; they form mildew in our bathroom showers and their spores cause allergies. The best way of describing the distribution of microorganisms is to say that they are ubiquitous, or found everywhere.

Intriguing questions to biologists are how are the various organisms related to one another and where do the individual organisms fit in an evolutionary scheme? Molecular biology techniques have provided a means to analyze the genetic relatedness of the organisms that comprise the biological world and determine where the various organisms fit into an evolutionary scheme. By comparing the sequence of ribosomal RNA molecules, coupled with biochemical data, investigators have developed a phylogenetic tree that illustrates the current thinking on the placement of the various organisms into such

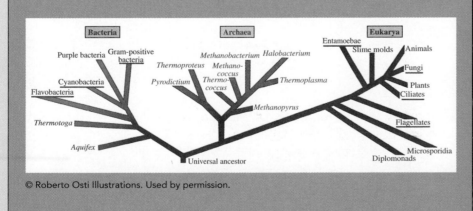

© Roberto Osti Illustrations. Used by permission.

a scheme. This evolutionary scheme divides the biological world into three domains.

Domain Bacteria   These organisms have a prokaryotic cell structure. They lack organelles such as mitochondria and chloroplasts, are devoid of an organized nucleus with a nuclear membrane, and possess 70S ribosomes that are inhibited by many broad-spectrum antibiotics. The vast majority of organisms are enclosed in a cell wall composed of peptidoglycan. The bacteria and cyanobacteria are members of this domain.

Domain Eukarya   Organisms in this domain have a eukaryotic cell structure. They contain membrane-bound organelles such as mitochondria and chloroplasts, an organized nucleus enclosed in a nuclear membrane, and 80S ribosomes that are not inhibited by broad spectrum antibiotics. Plants, animals, and microorganisms such as protozoa, algae, and fungi belong in this domain. Plants have cell walls composed of cellulose and fungi have cell walls composed of chitin. In contrast, animal cells lack a cell wall structure.

Domain Archaea   The Archaea exhibit the characteristics of both the bacteria and Eukarya. These organisms are considered to be the relatives of ancient microbes that existed during Archean times. Like their bacterial counterparts, they possess a simple cell structure that lacks organelles and an organized nucleus. They have 70S ribosomes like bacteria but the protein makeup and morphology of their ribosomes are more similar to eukaryotic ribosomes. Like eukaryotes, the ribosomes in Archaea are not sensitive to antibiotics. They have a cell wall but its structure is not composed of peptidoglycan. The principal habitats of these organisms are extreme environments such as volcanic hot springs, environments with excessively high salt, and environments devoid of oxygen. Thus, they are referred to as "extremophiles." The acido-thermophiles, the halobacteria, and the methanogens (methane bacteria) are examples of the Archaea.

In the exercises of Part 2, you will have the opportunity to study some of these organisms. In pond water, you may see amoebas, protozoans, various algae, diatoms, and cyanobacteria. You will sample for the presence of bacteria by exposing growth media to various environments. The fungi will be studied by looking at cultures and preparing slides of these organisms. Because the Archaea occur in extreme conditions and also require specialized culture techniques, it is unlikely that you will encounter any of these organisms.

Protozoa, Algae, and Cyanobacteria

In this exercise, a study will be made of protists and cyanobacteria that live in pond water. Containers of water and bottom debris from various ponds will be available for study. Illustrations and text provided in this exercise will be used to assist you in your attempt to identify the various organisms. Organisms that are bluish-green will probably be cyanobacteria (once categorized as blue-green algae). Most others will be protists. However, small nematodes, insect larvae, microcrustaceans, rotifers, and other invertebrates could be present. Supplementary books on the laboratory bookshelf may also be available to help you with identification of the organisms that are not described or diagrammed in the short text of this exercise.

The purpose of this exercise is to provide an opportunity for you to become familiar with common pond-dwelling microorganisms and to appreciate the vast diversity that exists in a drop of pond water. You will also become familiar with the differences among several groups by comparing their major characteristics. The extent to which you will be held accountable for the names of various organisms will be determined by your instructor. The amount of time available for this laboratory exercise will determine the depth and scope to be pursued.

To study the microorganisms of pond water, it will be necessary to make wet mount slides. The procedure for making such slides is very simple. All that is necessary is to place a drop of suspended organisms on a microscope slide and cover it with a cover glass. If several different samples are available, you should record the number of the container (from which you took your sample) on your slide with a marking pen. As you prepare your slides, observe the following guidelines below.

Materials

- bottles of pond-water samples
- microscope slides and cover glasses
- rubber-bulbed pipettes and forceps
- marking pen
- reference books

1. Clean a slide and cover glass with soap and water, rinse thoroughly, and dry. Do not attempt to study a slide that lacks a cover glass.

2. When using a pipette, insert it into the bottom of the sample bottle to get a maximum number of organisms. Very few organisms will be found swimming around in middepth of the bottle.
3. To remove filamentous algae from a sample bottle, use forceps. Avoid putting too much material on the slides.
4. Explore the slide first with the low-power objective. Reduce the lighting with the iris diaphragm; this is very important. Keep the condenser at its highest point.
5. When you find an organism of interest, swing the high-dry objective into position and adjust the lighting to get optimum contrast. If your microscope has phase-contrast elements, use them.
6. Refer to figures 6.1 through 6.6 and the text on these pages to identify the various organisms that you encounter.
7. Record your observations on the Laboratory Report.

Survey of Organisms

An impressive variety of protists and cyanobacteria will likely be encountered during this laboratory exercise. You will be asked to identify and categorize these organisms based on their morphological characteristics. Traditionally, such morphological characteristics were used to construct formal classification schemes; however, genetic analyses have demonstrated that such classification schemes do not necessarily represent evolutionary relationships between organisms. Presently, there is a lively debate among protistologists as to which taxonomic scheme (and there are many) of protists should be accepted worldwide. Because of this lack of consensus among scientists about how to classify protists, and because true evolutionary relationships cannot be determined simply by observing organisms, you will use an informal system based on morphology to categorize the organisms that you encounter in this exercise. The following table will help you understand some of the major morphological groups of the organisms that you may see. **Please keep in mind that these are not formally recognized taxonomic groups,** but are useful for identifying and categorizing organisms in the laboratory based on specific physical traits.

Table 6.1   Classification of Organisms

PROKARYOTES (DOMAIN PROKARYA)	EUKARYOTES (DOMAIN EUKARYA)

Cyanobacteria (Figure 6.6)	Protists:
	    Protozoa:
	       Flagellates (figure 6.1, illustrations 1–4)
	       Amoebae (figure 6.1, illustrations 5–8)
	       Ciliates (figure 6.1, illustrations 9–24)
	    Algae:
	       Euglenoids (euglenozoa) (figure 6.2, illustrations 1–6)
	       Green algae (chlorophytes) (figure 6.2, illustrations 8, 14, 15, 19, 20; figure 6.3; figure 6.4)
	       Golden-brown algae (chrysophytes) (figure 6.2, illustration 16)
	       Synurales (figure 6.2, illustration 13)
	       Yellow-green algae (xanthophytes) (figure 6.3, illustrations 5, 6)
	       Diatoms (bacillariophytes) (figure 6.5)
	       Dinoflagellates (figure 6.2, illustrations 17, 18)

Eukaryotes

The Protists

The protists are a large, paraphyletic group of organisms. Generally, protists are eukaryotic organisms that cannot be classified as plants, fungi, or animals. The majority of protists are unicellular, although some are colonial and some are multicellular. None have differentiated tissues.

Protozoa

Generally, moving, nonpigmented, single-celled organisms are referred to as "protozoa." Protozoa include heterotrophic flagellates, amoebae, ciliates, and apicomplexans. Either the entire cell will move, or the cell will be attached to a substrate and only parts of the cell will move. With respect to movement, there are three major means of locomotion of free-living protozoa; pseudopodia (found in amoeboid cells), flagella (found in flagellates), cilia (found in ciliates), and gliding (diatoms). These eukaryotic cells are bound by a plasma membrane that may have additional surface modifications depending on the species. All have a distinct nucleus, ribosomes, and mitochondria. Some possess a cytostome (for ingestion of food) and one or more contractile vacuoles (for osmoregulation). All can reproduce asexually, and some species can also reproduce sexually.

1. Flagellates

Flagellates contain one or several flagella, which are long whip-like structures that, internally, have a 9 + 2 arrangement of microtubules. During asexual reproduction, most flagellates divide longitudinally. Flagellated cells can be colorless and, thus, are heterotrophic (e.g., *Heteronema*) or pigmented (shades of green or golden brown). Pigmented, flagellated cells will be dis-cussed later. Illustrations 1 through 4 in figure 6.1 show colorless flagellates.

2. Amoeboid Cells

These organisms are sometimes simply called "amoeba." Most move by formation and extension of transitory pseudopodia, which are cytoplasmic extensions. Movement is generally slow. Amoebae are predators and usually feed on bacteria and protists by engulfing them and forming a food vacuole around the prey. Some amoebae that you might see today have a "test," a hard outer covering (e.g., *Arcella*, *Difflugia*). Illustrations 5 through 8 in figure 6.1 depict amoebae.

3. Ciliates

Ciliates are complex cells that possess two type of nuclei—a diploid micronucleus and a larger polyploid macronucleus. Ciliates are usually covered with many cilia, which are short, hairlike projections that beat in coordinated fashion to propel the cell forwards or backwards. Other than length, cilia are structurally very similar to flagella and have a 9 + 2 arrangement of internal microtubules; both cilia and flagella are covered by the plasma membrane. Some species are attached to a substrate (e.g., *Vorticella*, *Zoothamnium*) and use cilia to create feeding currents around the cytostome (mouth). The cytostome can be quite large for ingestion of food, which is usually bacteria or small protozoa. Ciliates can be colorless (e.g., *Stylonychia*), blue (*Stentor*), pink (*Blepharisma*), or green (*Paramecium bursaria*). The green color is due to the presence of endosymbiotic algal cells. Illustrations 9 through 24 in figure 6.1 depict representative ciliates.

4. Apicomplexa

Nearly all species in this group are parasitic. Most motile forms move by gliding. The genus *Plas-*

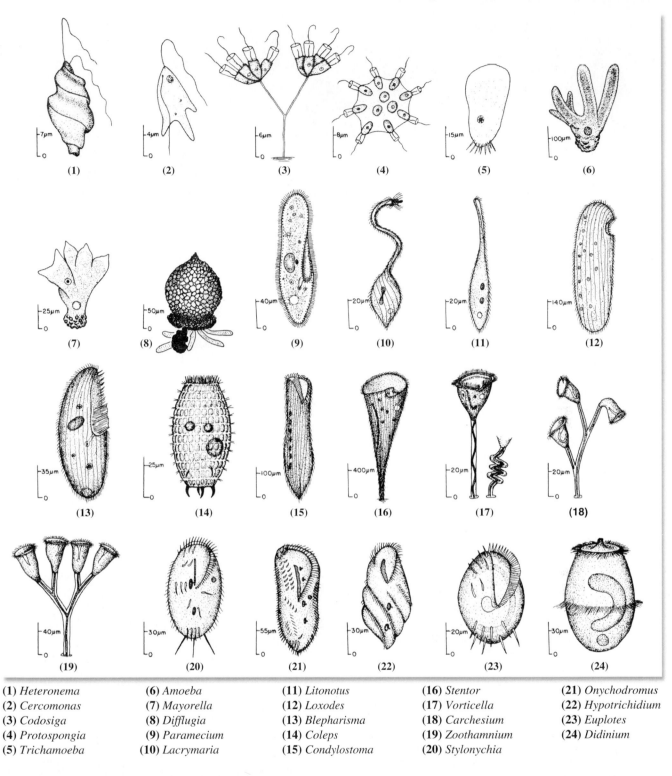

(1) Heteronema   **(6)** Amoeba   **(11)** Litonotus   **(16)** Stentor   **(21)** Onychodromus
(2) Cercomonas   **(7)** Mayorella   **(12)** Loxodes   **(17)** Vorticella   **(22)** Hypotrichidium
(3) Codosiga   **(8)** Difflugia   **(13)** Blepharisma   **(18)** Carchesium   **(23)** Euplotes
(4) Protospongia   **(9)** Paramecium   **(14)** Coleps   **(19)** Zoothamnium   **(24)** Didinium
(5) Trichamoeba   **(10)** Lacrymaria   **(15)** Condylostoma   **(20)** Stylonychia

Figure 6.1 Protozoans.

modium is an extremely important member of this group because several of its species are the causative agents of malaria in humans. Malaria is responsible for more deaths of humans per year than any other disease caused by a eukaryotic parasite. You will not see apicomplexans today. Parasitic, unicellular, eukaryotic cells will not be discussed further in this exercise but realize that they represent thousands of species.

The Algae—Photosynthetic Protists

The majority of eukaryotic algae are placed in the subgroups (a) Chlorophyta/Chloroplastida if they

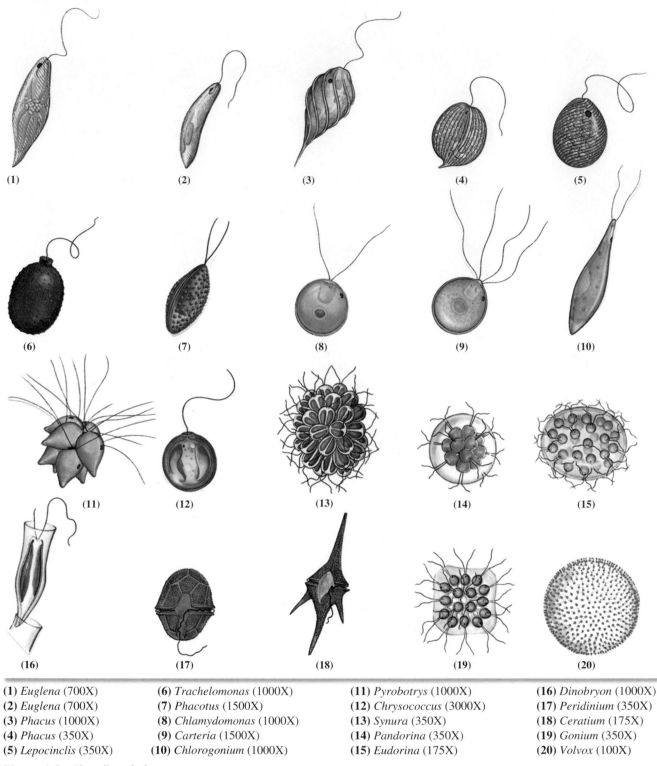

Figure 6.2 Flagellated algae.
(Courtesy of the U.S. Environmental Protection Agency, Office of Research & Development, Cincinnati, OH 45268)

(1) *Euglena* (700X) (6) *Trachelomonas* (1000X) (11) *Pyrobotrys* (1000X) (16) *Dinobryon* (1000X)
(2) *Euglena* (700X) (7) *Phacotus* (1500X) (12) *Chrysococcus* (3000X) (17) *Peridinium* (350X)
(3) *Phacus* (1000X) (8) *Chlamydomonas* (1000X) (13) *Synura* (350X) (18) *Ceratium* (175X)
(4) *Phacus* (350X) (9) *Carteria* (1500X) (14) *Pandorina* (350X) (19) *Gonium* (350X)
(5) *Lepocinclis* (350X) (10) *Chlorogonium* (1000X) (15) *Eudorina* (175X) (20) *Volvox* (100X)

are green; (b) Rhodophyceae if they are red (both of these groups are within the cluster Archaeplastida); (c) Euglenozoa if they are euglenoid (within the cluster Excavata); (d) Alveolata if they are dino-flagellates (within the cluster Chromalveolata); or (e) Straemenopile if they are brown, golden-brown, or a diatom (within the cluster Chromalveolata). Algae can be unicelluar (see top two rows of figure 6.2), colonial (see the four illustrations in the lower-right corner of figure 6.2), or filamentous

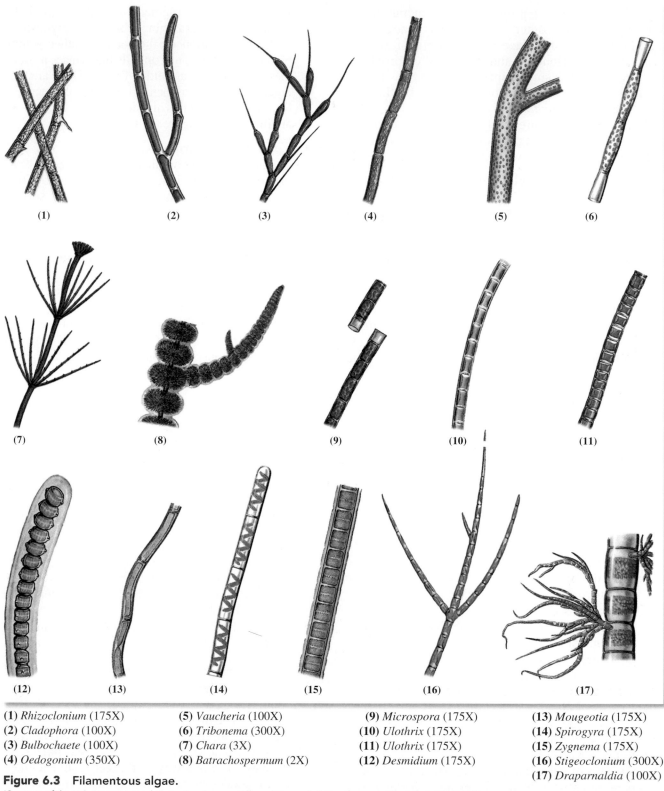

(1)   (2)   (3)   (4)   (5)   (6)

(7)   (8)   (9)   (10)   (11)

(12)   (13)   (14)   (15)   (16)   (17)

(1) *Rhizoclonium* (175X)   **(5)** *Vaucheria* (100X)   **(9)** *Microspora* (175X)   **(13)** *Mougeotia* (175X)
(2) *Cladophora* (100X)   **(6)** *Tribonema* (300X)   **(10)** *Ulothrix* (175X)   **(14)** *Spirogyra* (175X)
(3) *Bulbochaete* (100X)   **(7)** *Chara* (3X)   **(11)** *Ulothrix* (175X)   **(15)** *Zygnema* (175X)
(4) *Oedogonium* (350X)   **(8)** *Batrachospermum* (2X)   **(12)** *Desmidium* (175X)   **(16)** *Stigeoclonium* (300X)
                                                                                    (17) *Draparnaldia* (100X)

Figure 6.3   **Filamentous algae.**
(Courtesy of the U.S. Environmental Protection Agency, Office of Research & Development, Cincinnati, OH 45268)

(figure 6.3). The undifferentiated algal structure is often referred to as a "thallus." It lacks the stem, root, and leaf structures that result from tissue specialization. These organisms are universally present where ample moisture, favorable temperature, and suffi- cient sunlight exist. Although a great many of them live submerged in water, some grow on soil or the surface of snow. Others grow on the bark of trees or the surface of rocks. Algae have distinct, visible nu- clei and chloroplasts. Chloroplasts are organelles that contain thylakoids (parallel arrays of membranes), chlorophyll a, and other pigments. The reactions of photosynthesis take place in these organelles. The size, shape, distribution, and number of chloroplasts vary considerably among species. In some instances, a single chloroplast may occupy most of the cell space (e.g., *Chlamydomonas*). Several groups of algae (e.g., red algae) are rarely encountered in freshwater ponds. Following are descriptions of some of the more familiar freshwater groups.

1. Euglenoids (Euglenozoa/Excavata)

Illustrations 1 through 6 in figure 6.2 show typical euglenoids and represent four genera within this group. These unicells are flagellated (usually two flagella) and have no cell wall. Some have the ability to ingest food and assimilate organic molecules. Photosynthetic species have a photosensitive, red eyespot and chloroplasts that contain chlorophylls a and b and have a secondary origin. Three outer membranes surround the chloroplast. The euglenoid pellicle is located beneath the cell membrane and is made of proteinaceous strips such that some cells are capable of active, eye-catching distortion known as "euglenoid movement." Eulgenoid flagellates store "paramylon," a polysaccharide storage molecule that is unique to them.

2. Green Algae (Chlorophyta/Chloroplastida/Archaeplastida)

The majority of algae in ponds belong to this group. They are grass-green in color and have chlorophylls a and b. They store starch as an energy reserve. The archetype of the group may be *Chlamydomonas* (illustration 8, figure 6.2), a small, green, flagellated unicell that has been extensively studied. Many colonial forms such as *Pandorina, Eudorina, Gonium,* and *Volvox* (illustrations 14, 15, 19, and 20 in figure 6.2) exist. Except for *Vaucheria* and *Tribonema* (Xanthophyceae) and *Batrachospermum* (illustration 8; Rhodophyceae) all of the filamentous forms illustrated in figure 6.3 belong to Chloroplastida. All of the nonfilamentous, nonflagellated algae shown in figure 6.4 also are green algae.

The "desmids" are a unique group of green algae (illustrations 12 in figure 6.3 and 16–20 in figure 6.4). With the exception of a few species, the cells of desmids consist of two similar halves, or semi-cells. The two halves are separated by a constriction, the isthmus.

3. Golden-Brown Algae (Chrysophyceae/Stramenopile/Chromalveolata)

This large diverse division contains more than 6000 species. These organisms store food in the form of oils and leucosin (chrysolaminarin), a storage polysaccharide. Plastid pigments include chlorophylls a and c and fucoxanthin, a brownish pigment. It is the combination of fucoxanthin, other yellow pigments, and the chlorophylls that causes most of these algae to appear golden-brown. These are four outer membranes associated with the chloroplast. Representatives of this division, such as *Dinobryon,* are illustrated in figure 6.2 (illustration 16).

4. Synurales (Chromalveolata)

This small group contains a familiar genus, *Synura* (illustration 13 in figure 6.2). Members of this group contain chlorophylls a and c and several types of xanthin pigments. Silica scales cover the cells, most of which are flagellated.

5. Xanthophyceae (yellow-green algae; Chromalveolata)

Members of this group contain chlorophylls a and c and several types of xanthin pigments but not fucoxanthin. Genera that you might see today include *Tribonema* and *Vaucheria,* both of which are filamentous. These genera are diagrammed in illustrations 5 and 6 in figure 6.3.

6. Phaeophyceae (Stramenopile/Chromalveolata)

Members of this group are commonly called "brown algae." With the exception of three freshwater species, all members of this group live in saltwater environments; thus, it is unlikely that you will encounter brown algae during this lab exercise. These algae have essentially the same pigments seen in the golden-brown algae, but they appear brown because of the masking effect of the greater amount of fucoxanthin. The chloroplasts have four outer membranes. Food storage in the brown algae is in the form of laminarin, a polysaccharide, and mannitol, a sugar alcohol. All species of brown algae are multicellular and sessile. Most seaweeds (e.g., kelp) are brown algae.

7. Bacillariophyta (Stramenopile/Chromalveolata)

Members of this group include the diatoms (figure 6.5). The diatoms are unique in that they have hard cell walls made of tightly integrated silicified elements that are constructed in two halves, or "valves." The two valves fit together like a lid on a box. Skeletons of dead diatoms accumulate on the ocean bottom to form *diatomite,* or "diatomaceous earth," which is commercially available as an excellent polishing compound. It is postulated by some that much of our petroleum reserves may have been formulated by the accumulation of oil from dead diatoms over millions of years. Plastid pigments in diatoms are similar to those in brown

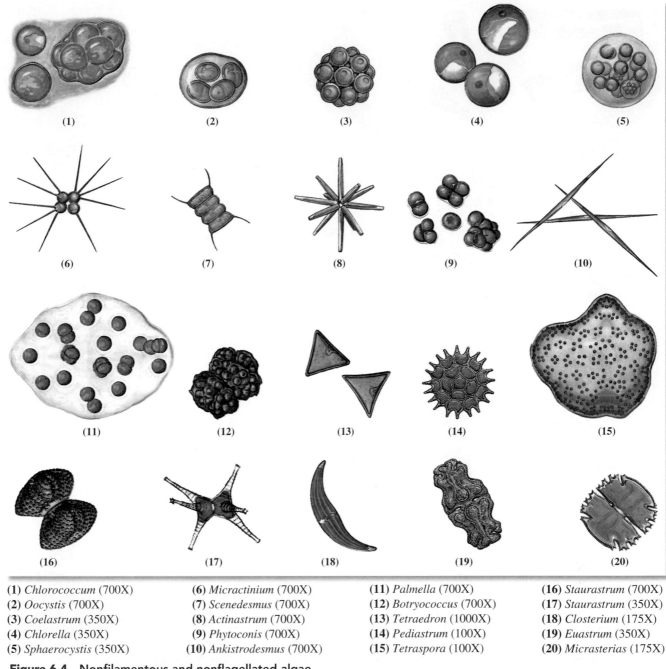

Figure 6.4 Nonfilamentous and nonflagellated algae.

(Courtesy of the U.S. Environmental Protection Agency, Office of Research & Development, Cincinnati, OH 45268)

(1) *Chlorococcum* (700X)　　**(6)** *Micractinium* (700X)　　**(11)** *Palmella* (700X)　　**(16)** *Staurastrum* (700X)
(2) *Oocystis* (700X)　　**(7)** *Scenedesmus* (700X)　　**(12)** *Botryococcus* (700X)　　**(17)** *Staurastrum* (350X)
(3) *Coelastrum* (350X)　　**(8)** *Actinastrum* (700X)　　**(13)** *Tetraedron* (1000X)　　**(18)** *Closterium* (175X)
(4) *Chlorella* (350X)　　**(9)** *Phytoconis* (700X)　　**(14)** *Pediastrum* (100X)　　**(19)** *Euastrum* (350X)
(5) *Sphaerocystis* (350X)　　**(10)** *Ankistrodesmus* (700X)　　**(15)** *Tetraspora* (100X)　　**(20)** *Micrasterias* (175X)

algae (i.e., chlorophylls a and c, fucoxanthin). The chloroplasts have four outer membranes. Chrysolaminarin, a type of laminarin, and sometimes oil are the storage molecules.

8. Dinozoa (Alveolata/Chromalveolata)

Members of this group are commonly called "dinoflagellates" or "fire algae." These unicellular, flagellated protists live in both marine and freshwater environments. You could possibly see *Peridinium* and *Ceratium* or others during this exercise (figure 6.2, illustrations 17 and 18). Most dinoflagellates have interlocking cellulose plates that form a theca. The thecal plates are within alveoil (membranous sacs) that are interior to the plasma membrane; therefore, these cellulose plates differ greatly from a plant's cell wall, which is exterior to the plasma membrane. Dinoflagellates have two flagella: a transverse flagellum sits in a groove and propels the cell and longitudinal flagellum appears to act as a steering device. Many species of marine dinoflagellates are bioluminescent and are easily seen at night in the wake of a moving boat. Some marine dinoflagellates are responsible

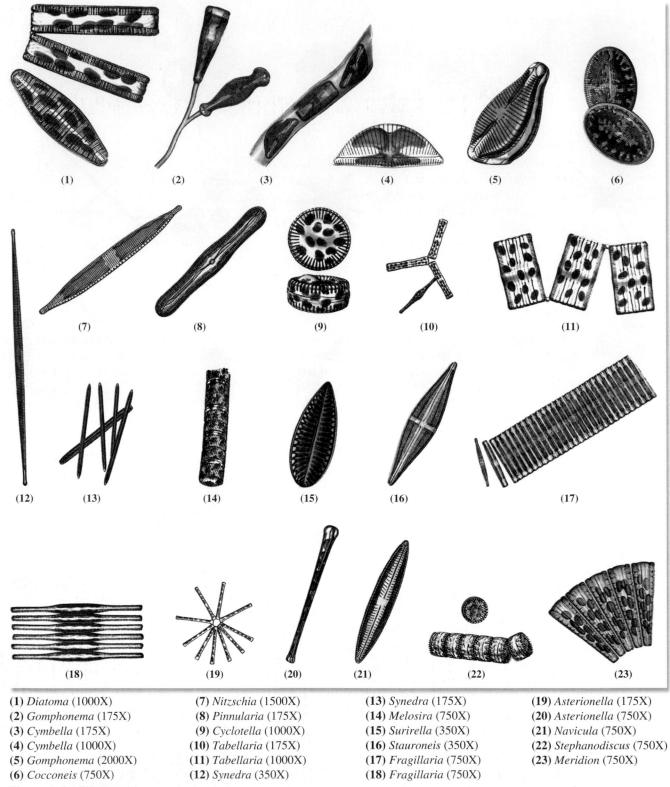

Figure 6.5 Diatoms.
(Courtesy of the U.S. Environmental Protection Agency, Office of Research & Development, Cincinnati, OH 45268)

(1) *Diatoma* (1000X)
(2) *Gomphonema* (175X)
(3) *Cymbella* (175X)
(4) *Cymbella* (1000X)
(5) *Gomphonema* (2000X)
(6) *Cocconeis* (750X)

(7) *Nitzschia* (1500X)
(8) *Pinnularia* (175X)
(9) *Cyclotella* (1000X)
(10) *Tabellaria* (175X)
(11) *Tabellaria* (1000X)
(12) *Synedra* (350X)

(13) *Synedra* (175X)
(14) *Melosira* (750X)
(15) *Surirella* (350X)
(16) *Stauroneis* (350X)
(17) *Fragillaria* (750X)
(18) *Fragillaria* (750X)

(19) *Asterionella* (175X)
(20) *Asterionella* (750X)
(21) *Navicula* (750X)
(22) *Stephanodiscus* (750X)
(23) *Meridion* (750X)

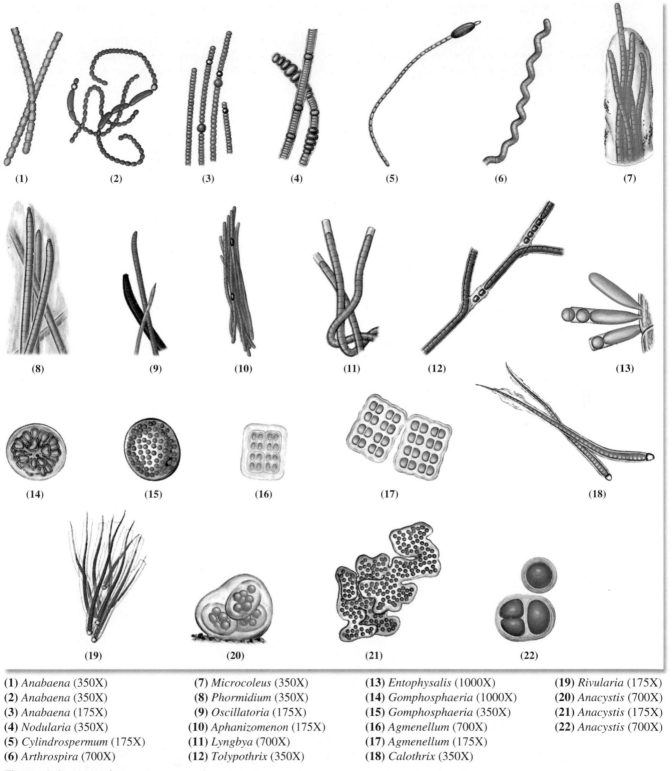

Figure 6.6 Cyanobacteria.

(Courtesy of the U.S. Environmental Protection Agency, Office of Research & Development, Cincinnati, OH 45268)

(1) *Anabaena* (350X)
(2) *Anabaena* (350X)
(3) *Anabaena* (175X)
(4) *Nodularia* (350X)
(5) *Cylindrospermum* (175X)
(6) *Arthrospira* (700X)
(7) *Microcoleus* (350X)
(8) *Phormidium* (350X)
(9) *Oscillatoria* (175X)
(10) *Aphanizomenon* (175X)
(11) *Lyngbya* (700X)
(12) *Tolypothrix* (350X)
(13) *Entophysalis* (1000X)
(14) *Gomphosphaeria* (1000X)
(15) *Gomphosphaeria* (350X)
(16) *Agmenellum* (700X)
(17) *Agmenellum* (175X)
(18) *Calothrix* (350X)
(19) *Rivularia* (175X)
(20) *Anacystis* (700X)
(21) *Anacystis* (175X)
(22) *Anacystis* (700X)

for "red tides" when great numbers of cells form spectacular blooms in the water, usually in response to a sudden influx of nutrients. The color of red tides can vary from intense red to orange to shades of brown depending on the number of cells present per volume of water and what plastid pigments are present. Red tides contain toxins that are produced by the dinoflagellates and stored in shellfish that ingest dinoflagellates and, thus, biologically magnify the toxin. These toxins can kill fish and other organisms, including humans, that eat the shellfish. Some dinoflagellates are heterotrophic; others are photosynthetic and contain chlorophylls a and c, fucoxanthin, peridinin, and several accessory pigments. Their chloroplasts are surrounded by three membranes and are thus very different from those of green plants. Dinoflagellates store a form of starch as an energy reserve. Zooxanthellae are certain species of dinoflagellates that are vital to healthy coral reefs in that they have an intimate endosymbiotic relationship with the coral. Storage molecules in dinoflagellates are diverse (e.g., oil, starch, and fat).

Prokaryotes

The Cyanobacteria

Cyanobacteria were once called "blue-green algae." Cyanobacteria have chlorophyll a and release oxygen during photosynthesis. They do not have an organized nucleus, chloroplasts, or other organelles. They do have peptidoglycan in their cell walls, hence they are prokaryotes. They differ from the green sulfur and the purple sulfur bacteria in that the latter use bacteriochlorophyll for photosynthesis and do not evolve oxygen because their photosynthetic metabolism is carried out under anaerobic conditions.

Over a thousand species of cyanobacteria have been described. Some representatives are depicted in figure 6.6. They are present in almost all moist environments from the tropics to the poles, in both marine and freshwater habitats. Cyanobacteria can be unicellular, colonial, or filamentous. Figure 6.6 illustrates only a random few that are frequently seen.

The designation of these bacteria as "blue-green" is somewhat misleading in that many are black, purple, red, and various shades of green, including blue-green. The varying colors are due to the varying proportions of photosynthetic pigments present, which include chlorophyll a, carotenoids, and phycobiliproteins. The latter pigments consist of allophycocyanin, phycocyanin, and phycoerythrin, that combine with protein molecules to form phycobilisomes, which serve as light harvesters for photosystems. These structures are also found in red algae. Because chloroplasts are not present, cyanobacteria use thylakoids as their photosynthetic apparatus. The phycobilisomes are attached to the thylakoids in parallel arrays.

Laboratory Report

Student: _____

Date: _____ Section: _____

6 Protozoa, Algae, and Cyanobacteria

A. Results

In this study of freshwater microorganisms, record your observations in the following tables. The number of organisms to be identified will depend on the availability of time and materials. Your instructor will indicate the number of each type that should be recorded.

Record the genus of each identifiable organism. Also, indicate the group to which the organism belongs. Microorganisms that you cannot identify should be sketched in the space provided. It is not necessary to draw those that are identified.

1.   **Protozoans**

GENUS	GROUP	BOTTLE NO.	SKETCHES OF UNIDENTIFIED

2.   **Algae**

GENUS	GROUP	BOTTLE NO.	SKETCHES OF UNIDENTIFIED

3.   **Cyanobacteria**

GENUS	BOTTLE NO.	SKETCHES OF UNIDENTIFIED

B. Short Answer Questions

1. In which domains are algae, protozoa, and cyanobacteria classified?

   _____

   _____

2. (a) Name one similarity between algae and plants. (b) Name one difference.

   _____

   _____

3. Compare and contrast the three mechanisms of motility displayed by protozoa.

   _____

   _____

4. (a) What organisms were formerly known as "blue-green algae"? (b) Why are these organisms not algae?

   _____

   _____

5. What makes "red tides" red?

   _____

   _____

6. What is the genus of the causative agent of malaria? In what group does it belong?

   _____

   _____

C. Fill-in-the-Blanks Questions

1. For each type of organism, place a check mark in the box to indicate whether the cellular characteristic or function is present.

CHARACTERISTIC OR FUNCTION	PROTOZOA	ALGAE	CYANOBACTERIA
Nucleus			
Flagella			
Pseudopodia			
Cilia			
Photosynthetic pigment(s)			
Chloroplasts			
Cell wall			

2. For each classification of protozoa, place a check mark in the box to indicate whether the cellular characteristic or function is present.

CHARACTERISTIC OR FUNCTION	AMOEBOID CELLS	FLAGELLATES	CILIATES	DIATOMS
Flagella				
Cilia				
Pseudopodia				
Gliding				
All members are parasitic				

3. For each group name, place a check mark in the box to indicate whether the cellular characteristic or function is present.

	CHLORO-PLASTIDA	CHRYSO-PHYCEAE	EUGLENO-ZOA	PHAEO-PHYCEAE	BACCILARIO-PHYTA	DINO-FLAGELLATES (ALVEOLATA)	CYANO-BACTERIA

Pigments

Chlorophyll a							
Chlorophyll b							
Chlorophyll c							
Fucoxanthin							
c-phycocyanin							
c-phycoerythrin							

Cell Features

Flagella							
Cell wall							
Chloroplasts							

Food Storage

Starch							
Paramylon							
Leucosin							
Laminarin							
Oils							
Mannitol							

Bacteria are the most widely distributed organisms in the biosphere. They occur as part of the normal flora of humans and animals, they are disease-causing parasites on many organisms, they abound in the soil and in water systems, and they are found deep within the earth's crust. We depend on them to degrade our sewage and solid wastes in landfills and to drive geochemical cycles involving nitrogen, carbon, and sulfur. Even though they are small in size and simple in their cell structure, they are considerably diverse in their metabolism and activities.

They are defined primarily by their cellular structure and small size. Bacteria have a simple cell structure that lacks a defined nucleus surrounded by nuclear membrane. Their nuclear genetic material is primarily supercoiled, circular DNA molecules that reside in the cell cytoplasm. Bacteria lack cellular organelles such as mitochondria and chloroplasts, but through modifications in their cell membrane, they carry out respiration and photosynthesis. Their ribosomes for synthesizing proteins are structurally different than higher cells and they are inhibited by many broad-spectrum antibiotics. For most bacteria, the cell is surrounded by a cell wall composed of a unique molecule called **peptidoglycan.** It is not found in any other kind of cell in the biological world. Most bacteria are small in size, averaging from 0.5 microns to 10 microns. Cyanobacteria tend to be larger, and some recently discovered bacteria are almost large enough to be seen without the aid of a microscope.

Figure 7.1 illustrates most of the common shapes of bacteria. They can be grouped into three morphological types: **rods** (bacilli); **cocci** (spherical); and **spirals** or curved rods. The rods or bacilli can have rounded, flat, or tapered ends and they can be motile or nonmotile. Cocci may occur singly, in chains, in a tetrad (packet of four cells), or in irregular masses. Most cocci are nonmotile because they lack flagella, the organelles of motility. The spiral bacteria can exist as slender spirochaetes, as a spirillum, or as a comma-shaped curved rod (vibrio). The spirochaetes are motile, using **axial filaments,** which are a type of flagella that originate from both ends of the cell and wrap around the cell body. Rotation of the axial filaments causes the spirochaete cell to move in a cork screwlike motion.

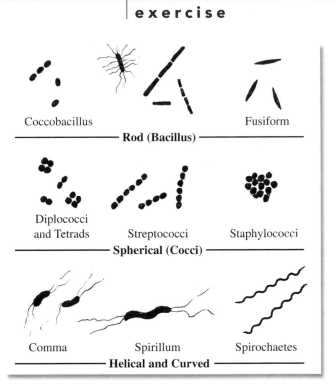

Figure 7.1   Bacterial morphology.

During this laboratory period, you will be provided with three kinds of sterile bacteriological media that you will expose to the environment in various ways. To ensure that these exposures cover as wide a spectrum as possible, specific assignments will be made for each student. In some instances, a moistened swab will be used to remove bacteria from some object; in other instances a petri plate of medium will be exposed to the air or a cough. You will be issued a number that will enable you to determine your specific assignment from the table on page 56.

Materials

per student:
- 1 tube of nutrient broth
- 1 petri plate of trypticase soy agar (TSA)
- 1 sterile cotton swab
- Sharpie marking pen

per two or more students:
- 1 petri plate of blood agar

1. Scrub down your desktop with a disinfectant (see Exercise 9, Aseptic Technique).

2. Expose your TSA plate according to your assignment in the table below. *Label the bottom* of your plate with your initials, your assignment number, and the date.

3. Moisten a sterile swab by immersing it into a tube of nutrient broth and expressing most of the broth out of it by pressing the swab against the inside wall of the tube.

4. Rub the moistened swab over a part of your body such as a finger or ear, or some object such as a doorknob or telephone mouthpiece, and return the swab to the tube of broth. It may be necessary to break off the stick end of the swab so that you can replace the cap on the tube.

5. Label the tube with your initials and the source of the bacteria.

6. Expose the blood agar plate by coughing onto it. Label the bottom of the plate with the initials of the individuals that cough onto it. Be sure to date the plate also.

7. Incubate the plates and tube at 37° C for 48 hours.

Evaluation

After 48 hours incubation, examine the tube of nutrient broth and two plates. Shake the tube vigorously without wetting the cap. Is it cloudy or clear? Compare it with an uncontaminated tube of broth. What is the significance of cloudiness? Do you see any colonies growing on the blood agar plate? Are the colonies all the same size and color? If not, what does this indicate? Group together a set of TSA plates representing all nine types of exposure. Record your results on the Laboratory Report.

Your instructor will indicate whether these tubes and plates are to be used for making slides in Exercise 12 (Simple Staining). If the plates and tubes are to be saved, containers will be provided for their storage in the refrigerator. Place the plates and tubes in the designated containers.

Laboratory Report

Record your results in Laboratory Report 7.

Table 7.1

EXPOSURE METHOD FOR TSA PLATE	STUDENT NUMBER
1. To the air in laboratory for 30 minutes	1, 10, 19, 28
2. To the air in room other than laboratory for 30 minutes	2, 11, 20, 29
3. To the air outside of building for 30 minutes	3, 12, 21, 30
4. Blow dust onto exposed medium	4, 13, 22, 31
5. Moist lips pressed against medium	5, 14, 23, 32
6. Fingertips pressed lightly on medium	6, 15, 24, 33
7. Several coins pressed temporarily on medium	7, 16, 25, 34
8. Hair combed over exposed medium (10 strokes)	8, 17, 26, 35
9. Optional: Any method not listed above	9, 18, 27, 36

Laboratory Report

Student: _____

Date: _____ Section: _____

7 The Bacteria

A. Results

1. After examining your TSA and blood agar plates, record your results in the following table and on a similar table that your instructor has drawn on the chalkboard. With respect to the plates, we are concerned with a quantitative evaluation of the degree of contamination and differentiation as to whether the organisms are bacteria or molds. Quantify your recording as follows:

   0 no growth
   + 1 to 10 colonies   + + + 51 to 100 colonies
   + + 11 to 50 colonies   + + + + over 100 colonies

   After shaking the tube of broth to disperse the organisms, look for cloudiness (turbidity). If the broth is clear, no bacterial growth occurred. Record no growth as 0. If tube is turbid, record + in the last column.

STUDENT INITIALS	PLATE EXPOSURE METHOD		COLONY COUNTS		BROTH	
	TSA	Blood Agar	Bacteria	Mold	Source	Result

Use the class results to answer the following questions.

2. Using the number of colonies as an indicator, which habitat sampled by the class appears to contain more bacteria? Why? _____

    _____

3. Why do you suppose this habitat contains such a high microbial count? _____

    _____

4. a. Were any plates completely lacking in colonies? _____

    b. Do you think that the habitat sampled was really sterile? _____

    c. If your answer to *b* is *no,* then how can you account for the lack of growth on the plate?

    _____

    d. If your answer to *b* is *yes,* defend it:

    _____

    _____

B. Short Answer Questions

1. In what ways do the macroscopic features of bacterial colonies differ from that of molds?

    _____

    _____

2. Why is the level of contamination measured as number of colonies rather than size of colonies?

    _____

    _____

3. Should one be concerned to find bacteria on the skin? How about molds? Explain.

    _____

    _____

4. How can microbial levels be controlled on the skin? On surfaces in the environment? In the air?

    _____

    _____

5. Compare the following features of bacteria to those of eukaryotic microorganisms:

    a. size.

    _____

    b. organization of genetic material.

    _____

    c. ribosomes.

    _____

      d. cell wall.

      _____

      e. respiration and photosynthesis.

      _____

      f. motility mechanisms.

      _____

The Fungi:
Molds and Yeasts

The fungi comprise large groups of **heterotrophic** (non-photosynthesizing) **eukaryotic** organisms that produce exoenzymes and absorb their nutrients (**osmotrophic**). Fungi may be saprophytic or parasitic. They may be unicellular or filamentous. The distinguishing characteristics of the group as a whole is that they are (1) eukaryotic, (2) heterotrophic, (3) lack tissue differentiation, (4) have cell walls of chitin or other polysaccharides, and (5) propagate by spores (sexual and/or asexual).

Many diverse organisms have been traditionally categorized as fungi, including water molds, mushrooms, puffballs, bracket fungi, yeasts and molds. These organisms do not have a uniform genetic background, and are believed to have evolved from at least two ancestral lineages. Classification of fungi is a complex and dynamic process. Traditional classification schemes rely primarily on morphological characteristics and reproductive mechanisms to categorize fungal groups. More modern schemes use genetic analysis to determine relatedness between species and subsequent categorization. Recently collected data from genetic analyses of fungi indicate that fungal classification based on morphology does not necessarily reflect evolutionary relationships between organisms. However, identification and categorization of fungal types are still most easily performed in the introductory laboratory using morphological characteristics. In this exercise, we will examine prepared stained slides and slides made from living cultures of yeasts and molds. Molds are normally present in the air and can be cultured and studied macroscopically and microscopically. In addition, an attempt will be made to identify the various fungal types that are cultured based on morphological characteristics.

Before attempting to identify the various molds, familiarize yourself with the basic differences between molds and yeasts. Note in figure 8.1 that yeasts are essentially unicellular and molds are multicellular.

Fungi

Fungal forms: molds and yeasts

Molds   Molds are fungi that contain microscopic intertwining filaments called *hyphae* (*hypha*, singular). A mass of hyphae forms the **mycelium** that is

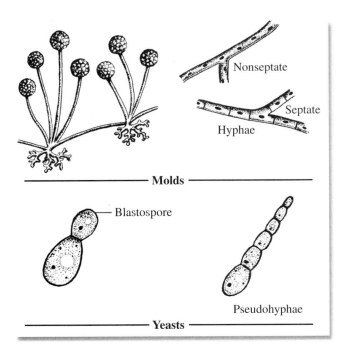

Figure 8.1   Structural differences between molds and yeasts.

seen macroscopically as the fungal colony. As shown in figure 8.1, **septa** or crosswalls separate the hyphae of some fungi into individual compartments with an organized nucleus and organelles. In most fungi, the septa have central openings or pores that allow for the streaming of cytoplasm between compartments. If no crosswalls are present, the hyphae are said to be nonseptate or **coenocytic.** A mass of intermeshed hyphae, as seen macroscopically, is a *mycelium.*

Yeasts   Yeasts are fungi that lack hyphae. These fungi multiply by budding or fission or a combination of these two processes. As shown in figure 8.1, yeast cells may form chains of buds called **pseudohyphae.**

Some species of fungi occur exclusively as molds, while other species occur exclusively as yeasts. A number of fungal species, known as **dimorphic** fungi, can occur as either molds or yeasts, depending on environmental conditions.

Fungal spores: asexual and sexual

Fungal spores are important in the reproductive cycle and can be produced through either asexual or sexual

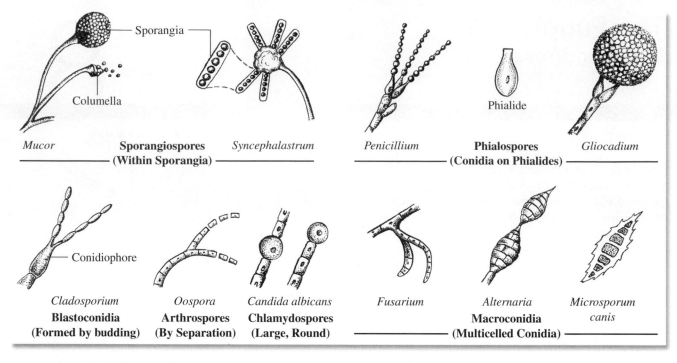

Figure 8.2 Types of asexual spores seen in fungi.

means. There is great variation in spore size, color, and appearance between species. Much of the traditional taxonomy of fungi is based on morphological characteristics of spores and spore-related structures. In this exercise, you will attempt to identify fungi based, in part, on these characteristics.

Asexual spores   Fungi produce asexual spores by mitotic division of a parental cell. Asexual spore formation does not involve the union of cell nuclei. Two types of asexual spores are sporangiospores and conidia. Both types are shown in figure 8.2.

  Sporangiospores: Sporangiospores are asexual spores that form *within* a thin-walled sac called a sporangium. Sporangiospores can be motile or non-motile.

  Conidia: Conidia are asexual non-motile spores that form on specialized hyphae called conidiophores. Several different categories of conidia exist. They include Phialospores, Blastospores, Arthrospores, and Chlamydospores.

  Phialospores: are asexual reproductive spores produced from a vase or flask-shaped cell called a phialide. *Penicillium* and *Gliocadium* produce this type.

  Blastospores: are asexual reproductive spores produced by budding from cells in yeast and some filamentous fungi (molds). *Cladosporium* and *Candida* are examples

of genera that reproduce asexually by blastospores.

  Arthrospores: are asexual reproductive spores produced by fragmentation of preexisting hyphae. *Geotrichum* and *Galactomyces* produce this type of spore.

  Chlamydospores: are thick-walled, round or irregular asexual spores whose function is survival. These are common to most fungi.

Sexual spores   Fungi produce sexual spores by the union of two parental nuclei followed by meiotic division. Three kinds of sexual spores are seen in fungi. These are the zygospores, ascospores, and basidiospores. Figure 8.3 illustrates these three types.

  Zygospores: are formed by the union of nuclear material from the hyphae of two different gametangia (+, −) which appear morpholically identical but are genetically different.

  Ascospores: are haploid sexual spores formed in the interior of an oval or elongated enclosure usually termed an ascus.

  Basidiospores: are sexual haploid spores produced externally on a club-shaped basidium.

Yeasts

Hyphae   Unicellular fungi multiply by budding or fission or by a combination of the two processes (figure 8.1).

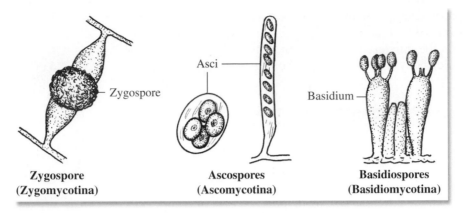

Figure 8.3 Types of sexual spores seen in the Amastigomycota.

Subdivision of the Kingdom Eumycota

The Kingdom Eumycota is made up of three phyla: Chytridiomycota, Zygomycota, and the Dikaryomycota.

The **Chytridiomycota** are aquatic organisms that vary morphologically and may be unicellular or filamentous. Asexual reproduction generates zoospores with one whiplash flagellum.

The **Zygomycota** prefer an atmosphere with high relative humidity and produce coencytic hyphae that are wide and thin-walled. They are coenocytic, having no crosswalls. This phylum reproduces sexually by fusion of gametangia, forming a zygospore, and asexually by non-motile sporangiospores.

The **Dikaryomycota** contain the most species diversity and are divided into the subphyla **Ascomycotina** and **Basidiomycotina.** The Dikaryomycota contain chitin in their cell walls, have hyphae with septa, and their hyphae fuse and exchange nuclei **(anastomosis).** In a phase of their life cycle, the hyphae can contain two haploid nuclei in each cell **(dikaryon).**

Ascomycotina produce ascospores in an ascus and they are commonly referred to as the "sac fungi," or ascomycetes. Ascomycotina produce ascospores in one of the following structures:

Apothecia: A cup or dish-shaped fruiting body that opens widely to expose the asci-containing ascospores.
Perithecia: Flask-shaped fruiting structures containing asci with ascospores.
Cleithothecia: Enclosed ascocarps that lack a pore-containing asci with ascospores.

Basidiomycotina produce basidiospores 1, 2, or 4 externally on a basidium. These are the puffballs, mushrooms, smuts, rusts, and shelf fungi.

In zygomycetes, the anamorph and telemorph stages often occur at the same time. However, in ascomycetes and basidiomycetes these stages do not occur simultaneously but appear at different times on different substrates. The International Code of Botanical Nomenclature thus stipulates that a different genus-species name for each stage is an accepted practice. However, as the anamorph and telemorph stages are identified as the same fungus, the organism is referred to by the telemorph or sexual stage binomial name. The anamorph stage fungi are often classed as the imperfect fungi, or deuteromycetes. There are more than 15,000 species in this group.

Laboratory Procedures

Several options are provided here for the study of fungi. The procedures to be followed will be outlined by your instructor.

Yeast Study

The organism *Saccharomyces cerevisiae,* which is used in bread making and alcohol fermentation, will be used for this study. Either prepared slides or living organisms may be used.

Materials

- prepared slides of *Saccharomyces cerevisiae*
- broth cultures of *Saccharomyces cerevisiae*
- methylene blue stain
- microscope slides and coverslips

Prepared Slides   If prepared slides are used, they may be examined under high-dry or oil immersion. Look for typical **blastospores** and **ascospores.** Space is provided on the Laboratory Report for drawing the organisms.

Living Materials   If broth cultures of *Saccharomyces cerevisiae* are available, they should be examined on a wet mount slide with phase-contrast or brightfield optics. Place two or three loopfuls of the organisms on the slide with a drop of methylene blue stain. Oil immersion will reveal the greatest amount of detail. Look for the **nucleus** and **vacuole.** The nucleus is the smaller body. Draw a few cells on the Laboratory Report.

Fungi Study

Examine a petri plate of Sabouraud's agar that has been exposed to the air for about an hour and incubated at room temperature for 3–5 days. This medium has a low pH, which makes it selective for fungi. A good plate will have many different-colored colonies. Note the characteristic "cottony" nature of the colonies. Also, look at the bottom of the plate and observe how the colonies differ in color here. Colony surface color, underside pigmentation, hyphal structure, and the type of spores produced are important phenotypic characteristics used in the identification of fungi.

Figure 8.4 reveals how some common molds appear when grown on Sabouraud's agar. Keep in mind that the appearance of a fungal colony can change appreciably as it gets older. The photographs in figure 8.4 are of colonies that are 10 to 21 days old.

Conclusive identification cannot be made unless a microscope slide is made to determine the type of hyphae and spores that are present. Figure 8.5 reveals, diagrammatically, the microscopic differences to look for when identifying fungal genera.

Two Options In making slides from fungal colonies, one can make either wet mounts directly from the colonies by the procedure outlined here or make cultured slides as outlined in Exercise 24. The following steps should be used for making stained slides directly from the colonies. Your instructor will indicate the number of identifications that are to be made.

Materials

- fungal cultures on Sabouraud's agar
- microscope slides and coverslips
- lactophenol cotton blue stain
- sharp-pointed scalpels or dissecting needles

1. Place an uncovered plate on a dissecting microscope and examine the colony. Look for hyphal structures and spore arrangement. Increase the magnification if necessary to more clearly see spores. Ignore white colonies as they are usually young and have not begun the sporulation process.

2. Consult figures 8.4 and 8.5 to make a preliminary identification based on colony characteris-

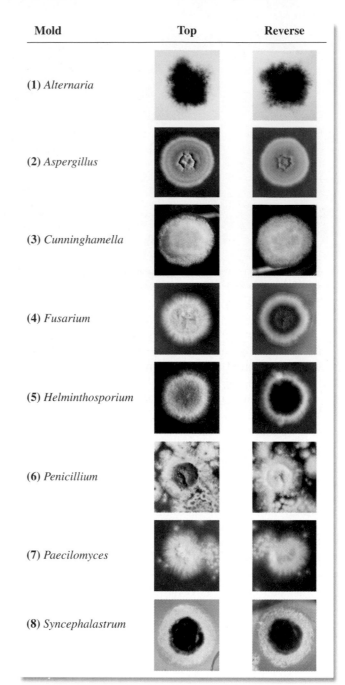

Mold	Top	Reverse
(1) *Alternaria*		
(2) *Aspergillus*		
(3) *Cunninghamella*		
(4) *Fusarium*		
(5) *Helminthosporium*		
(6) *Penicillium*		
(7) *Paecilomyces*		
(8) *Syncephalastrum*		

Figure 8.4 Colony characteristics of some of the more common molds.

tics and low-power magnification of hyphae and spores.

3. Make a wet mount slide by transferring a small amount of mycelium with a scalpel, dissecting needle, or toothpick to a drop of lactophenol cotton blue. Gently tease apart the mycelium with the dissecting needles. Cover the specimen with a coverslip and examine with the low-power objective. Look for hyphae that have spore structures.

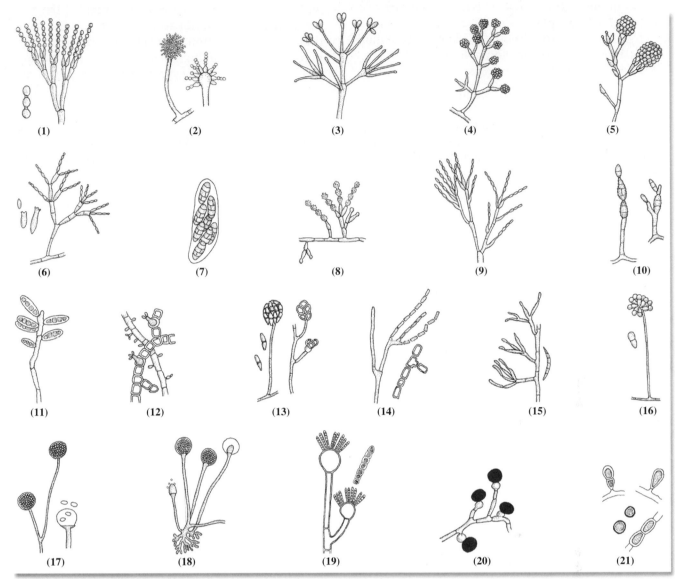

(1) (2) (3) (4) (5)

(6) (7) (8) (9) (10)

(11) (12) (13) (14) (15) (16)

(17) (18) (19) (20) (21)

(1) *Penicillium*– bluish-green; brush arrangement of phialospores.

(2) *Aspergillus*– bluish-green with sulfur-yellow areas on the surface. *Aspergillus niger* is black.

(3) *Verticillium*– pinkish-brown, elliptical microconidia.

(4) *Trichoderma*– green, resemble *Penicillium* macroscopically.

(5) *Gliocadium*– dark-green; conidia (phialospores) borne on phialides, similar to *Penicillium*; grows faster than *Penicillium*.

(6) *Cladosporium (Hormodendrum)*– light green to grayish surface; gray to black back surface; blastoconidia.

(7) *Pleospora*– tan to green surface with brown to black back; ascospores shown are produced in sacs borne within brown, flask-shaped fruiting bodies called pseudothecia.

(8) *Scopulariopsis*– light-brown; rough-walled microconidia.

(9) *Paecilomyces*– yellowish-brown, elliptical microconidia.

(10) *Alternaria*– dark greenish-black surface with gray periphery; black on reverse side; chains of macroconidia.

(11) *Bipolaris*– black surface with grayish periphery; macroconidia shown.

(12) *Pullularia*– black, shiny, leathery surface; thick-walled; budding spores.

(13) *Diplosporium*– buff-colored wooly surface; reverse side has red center surrounded by brown.

(14) *Oospora (Geotrichum)*– buff-colored surface; hyphae break up into thin-walled rectangular arthrospores.

(15) *Fusarium*– variants, of yellow, orange, red, and purple colonies; sickle-shaped macroconidia.

(16) *Trichothecium*– white to pink surface; two-celled conidia.

(17) *Mucor*– a zygomycete; sporangia with a slimy texture; spores with dark pigment.

(18) *Rhizopus*– a zygomycete; spores with dark pigment.

(19) *Syncephalastrum*– a zygomycete; sporangiophores bear rod-shaped sporangioles, each containing a row of spherical spores.

(20) *Nigrospora*– conidia black, globose, one-celled, borne on a flattened, colorless vesicle at the end of a conidiophore.

(21) *Montospora*– dark gray center with light gray periphery; yellow-brown conidia.

Figure 8.5 Microscopic appearance of some of the more common molds.

Go to the high-dry objective to discern more detail about the spores. Compare your specimen to figure 8.5 and see if you can identify the culture based on microscopic morphology.

4. Repeat the procedure for each colony.

Note: An alternative procedure that preserves the fruiting structures is the **cellophane tape method.** Place 1–2 drops of lactophenol cotton blue on a microscope slide. Using a piece of *clear* cellophane tape slightly smaller than the length of the slide, gently touch the surface of a fungal colony with the sticky side of the tape. Transfer the tape containing the material from the fungal colony to the lactophenol cotton blue stain and press the tape onto the slide, making sure that the culture material is in the stain. Observe with the low-power and high-dry lens.

Laboratory Report

After recording your results on the Laboratory Report, answer all the questions.

Laboratory Report

Student: _____

Date: _____ Section: _____

8   The Fungi: Molds and Yeasts

A. Results

1. **Yeast Study**

    Draw a few representative cells of *Saccharomyces cerevisiae* in the appropriate circles below. Blastospores (buds) and ascospores, if seen, should be shown and labeled.

Prepared Slide	**Living Cells**

2. **Mold Study**

    In the following table, list the genera of molds identified in this exercise. Under colony description, give the approximate diameter of the colony, its topside color and backside (bottom) color. For microscopic appearance, make a sketch of the organism as it appears on slide preparation.

GENUS	COLONY DESCRIPTION	MICROSCOPIC APPEARANCE (DRAWING)

B. Short Answer Questions

1. What are the five distinguishing characteristics of fungi?

   _____

   _____

2. Describe how modern fungal classification schemes differ from more traditional ones. Why are we using traditional methods to characterize fungi in this laboratory exercise?

   _____

   _____

3. Name important phenotypic characteristics used in the identification of fungi.

   _____

   _____

4. Differentiate between molds and yeasts. What term is given to fungal species that have both a mold and a yeast phase?

   _____

   _____

5. Differentiate between hyphae and mycelia.

   _____

   _____

6. What characteristic determines that fungal hyphae are coenocytic?

   _____

   _____

7. What are the types of asexual spores of fungi?

   _____

   _____

8. What are the types of sexual spores of fungi?

   _____

   _____

9. Describe how sporangiospores differ from conidia.

   _____

   _____

10. If you have previously examined bacteria in the laboratory, use your experience from the current exercise to compare the morphological characteristices of bacteria and fungi.

    _____

    _____

Manipulation of Microorganisms

One of the most critical techniques that any beginning student in microbiology must learn is aseptic technique. This technique insures that an aseptic environment is maintained when handling microorganisms. This means two things:

1. no contaminating microorganisms are introduced into cultures or culture materials and
2. the microbiologist is not contaminated by cultures that are being manipulated.

Aseptic technique is crucial in characterizing an unknown organism. Oftentimes, multiple transfers must be made from a stock culture to various test media. It is imperative that only the desired organism is transferred each time and that no foreign bacteria are introduced during the transfer. Aseptic technique is also obligatory in isolating and purifying bacteria from a mixed source of organisms. The streak-plate and pour-plate techniques provide a means to isolate an individual species. And once an organism is in pure culture and stored as a stock culture, aseptic technique insures that the culture remains pure when it is necessary to retrieve the organism.

Individuals who work with pathogenic bacteria must be sure that any pathogen that is being handled is not accidently released causing harm to themselves or to coworkers. Failure to observe aseptic technique can obviously pose a serious threat to many.

In the following exercises, you will learn the techniques that allow you to handle and manipulate cultures of microorganisms. Once you have mastered these procedures, you will be able to make transfers of microorganisms from one kind of medium to another with confidence. You will also be able to isolate an organism from a mixed culture to obtain a pure isolate. It is imperative that you have a good grasp of these procedures, as they will be required over and over in the exercises in this manual.

Aseptic Technique

The use of aseptic technique insures that no contaminating organisms are introduced into culture materials when the latter are inoculated or handled in some manner. It also insures that organisms that are being handled do not contaminate the handler or others who may be present. And its use means that no contamination remains after you have worked with cultures.

As you work with these procedures, with time, they will become routine and second nature to you. You will automatically know that a set of procedures outlined below will be used when dealing with cultures of microorganisms. This may involve the transfer of a broth culture to a plate for streaking, or inoculating an isolated colony from a plate onto a slant culture to prepare a stock culture. It may also involve inoculating many tubes of media and agar plates from a stock culture in order to characterize and identify an unknown bacterium. Ensuring that only the desired organism is transferred in each inoculation is of paramount importance in the identification process. The general procedure for aseptic technique follows.

Work Area Disinfection The work area is first treated with a disinfectant to kill any microorganisms that may be present. This process destroys vegetative cells and viruses but may not destroy endospores.

Loops and Needles The transfer of cultures will be achieved using inoculating loops and needles. These implements must be sterilized before transferring any culture. A loop or needle is sterilized by inserting it into a Bunsen burner flame until it is red-hot. This will incinerate any contaminating organisms that may be present. Allow the loop to cool completely before picking up inoculum. This will ensure that viable cells are transferred.

Culture Tube Flaming and Inoculation Prior to inserting a cooled loop or needle into a culture tube, the cap is removed and the mouth of the tube is flamed. If the tube is a broth tube, the loop is inserted into the tube and twisted several times to ensure that the organisms on the loop are delivered to the liquid. If the tube is an agar slant, the surface of the slant is inoculated by drawing the loop up the surface of the slant from the bottom of the slant to its top. For stab

cultures, a needle is inserted into the agar medium by stabbing it into the agar. After the culture is inoculated, the mouth of the tube is reflamed and the tube is recapped.

Final Flaming of the Loop or Needle After the inoculation is complete, the loop or needle is flamed in the Bunsen burner to destroy any organisms that remain on these implements. The loop or needle is then returned to its receptacle for storage. It should never be placed on the desk surface.

Petri Plate Inoculations Loops are used to inoculate or streak petri plates. The plate cover is raised and held diagonally over the plate to protect the surface from any contamination in the air. The loop containing the inoculum is then streaked gently over the surface of the agar. It is important not to gouge or disturb the surface of the agar with the loop. The cover is replaced and the loop is flamed in a Bunsen burner.

Final Disinfection of the Work Area When all work for the day is complete, the work area is treated with disinfectant to insure that any organism that might have been deposited during any of the procedures is killed.

To gain some practice in aseptic transfer of bacterial cultures, three simple transfers will be performed in this exercise:
1. broth culture to broth tube
2. agar slant culture to an agar slant and
3. agar plate to an agar slant.

Transfer from Broth Culture to Another Broth

Do a broth tube to broth tube inoculation using the following technique. Figure 9.1 illustrates the procedure for removing organisms from a culture, and figure 9.2 shows how to inoculate a tube of sterile broth.

Materials
- broth culture of *Escherichia coli*
- tubes of sterile nutrient broth
- inoculating loop

- Bunsen burner
- disinfectant for desktop and sponge
- Sharpie marking pen

1. Prepare your desktop by swabbing down its surface with a disinfectant. Use a sponge or paper towels.
2. With a marking pen, label a tube of sterile nutrient broth with your initials and *E. coli.*
3. Sterilize your inoculating loop by holding it over the flame of a Bunsen burner **until it becomes bright red.** The entire wire must be heated. See illustration 1, figure 9.1.
4. Using your free hand, gently shake the tube to disperse the culture (illustration 2, figure 9.1).
5. Grasp the tube cap with the little finger of your hand holding the inoculating loop and remove it from the tube. Flame the mouth of the tube as shown in illustration 3, figure 9.1.
6. Insert the inoculating loop into the culture (illustration 4, figure 9.1).

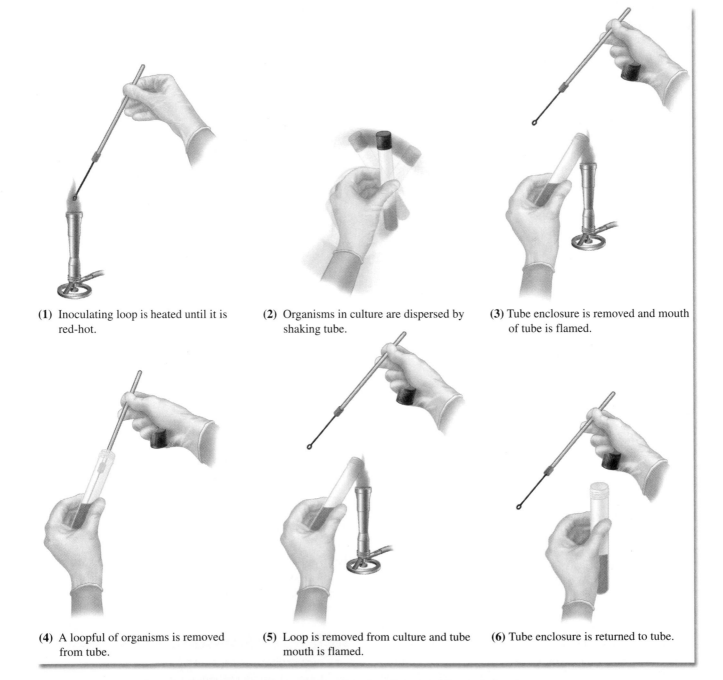

(1) Inoculating loop is heated until it is red-hot.

(2) Organisms in culture are dispersed by shaking tube.

(3) Tube enclosure is removed and mouth of tube is flamed.

(4) A loopful of organisms is removed from tube.

(5) Loop is removed from culture and tube mouth is flamed.

(6) Tube enclosure is returned to tube.

Figure 9.1 **Procedure for removing organisms from a broth culture with inoculating loop.**

7. Remove the loop containing the culture, flame the mouth of the tube again (illustration 5, figure 9.1), and recap the tube (illustration 6). Place the culture tube back on the test-tube rack.

8. Grasp a tube of sterile nutrient broth with your free hand, carefully remove the cap with your little finger, and flame the mouth of this tube (illustration 1, figure 9.2).

9. Without flaming the loop, insert it into the sterile broth, inoculating it (illustration 2, figure 9.2). To disperse the organisms into the medium, move the loop back and forth in the tube.

10. Remove the loop from the tube and flame the mouth (illustration 3, figure 9.2). Replace the cap on the tube (illustration 4, figure 9.2).

11. Sterilize the loop by flaming it (illustration 5, figure 9.2). Return the loop to its container.

12. Incubate the culture you just inoculated at 37° C for 24–48 hours.

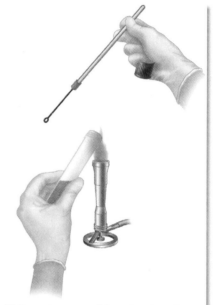

(1) Cap is removed from sterile broth and tube mouth is flamed.

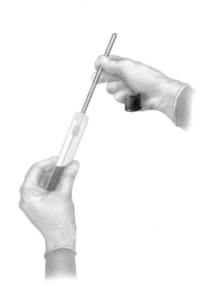

(2) Unheated loop is inserted into tube of sterile broth.

(3) Loop is removed from broth and tube mouth is flamed.

(4) Tube enclosure is returned to tube.

(5) Loop is flamed and returned to receptacle.

Figure 9.2   Procedure for inoculating a nutrient broth.

Transfer of Bacteria from a Slant

To inoculate a sterile nutrient agar slant from an agar slant culture, use the following procedure. Figure 9.3 illustrates the entire process.

Materials

- agar slant culture of *E. coli*
- sterile nutrient agar slant
- inoculating loop
- Bunsen burner
- Sharpie marking pen

1. If you have not already done so, prepare your desktop by swabbing down its surface with a disinfectant.
2. With a marking pen label a tube of nutrient agar slant with your initials and *E. coli*.
3. Sterilize your inoculating loop by holding it over the flame of a Bunsen burner *until it become bright red* (illustration 1, figure 9.3). The entire wire must be heated. Allow the loop to cool completely.
4. Using your free hand, pick up the slant culture of *E. coli* and remove the cap using the little finger of the hand that is holding the loop (illustration 2, figure 9.3).
5. Flame the mouth of the tube and insert the cooled loop into the tube. Pick up some of the culture on the loop (illustration 3, figure 9.3) and remove the loop from the tube.

6. Flame the mouth of the tube (illustrations 4 and 5, figure 9.3) and replace the cap, being careful not to burn your hand. Return tube to rack.
7. Pick up a sterile nutrient agar slant with your free hand, remove the cap with your little finger as before, and flame the mouth of the tube (illustration 6, figure 9.3).
8. Without flaming the loop containing the culture, insert the loop into the tube and gently inoculate the surface of the slant by moving the loop back and forth over the agar surface, while moving up the surface of the slant (illustration 7, figure 9.3). This should involve a type of serpentine or zig-zag motion.
9. Remove the loop, flame the mouth of the tube, and recap the tube (illustration 8, figure 9.3). Replace the tube in the rack.
10. Flame the loop, heating the entire wire to red-hot (illustration 9, figure 9.3), allow to cool, and place the loop in its container.
11. Incubate the inoculated agar slant at 30° C for 24–48 hours.

Working with Agar Plates

(Inoculating a slant from a petri plate)

The transfer of organisms from colonies on agar plates to slants or broth tubes is very similar to the procedures used in the last two transfers (broth to broth and slant to slant). The following rules should be observed.

(1) Inoculating loop is heated until it is red-hot.

(2) Cap is removed from slant culture and tube mouth is heated.

(3) Organism is picked up from slant with inoculating loop.

continued

Figure 9.3 Procedure for inoculating a nutrient agar slant from a slant culture.

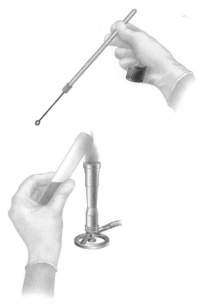

(4) Mouth of tube is flamed. Inoculating loop is not flamed.

(5) Slant culture is recapped and returned to test-tube rack.

(6) Tube of sterile agar slant is uncapped and mouth is flamed.

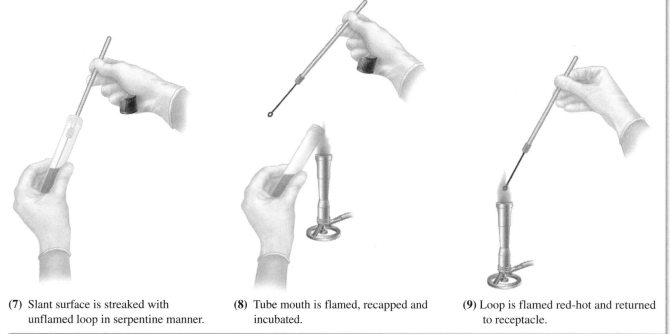

(7) Slant surface is streaked with unflamed loop in serpentine manner.

(8) Tube mouth is flamed, recapped and incubated.

(9) Loop is flamed red-hot and returned to receptacle.

Figure 9.3 *(continued)*

Loops and Needles   Loops are routinely used when streaking agar plates and slants. When used properly, a loop will not gouge or tear the agar surface. Needles are used in transfers involving stab cultures.

Plate Handling   Media in plates must always be protected against contamination. To prevent exposure to air contamination, covers should always be left closed. When organisms are removed from a plate culture, the cover should be only partially opened as shown in illustration 2, figure 9.4.

Flaming Procedures   Inoculating loops or needles must be flamed in the same manner that you used when working with previous tubes. One difference when working with plates is that plates are never flamed!

Plate Labeling and Incubation   Petri plates containing inoculated media are labeled on the bottom of the plate. Inoculated plates are almost always incubated upside down. This prevents moisture from condensing on the agar surface and spreading the inoculated organisms.

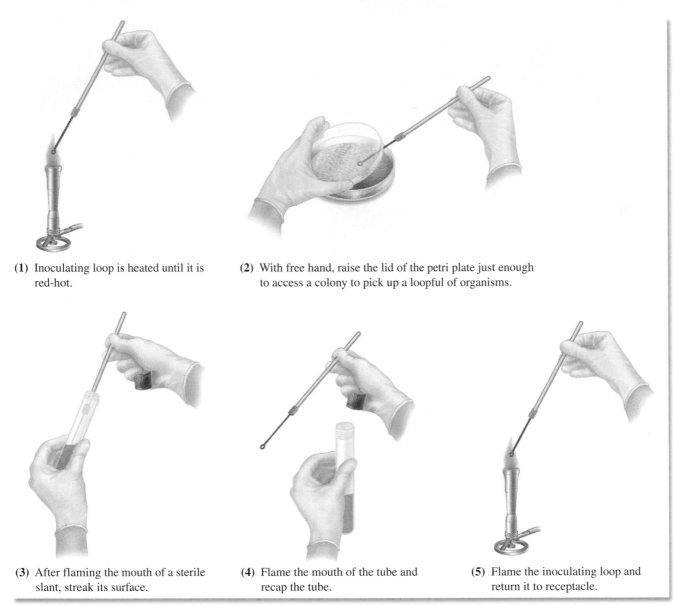

(1) Inoculating loop is heated until it is red-hot.

(2) With free hand, raise the lid of the petri plate just enough to access a colony to pick up a loopful of organisms.

(3) After flaming the mouth of a sterile slant, streak its surface.

(4) Flame the mouth of the tube and recap the tube.

(5) Flame the inoculating loop and return it to receptacle.

Figure 9.4 Procedure for inoculating a nutrient agar slant from an agar plate.

To transfer organisms from a petri plate to an agar slant, use the following procedure:

Materials

- nutrient agar plate with bacterial colonies
- sterile nutrient agar slant
- inoculating loop
- Sharpie marking pen

1. If you have not done so, swab your work area with disinfectant. Allow area to dry.
2. Label a sterile nutrient agar slant with your name and organism to be transferred.
3. Flame an inoculating loop until it is red-hot (illustration 1, figure 9.4). Allow the loop to cool.
4. As shown in illustration 2, figure 9.4, raise the lid of a petri plate sufficiently to access a colony with your sterile loop.

   Do not gouge into the agar with your loop as you pick up organisms, and do not completely remove the lid, exposing the surface to the air. Close the lid once you have picked up the organisms.
5. With your free hand, pick up the sterile nutrient agar slant tube. Remove the cap by grasping the cap with the little finger of the hand that is holding the loop.

6. Flame the mouth of the tube and insert the loop into the tube to inoculate the surface of the slant, using a serpentine motion (illustration 3, figure 9.4). Avoid disrupting the agar surface with the loop.
7. Remove the loop from the tube and flame the mouth of the tube. Replace the cap on the tube (illustration 4, figure 9.4).

8. Flame the loop (illustration 5, figure 9.4) and place it in its container.
9. Incubate the nutrient agar slant at 37° C for 24–48 hours.

Second Period

Examine all three tubes and record your results in Laboratory Report 9.

Laboratory Report

Student: _____

Date: _____ Section: _____

9 Aseptic Technique

A. Results

1. Were all your transfers successful? _____

2. What evidence do you have that they were successful? _____
   _____

3. What evidence do you have that a transfer is unsuccessful? _____
   _____

B. Short Answer Questions

1. Provide three reasons why the use of aseptic technique is essential when handling microbial cultures in the laboratory.
   _____
   _____

2. Provide two examples of how the Bunsen burner is used during inoculation of a tube culture.
   _____
   _____

3. How is air contamination prevented when an inoculating loop is used to introduce or take a bacterial sample to/from an agar plate?
   _____
   _____

4. Where should a label be written on an agar plate?
   _____
   _____

5. How should agar plates be incubated? Why?
   _____
   _____

6. Against which organisms are disinfectants effective? Against which type of organism may they not be effective? What disinfectant(s) is used in your laboratory?
   _____
   _____

C. Multiple Choice Questions

Select the answer that best completes the following statements.

1. A disinfectant is used on your work surface
   a. before the beginning of laboratory procedures.
   b. after all work is complete.
   c. after any spill of live microorganisms.
   d. Both (b) and (c) are correct.
   e. All of the above are correct.

2. To retrieve a sample from a culture tube with an inoculating loop, the cap of the tube is
   a. removed and held in one's teeth.
   b. removed and held with the fingers of the loop hand.
   c. removed with the fingers of the loop hand and placed in the fingers of the tube hand.
   d. removed with the fingers of the loop hand and placed on the laboratory bench.
   e. Any of these methods can be used.

3. An inoculating loop or needle is sterilized in a flame
   a. by one brief passage.
   b. for exactly 5 minutes.
   c. until it is entirely bright red.
   d. until the handle is bright red.
   e. until the tip is bright red.

ANSWERS

Multiple Choice

1. _____

2. _____

3. _____

Pure Culture Techniques

When we try to study the bacterial flora of the body, soil, water, or just about any environment, we realize quickly that bacteria exist in natural environments as mixed populations. It is only in very rare instances that they occur as a single species. Robert Koch, the father of medical microbiology, was one of the first to recognize that if he was going to prove that a particular bacterium causes a specific disease, it would be necessary to isolate the agent from all other bacteria and characterize the pathogen. From his studies on pathogenic bacteria, his laboratory contributed many techniques to the science of microbiology, including the method for obtaining **pure cultures** of bacteria. A pure culture contains only a single kind of an organism whereas a mixed culture contains more than one kind of organism. A contaminated culture contains a desired organism but also unwanted organisms. With a pure culture, we are able to study the cultural, morphological, and physiological characteristics of an individual organism.

Several methods for obtaining pure cultures are available to the microbiologist. Two commonly used procedures are the **streak plate** and the **pour plate.** Both procedures involve diluting the bacterial cells in a sample to an end point where a single cell divides giving rise to single **pure colony.** The colony is therefore assumed to be the identical progeny of the original cell and can be picked and used for further study of the bacterium.

In this exercise, you will use both the streak-plate and pour-plate methods to separate a mixed culture of bacteria. The bacteria may be differentiated by the characteristics of the colony, such as color, shape, and other colony characteristics. Isolated colonies can then be subcultured and stains prepared to check for purity.

Streak-Plate Method

For economy of materials and time, the streak-plate method is best. It requires a certain amount of skill, however, which is forthcoming with experience. A properly executed streak plate will give as good an

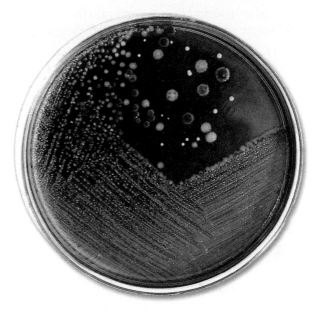

Figure 10.1   A streak plate demonstrating well-isolated colonies of three different bacteria.

isolation as is desired for most work. Figure 10.1 illustrates how colonies of a mixed culture should be spread out on a properly made streak plate. The important thing is to produce good spacing between colonies.

Materials

- electric hot plate (or tripod and wire gauze)
- Bunsen burner and beaker of water
- wire loop, thermometer, and Sharpie marking pen
- nutrient agar pour and 1 sterile petri plate
- mixed culture of *Serratia marcescens* or *Escherichia coli*, and *Micrococcus luteus* or *Chromobacterium violaceum*

1. Prepare your tabletop by disinfecting its surface with the disinfectant that is available in the laboratory (Roccal, Zephiran, Betadine, etc.). Use a sponge or paper towels to scrub it clean.

2. Label the bottom surface of a sterile petri plate with your name and date. Use a marking pen such as a Sharpie.

3. Liquefy a tube of nutrient agar, cool to 50° C, and pour the medium into the bottom of the plate, following the procedure illustrated in figure 10.2. Be sure to flame the neck of the tube prior to pouring to destroy any bacteria around the end of the tube.

After pouring the medium into the plate, gently rotate the plate so that it becomes evenly distributed, but do not splash any medium up over the sides.

Agar-agar, the solidifying agent in this medium, becomes liquid when boiled and resolidifies at around 42° C. Failure to cool it prior to pouring into the plate will result in condensation of moisture on the cover. Any moisture on the cover is undesirable because it can become deposited on the agar surface, causing the organisms to spread over the surface and thereby defeating the entire isolation procedure.

(1) Liquefy a nutrient agar pour by boiling for 5 minutes.

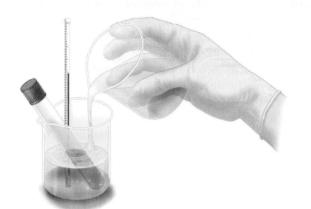

(2) Cool down the nutrient agar pour to 50° C by pouring off some of the hot water and adding cold water to the beaker. Hold at 50° C for 5 minutes.

(3) Remove the cap from the tube and flame the open end of the tube.

(4) Pour the contents of the tube into the bottom of the petri plate and allow it to solidify.

Figure 10.2 Procedure for pouring an agar plate for streaking.

4. Streak the plate by one of the methods shown in figure 10.4. Your instructor will indicate which technique you should use.

Caution: Be sure to follow the routine in figure 10.3 for getting the organism out of culture.

5. Incubate the plate in an *inverted position* at 25° C for 24–48 hours.

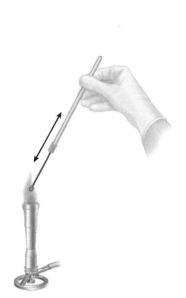

(1) Shake the culture tube from side to side to suspend organisms. Do not moisten cap on tube.

(2) Heat loop and wire to red-hot. Flame the handle slightly also.

(3) Remove the cap and flame the neck of the tube. Do not place the cap down on the table.

(4) After allowing the loop to cool for at least 5 seconds, remove a loopful of organisms. Avoid touching the side of the tube.

(5) Flame the mouth of the culture tube again.

(6) Return the cap to the tube and place the tube in a test-tube rack.

continued

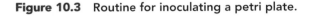

Figure 10.3 Routine for inoculating a petri plate.

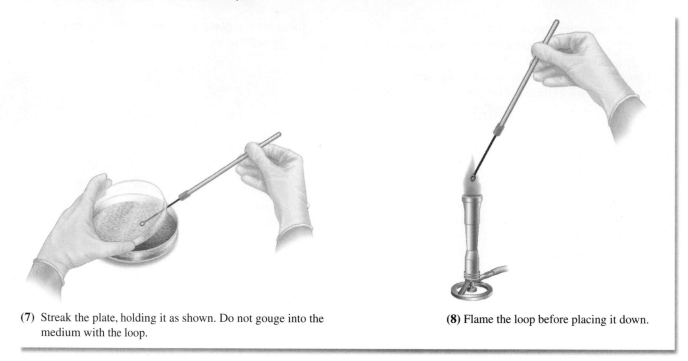

(7) Streak the plate, holding it as shown. Do not gouge into the medium with the loop.

(8) Flame the loop before placing it down.

Figure 10.3 *(continued)*

Quadrant Streak
(Method A)

(1) Streak one loopful of organisms over Area 1 near edge of the plate. Apply the loop lightly. Don't gouge into the medium.
(2) Flame the loop, cool 5 seconds, and make 5 or 6 streaks from Area 1 through Area 2. Momentarily touching the loop to a sterile area of the medium before streaking insures a cool loop.
(3) Flame the loop again, cool it, and make 6 or 7 streaks from Area 2 through Area 3.
(4) Flame the loop again, and make as many streaks as possible from Area 3 through Area 4, using up the remainder of the plate surface.
(5) Flame the loop before putting it aside.

Quadrant Streak
(Method B)

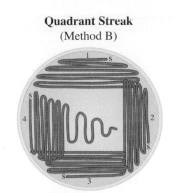

(1) Streak one loopful of organisms back and forth over Area 1, starting at point designated by "s". Apply loop lightly. Don't gouge into the medium.
(2) Flame the loop, cool 5 seconds, and touch the medium in sterile area momentarily to insure coolness.
(3) Rotate dish 90 degrees while keeping the dish closed. Streak Area 2 with several back and forth strokes, hitting the original streak a few times.
(4) Flame the loop again. Rotate the dish and streak Area 3 several times, hitting last area several times.
(5) Flame the loop, cool it, and rotate the dish 90 degrees again. Streak Area 4, contacting Area 3 several times and drag out the culture as illustrated.
(6) Flame the loop before putting it aside.

Radiant Streak

(1) Spread a loopful of organisms in small area near the edge of the plate in Area 1. Apply the loop lightly. Don't gouge medium.
(2) Flame the loop and allow it to cool for 5 seconds. Touching a sterile area will insure coolness.
(3) **From the edge** of Area 1 make 7 or 8 straight streaks to the opposite side of the plate.
(4) Flame the loop again, cool it sufficiently, and cross streak over the last streaks, **starting near Area 1.**
(5) Flame the loop before putting it aside.

Continuous Streak

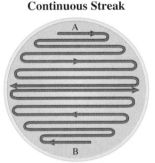

(1) Starting at the edge of the plate (Area A) with a loopful of organisms, spread the organisms in a single continuous movement to the center of the plate. Use light pressure and avoid gouging the medium.
(2) Rotate the plate 180 degrees so that the uninoculated portion of the plate is away from you.
(3) Without flaming the loop again, and using the same face of the loop, continue streaking the other half of the plate by starting at Area B and working toward the center.
(4) Flame the loop before putting it aside.

Figure 10.4 Four different streak techniques.

Pour-Plate Method
(Loop Dilution)

This method of separating one species of bacteria from another consists of diluting out one loopful of organisms with three tubes of liquefied nutrient agar in such a manner that one of the plates poured will have an optimum number of organisms to provide good isolation. Figure 10.5 illustrates the general procedure. One advantage of this method is that it requires somewhat less skill than that required for a good streak plate; a disadvantage, however, is that it requires more media, tubes, and plates. Proceed as follows to make three dilution pour plates, using the same mixed culture you used for your streak plate.

Materials

- mixed culture of bacteria
- 3 nutrient agar pours
- 3 sterile petri plates
- electric hot plate
- beaker of water
- thermometer
- inoculating loop and Sharpie marking pen

1. Label the three nutrient agar pours **I, II,** and **III** with a marking pen and place them in a beaker of water on an electric hot plate to be liquefied. To save time, start with hot tap water if it is available.
2. While the tubes of media are being heated, label the bottoms of the three petri plates **I, II,** and **III.**
3. Cool down the tubes of media to 50° C, using the same method that was used for the streak plate.
4. Following the routine in figure 10.5, inoculate tube I with one loopful of organisms from the mixed culture. Note the sequence and manner of handling the tubes in figure 10.6.
5. Inoculate tube II with one loopful from tube I after thoroughly mixing the organisms in tube I by shaking the tube from side to side or by rolling the tube vigorously between the palms of both hands. ***Do not splash any of the medium up onto the tube closure.*** Return tube I to the water bath.
6. Agitate tube II to completely disperse the organisms and inoculate tube III with one loopful from tube II. Return tube II to the water bath.
7. Agitate tube III, flame its neck, and pour its contents into plate III.
8. Flame the necks of tubes I and II and pour their contents into their respective plates.
9. After the medium has completely solidified, incubate the *inverted* plates at 25° C for 24–48 hours.

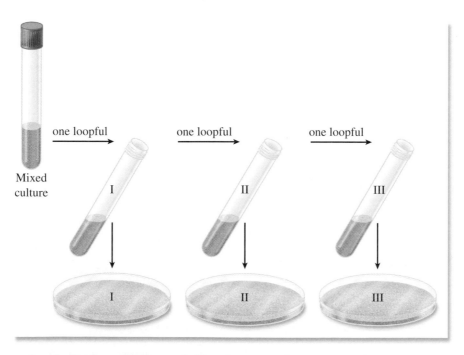

one loopful   one loopful   one loopful

Mixed culture   I   II   III

I   II   III

Figure 10.5 Three steps in the loop dilution technique for separating out organisms.

Evaluation of the Two Methods

After 24 to 48 hours of incubation examine all four petri plates. Look for colonies that are well isolated from the others. Note how crowded the colonies appear on plate I as compared with plates II and III. Plate I will be unusable. Either plate II or III will have the most favorable isolation of colonies. Can you pick out well-isolated colonies on your best pour plate that are distinct from one another?

Draw the appearance of your streak plate and pour plates on the Laboratory Report.

(1) Liquefy three nutrient agar pours, cool to 50° C, and let stand for 10 minutes.

(2) After shaking the culture to disperse the organisms, flame the loop and necks of the tubes.

(3) Transfer one loopful of the culture to tube I.

(4) Flame the loop and the necks of both tubes.

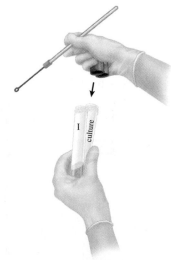

(5) Replace the caps on the tubes and return culture to the test-tube rack.

(6) Disperse the organisms in tube I by shaking the tube or rolling it between the palms.

continued

Figure 10.6   Tube-handling procedure in making inoculations for pour plates.

(7) Transfer one loopful from tube I to tube II. Return tube I to the water bath.

(8) After shaking tube II and transferring one loopful to tube III, flame the necks of each tube.

(9) Pour the inoculated pours into their respective petri plates.

Figure 10.6 *(continued)*

Subculturing Techniques

The next step in the development of a pure culture is the transfer of an isolated colony from the petri plate to a tube of nutrient broth or a slant of nutrient agar. Use your loop to carefully pick an isolated colony and aseptically transfer a colony to the broth tube or slant. To insure that the broth is inoculated, rotate the loop in the broth several times before withdrawing it from the broth tube. For the slant, make an "S" motion by drawing the loop from the bottom of the tube up the surface of the slant. Use the following routine to subculture the different organisms that you have isolated.

Materials

- nutrient agar slants
- inoculating loops
- Bunsen burners

1. Label one tube *Micrococcus luteus* and a second *Serratia marcescens*.
2. Select a well-isolated red colony on either the streak plate or the pour plate. Use your inoculating loop to pick a well-isolated colony and transfer it to the tube labeled *S. marcescens*.
3. Repeat this procedure for a yellow/cream-colored colony and transfer the colony to the tube labeled *M. luteus*.
4. Incubate the tubes at 25° C for 24–48 hours.

Evaluation of Slants

After incubation, examine the slants. Is the *S. marcescens* culture red? If the culture was incubated at a temperature higher than 25° C it may not be red because higher temperatures inhibit the formation of the organism's red pigment. Draw the appearance of the slant. What color is the *M. luteus* culture? Draw the slant.

You cannot be sure that your cultures are pure until you have made a microscopic examination of the respective cultures. It is entirely possible that the *S. marcescens* culture harbors some contaminating *M. luteus* and vice versa. Prepare smears for each culture and Gram stain the smears (see Exercise 15). *S. marcescens* is a gram-negative short rod while *M. luteus* is a gram-positive coccus. Draw the Gram-stained smears on the Laboratory Report.

Laboratory Report

Complete the Laboratory Report for this exercise.

Laboratory Report

Student: _____

Date: _____ Section: _____

10 Pure Culture Techniques

A. Results

1. **Evaluation of Streak Plate**

   Show within the circle the distribution of the colonies on your streak plate. To identify the colonies, use red for *Serratia marcescens*, yellow for *Micrococcus luteus,* and purple for *Chromobacterium violaceum*. If time permits, your instructor may inspect your plate and enter a grade where indicated.

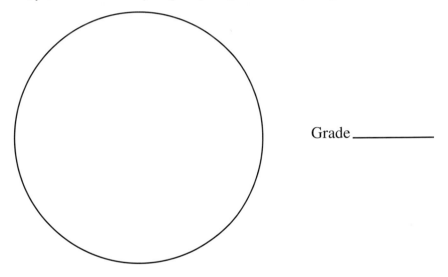

   Grade _____

2. **Evaluation of Pour Plates**

   Show the distribution of colonies on plates II and III, using only the quadrant section for plate II. If plate III has too many colonies, follow the same procedure. Use colors.

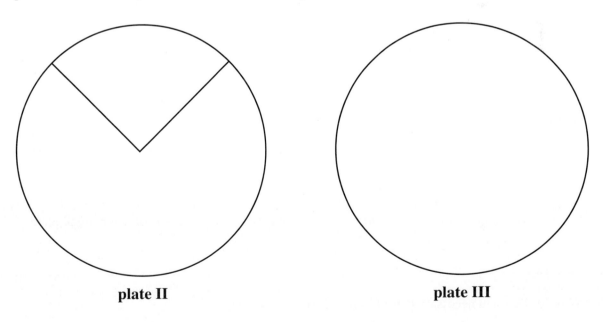

   plate II   **plate III**

3. **Subculture Evaluation**

With colored pencils, sketch the appearance of the growth on the slant diagrams below. Also, draw a few cells of each organism as revealed by Gram staining in the adjacent circle.

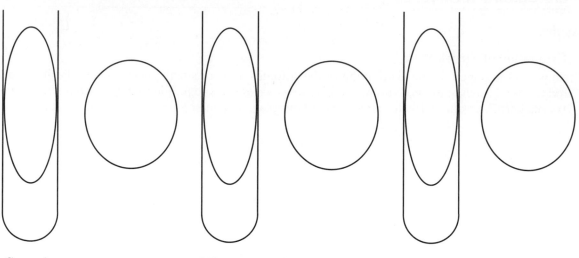

Serratia marcescens

Micrococcus luteus
or
Chromobacterium violaceum

Escherichia coli

4. Compare the results of your streak and pour plates. Which method achieved the best separation of species?

_____

_____

5. Do your slants contain pure cultures? How would you confirm their purities?

_____

_____

B. Short Answer Questions

1. In regards to bacterial growth on solid media, define the term "colony."

_____

_____

2. What colony characteristics can be used for differentiation of bacterial species? As an example, compare the properties of colonies of *Serratia marcescens* and *Micrococcus luteus* on your streak plate.

_____

_____

3. Why is dilution a necessary part of pure culture preparation?

_____

_____

4. What advantage(s) does the streak-plate method have over the pour-plate method?

_____

_____

5. What advantage(s) does the pour-plate method have over the streak-plate method?

_____

_____

6. Why is the loop flamed before it is placed in a culture tube? Why is it flamed after completing the inoculation?

_____

_____

7. Before inoculating and pouring molten nutrient agar into a plate, why must the agar first be cooled to 50° C?

_____

_____

8. Provide two reasons why plates should be inverted during incubation.

_____

_____

Staining and Observation of Microorganisms

The eight exercises in this unit include the procedures for ten slide techniques that one might employ in morphological studies of bacteria. A culture method in Exercise 18 also is included as a substitute for slide techniques when pathogens are encountered.

These exercises are intended to serve two equally important functions: (1) to help you to develop the necessary skills in making slides and (2) to introduce you to the morphology of bacteria. Although the title of each exercise pertains to a specific technique, the organisms chosen for each method have been carefully selected so that you can learn to recognize certain morphological features. For example, in the exercise on simple staining (Exercise 12), a single staining procedure applied to the selected organism can be used to demonstrate cell morphology, cell arrangement, and internal storage materials such as metachromatic granules. In Exercise 15 (Gram Staining), you will learn how to perform an important differential stain that employs more than one stain. This procedure allows you to taxonomically differentiate between two different kinds of bacteria as well as distinguish their cell morphology—cocci or rods.

The importance of the mastery of these techniques cannot be overemphasized. Although one is seldom able to make species identification on the basis of morphological characteristics alone, it is a very significant starting point. This fact will become increasingly clear with subsequent experiments.

Although the steps in the various staining procedures may seem relatively simple, student success is often quite unpredictable. Unless your instructor suggests a variation in the procedure, try to follow the procedures exactly as stated, without improvisation. Photomicrographs in color have been provided for many of the techniques; use them as a guide to evaluate the slides you have prepared. Once you have mastered a specific technique, feel free to experiment.

Smear Preparation

The success for most staining procedures depends upon the preparation of a good **smear.** There are several goals in preparing a smear. The first goal is to cause the cells to adhere to the microscope slide so that they are not washed off during subsequent staining and washing procedures. Second, it is important to insure that shrinkage of cells does not occur during staining, otherwise distortion and artifacts can result. A third goal is to prepare thin smears because the thickness of the smear will determine if you can visualize individual cells, their arrangement, or details regarding microstructures associated with cells. Thick smears of cells with large clumps obscure details about individual cells and, furthermore, the smear can entrap stain, keeping it from being removed by washing or destaining, leading to erroneous results. The procedure for making a smear is illustrated in figure 11.1.

The first step in preparing a bacteriological smear differs according to the source of the organisms. If the bacteria are growing in a liquid medium (broths, milk, saliva, urine, etc.), one starts by placing two or more loopfuls of the liquid medium directly on the slide.

From solid media such as nutrient agar, blood agar, or some part of the body, one starts by placing one or two loopfuls of water on the slide and then using an inoculating loop to disperse the organisms in the water. Bacteria growing on solid media tend to cling to each other and must be dispersed sufficiently by dilution in water; unless this is done, the smear will be too thick. *The most difficult concept for students to understand about making slides from solid media is that it takes only a very small amount of material to make a good smear.* When your instructor demonstrates this step, pay very careful attention to the amount of material that is placed on the slide.

The organisms to be used for your first slides may be from several different sources. If the plates from Exercise 7 were saved, some slides may be made from them. If they were discarded, the first slides may be made for Exercise 12, which pertains to simple staining. Your instructor will indicate which cultures to use.

From Liquid Media

(Broths, saliva, milk, etc.)

If you are preparing a bacterial smear from liquid media, follow this routine, which is depicted on the left side of figure 11.1.

Materials

- microscope slides
- Bunsen burner
- wire loop
- Sharpie marking pen
- slide holder (clothespin)

1. Wash a slide with soap or Bon Ami and hot water, removing all dirt and grease. Handle the clean slide by its edges.
2. Write the initials of the organism or organisms on the left-hand side of the slide with a marking pen.
3. To provide a target on which to place the organisms, make a $\frac{1}{2}''$ circle on the *bottom* side of the slide, centrally located, with a marking pen. Later on, when you become more skilled, you may wish to omit the use of this "target circle."
4. Shake the culture vigorously and transfer two loopfuls of organisms to the center of the slide over the target circle. Follow the routine for inoculations shown in figure 11.2. *Be sure to flame the loop after it has touched the slide.*

> **CAUTION:** Be sure to cool the loop completely before inserting it into a medium. A loop that is too hot will spatter the medium and move bacteria into the air.

5. Spread the organisms over the area of the target circle.
6. Allow the slide to dry by normal evaporation of the water. Don't apply heat.

7. After the smear has become completely dry, place the slide in a clothespin and pass it several times through the flame of a Bunsen burner. Avoid prolonged heating of the slide as it can shatter from excessive exposure to heat. The underside of the slide should feel warm to the touch.

Note that in this step one has the option of using or not using a clothespin to hold the slide. *Use the option preferred by your instructor.*

From Solid Media

When preparing a bacterial smear from solid media, such as nutrient agar or a part of the body, follow this routine, which is depicted on the right side of figure 11.1.

Materials

- microscope slides
- inoculating needle and loop

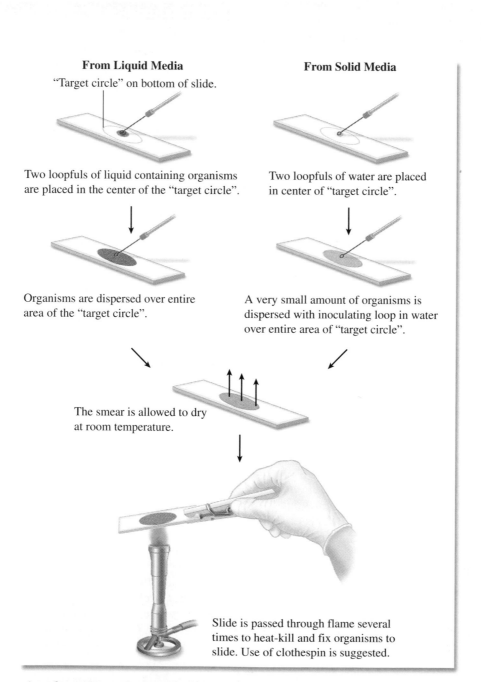

From Liquid Media

"Target circle" on bottom of slide.

Two loopfuls of liquid containing organisms are placed in the center of the "target circle".

Organisms are dispersed over entire area of the "target circle".

The smear is allowed to dry at room temperature.

From Solid Media

Two loopfuls of water are placed in center of "target circle".

A very small amount of organisms is dispersed with inoculating loop in water over entire area of "target circle".

Slide is passed through flame several times to heat-kill and fix organisms to slide. Use of clothespin is suggested.

Figure 11.1 Procedure for making a bacterial smear.

- Sharpie marking pen
- slide holder (clothespin)
- Bunsen burner

1. Wash a slide with soap or Bon Ami and hot water, removing all dirt and grease. Handle the clean slide by its edges.

2. Write the initials of the organism or organisms on the left-hand side of the slide with a marking pen.

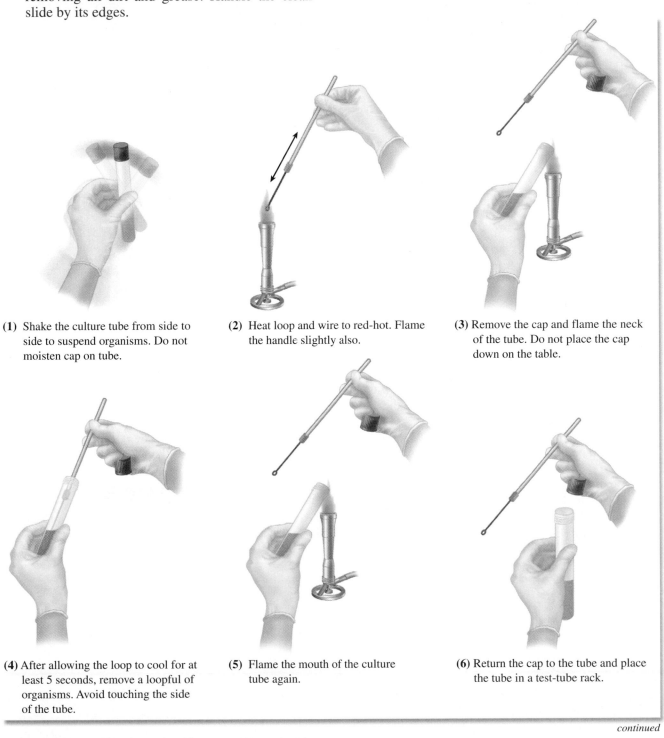

(1) Shake the culture tube from side to side to suspend organisms. Do not moisten cap on tube.

(2) Heat loop and wire to red-hot. Flame the handle slightly also.

(3) Remove the cap and flame the neck of the tube. Do not place the cap down on the table.

(4) After allowing the loop to cool for at least 5 seconds, remove a loopful of organisms. Avoid touching the side of the tube.

(5) Flame the mouth of the culture tube again.

(6) Return the cap to the tube and place the tube in a test-tube rack.

continued

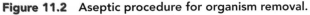

Figure 11.2 **Aseptic procedure for organism removal.**

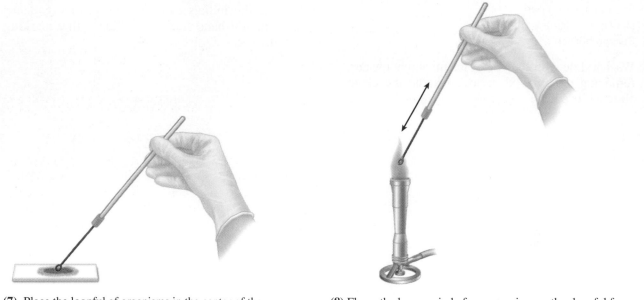

(7) Place the loopful of organisms in the center of the target circle on the slide.

(8) Flame the loop again before removing another loopful from the culture or setting the inoculating loop aside.

Figure 11.2 *(continued)*

3. Mark a "target circle" on the bottom side of the slide with a marking pen. (See comments in step 3 on page 95.)
4. Flame an inoculating loop, let it cool, and transfer two loopfuls of water to the center of the target circle.
5. Flame an inoculating needle then let it cool. Pick up *a very small amount of the organisms,* and mix it into the water on the slide. Disperse the mixture over the area of the target circle. Be certain that the organisms have been well emulsified in the liquid. *Be sure to flame the inoculating needle before placing it in its holder.*

6. Allow the slide to dry by normal evaporation of the water. Don't apply heat.
7. After the slide has become completely dry, place it in a clothespin and pass it several times through the flame of a Bunsen burner. Avoid prolonged heating of the slide as it can shatter from excessive exposure to heat. The underside of the slide should feel warm to the touch.

Laboratory Report

Answer the questions on Laboratory Report 11–14 that relate to this exercise.

Simple Staining

The use of a single stain to color a bacterial cell is commonly referred to as **simple staining.** Some of the most commonly used dyes for simple staining are methylene blue, basic fuchsin, and crystal violet. All of these dyes work well on bacteria because they have color-bearing ions (*chromophores*) that are positively charged (cationic).

The fact that bacteria are negatively charged produces a pronounced attraction between these cationic chromophores and the organism. Such dyes are classified as **basic dyes.** The basic dye methylene blue (methylene$^+$ chloride$^-$) will be used in this exercise. Those dyes that have anionic chromophores are called **acidic dyes.** Eosin (sodium$^+$ eosinate$^-$) is such a dye. The anionic chromophore, eosinate$^-$, will not stain bacteria because of the electrostatic repelling forces that are involved.

The staining times for most simple stains are relatively short, usually from 30 seconds to 2 minutes, depending on the affinity of the dye. After a smear has been stained for the required time, it is washed off gently, blotted dry, and examined directly under oil immersion. Such a slide is useful in determining basic morphology and the presence or absence of certain kinds of granules.

An avirulent strain of *Corynebacterium diphtheriae* will be used here for simple staining. In its pathogenic form, this organism is the cause of diphtheria, a very serious disease. One of the steps in identifying this pathogen is to do a simple stain of it to demonstrate the following unique characteristics: pleomorphism, metachromatic granules, and palisade arrangement of cells.

Pleomorphism pertains to irregularity of form: that is, demonstrating several different shapes. While *C. diphtheriae* is basically rod-shaped, it also appears club-shaped, spermlike, or needle-shaped. *Bergey's Manual* uses the terms "pleomorphic" and "irregular" interchangeably.

Metachromatic granules are distinct reddish-purple granules within cells that show up when the organisms are stained with methylene blue. These granules are masses of *volutin*, a polymetaphosphate.

Palisade arrangement pertains to parallel arrangement of rod-shaped cells. This characteristic, also called "picket fence" arrangement, is common to many corynebacteria.

Procedure

Prepare a slide of *C. diphtheriae*, using the procedure outlined in figure 12.1. It will be necessary to refer back to Exercise 11 for the smear preparation procedure.

Materials

- slant culture of avirulent strain of *Corynebacterium diphtheriae*
- methylene blue (Loeffler's)
- wash bottle
- bibulous paper

After examining the slide, compare it with the photomicrograph in illustration 1, figure 15.4 (page 112). Record your observations on Laboratory Report 11–14.

(1) A bacterial smear is stained with methylene blue for one minute.

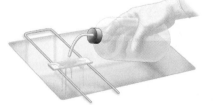

(2) Stain is briefly washed off slide with water.

(3) Water drops are carefully blotted off slide with bibulous paper.

Figure 12.1   Procedure for simple staining.

Negative Staining

Negative stains can be useful in studying the morphology of bacterial cells and characterizing some of the external structures, such as capsules, that are associated with bacterial cells. Negative stains are acidic and thus have a negatively charged chromophore that does not penetrate the cell but rather is repelled by the similarly charged bacterial cell. The background surrounding the cell is colored by a negative stain, resulting in a negative or indirect staining of the cell. Usually cells appear as transparent objects against a dark background (see figure 15.4, photo 7). Examples of negative stains are india ink and nigrosin. The negative stain procedure consists of mixing the organism with a small amount of stain and spreading a very thin film over the surface of a microscope slide. For negative stains, cells are not usually heat fixed prior to the application of the negative stain. Sometimes negative staining can be combined with positive staining to better demonstrate structures such as capsules. In this case, the capsule can be seen as a halo surrounding a positively stained cell against a dark background.

Negative staining can also be useful for accurately determining cell dimensions. Because heat fixation is not performed, no shrinkage of cells occurs and size determinations are more accurate than those determined on fixed material. Avoiding heat fixation is also important if the capsule surrounding the cell is to be observed because heat fixation will severely shrink this structure. The negative stain is also useful for observing spirochaetes, which tend to be very thin cells that do not readily stain with positive stains.

Three Methods

Negative staining can be done by one of three different methods. Figure 13.1 illustrates the more commonly used method in which the organisms are mixed in a drop of nigrosin and spread over the slide with another slide. The goal is to produce a smear that is thick at one end and feather-thin at the other end. Somewhere between the too thick and too thin areas will be an ideal spot to study the organisms.

Figure 13.2 illustrates a second method, in which organisms are mixed in only a loopful of nigrosin instead of a full drop. In this method, the organisms are spread over

(1) Organisms are dispersed into a small drop of nigrosin or india ink. Drop should not exceed ⅛" diameter and should be near the end of the slide.

(2) Spreader slide is moved toward drop of suspension until it contacts the drop causing the liquid to be spread along it's spreading edge.

(3) Once spreader slide contacts the drop on the bottom slide, the suspension will spread out along the spreading edge as shown.

(4) Spreader slide is pushed to the left, dragging the suspension over the bottom slide. After the slide has air-dried, it may be examined under oil immersion.

Figure 13.1 Negative staining technique, using a spreader slide.

(1) A loopful of nigrosin or india ink is placed in the center of a clean microscope slide.

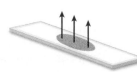

(2) A sterile inoculating wire is used to transfer the organisms to the liquid and mix the organisms into the stain.

(3) Suspension of bacteria is spread evenly over an area of one or two centimeters with the straight wire.

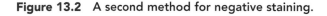

(4) Once the preparation has completely air-dried, it can be examined under oil immersion. No heat should be used to hasten drying.

Figure 13.2 A second method for negative staining.

a smaller area in the center of the slide with an inoculating needle. No spreader slide is used in this method.

The third procedure (Woeste-Demchick's method), which is not illustrated here, involves applying ink to a conventional smear with a black felt-tip marking pen. If this method is used, it should be done on a smear prepared in the manner described in Exercise 14. Simply put, the technique involves applying a *single coat* of marking-pen ink over a smear.

Note in the procedure below that slides may be made from organisms between your teeth or from specific bacterial cultures. Your instructor will indicate which method or methods you should use and demonstrate some basic aseptic techniques. Various options are provided here to ensure success.

Materials

- microscope slides (with polished edges)
- nigrosin solution or india ink
- slant cultures of *S. aureus* and *B. megaterium*
- inoculating straight wire and loop
- sterile toothpicks
- Bunsen burner
- Sharpie marking pen
- felt-tip marking pen (see Instructor's Handbook)

1. Swab down your tabletop with disinfectant in preparation for making slides.

2. Clean two or three microscope slides with Bon Ami to rid them of all dirt and grease.
3. By referring to figure 13.1 or 13.2, place the proper amount of stain on the slide.
4. **Oral Organisms:** Remove a small amount of material from between your teeth with a sterile straight toothpick and mix it into the stain on the slide. Be sure to break up any clumps of organisms with the toothpick or a sterile inoculating loop. When using a loop, *be sure to flame it first to make it sterile.*

> **CAUTION:** If you use a toothpick, discard it into a beaker of disinfectant.

5. **From Cultures:** With a *sterile* straight wire, transfer a very small amount of bacteria from the slant to the center of the stain on the slide.
6. Spread the mixture over the slide according to the procedure used in figure 13.1 or 13.2.
7. Allow the slide to air-dry and examine with an oil immersion objective.

Laboratory Report

Draw a few representative types of organisms on Laboratory Report 11–14. If the slide is of oral organisms, look for yeasts and hyphae as well as bacteria. Spirochaetes may also be present.

Some bacterial cells are surrounded by an extracellular slime layer called a **capsule** or **glycocalyx.** This structure can play a protective role for certain pathogenic bacteria such as *Streptococcus pneumoniae.* The capsule prevents phagocytic white blood cells from engulfing and destroying this bacterial pathogen, enabling the organism to invade the lungs and cause pneumonia. The capsule is also a means for many bacteria to attach to solid surfaces in the environment. For example, *Streptococcus mutans* can attach to the surface of a tooth by its capsular material resulting in the formation of dental plaque, which contributes to the process of tooth decay in humans. Most capsules are usually composed of polysaccharides but in some cases a capsule can consist of polypeptides with unique amino acids. Evidence supports the view that probably all bacterial cells have some amount of slime layer, but in most cases the amount is not enough to be readily discernible.

Staining of the bacterial capsule cannot be accomplished by ordinary staining procedures. If smears are heat fixed prior to staining, the capsule shrinks or is destroyed and cannot be seen in stains. However, the capsule can be demonstrated by combining the simple stain with the negative stain as shown in figure 14.1. First, a negative stain is prepared that will provide a dark background and will outline the capsule surrounding the cell. Next, cells will be gently heat fixed to ensure that they will adhere to the microscope slide but will not shrink or destroy the capsule. A positive stain consisting of crystal violet is then applied to specifically stain the cells on the slide. The final result is that capsules will appear as a halo-like structures surrounding purple cells against a dark background.

(1) Two loopfuls of the organism are mixed in a small drop of india ink.

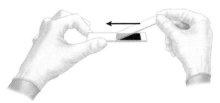

(2) The ink suspension of bacteria is spread over slide and air-dried.

(3) The slide is *gently* heat-dried to fix the organisms to the slide.

(4) Smear is stained with crystal violet for one minute.

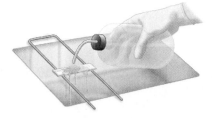

(5) Crystal violet is *gently* washed off with water.

(6) Slide is blotted dry with bibulous paper, and examined with oil immersion objective.

Figure 14.1   Procedure for demonstration of a capsule.

You will use this procedure to stain the capsules of *Klebsiella pneumoniae*.

Materials

- 36–48 hour milk culture of *Klebsiella pneumoniae*
- india ink
- crystal violet

Observation Examine the slide under oil immersion and compare your slide with illustration 2, figure 15.4 on page 112. Record your results on Laboratory Report 11–14.

Laboratory Report

Student: _____

Date: _____ Section: _____

11–14   Negative Staining, Smear Preparation, Simple and Capsular Staining

A. Results

Simple Staining (Exercise 12)

1. What three noteworthy physical characteristics of *Corynebacterium diphtheriae* are visible after performing a simple stain? Draw cells from your slide to demonstrate these characteristics.

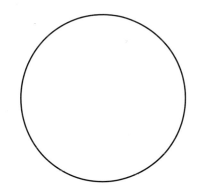

Corynebacterium diphtheriae

Negative Staining (Exercise 13)

2. Draw the different types of microorganisms that were found in the negative stain of the oral sample. How would you differentiate between oral streptococci, yeasts, and spirochaetes in your sample?

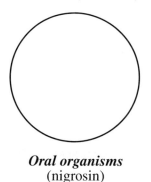

Oral organisms
(nigrosin)

Capsular Staining (Exercise 14)

3. Draw cells that display a capsule from your stained slide of *Klebsiella pneumoniae.* Explain how the capsule is visualized without the use of dyes that adhere to a capsule.

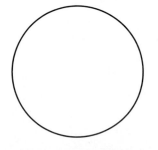

Klebsiella pneumoniae
(capsular stain)

B. Short Answer Questions

1. How does smear preparation of cells from a liquid medium differ from preparation of cells from a solid medium?

   _____

   _____

2. Why is it important to limit the quantity of cells used to prepare a smear?

   _____

   _____

3. For preparation of a smear on a slide, what is the purpose of heat fixation? What problems can arise when the slide is heated in a flame?

   _____

   _____

4. What causes a stain to adhere to bacterial cells? Why are all colored dyes not necessarily useful for simple staining?

   _____

   _____

5. Which type of microscopy produces an image of unstained cells that is most similar to the one achieved by negative staining?

   _____

   _____

6. How might fluorescence microscopy be used to visualize the bacterial capsule?

   _____

   _____

7. Why are encapsulated strains of *Streptococcus pneumoniae* much more likely to cause disease than strains that do not produce a capsule?

   _____

   _____

C. Matching Questions

Match the bacterial staining technique (simple, negative, capsule) to its description. Some descriptions may require more than one choice.

1. A basic dye is utilized to stain bacterial cells.
2. A stain that does not penetrate cells is used to color the background.
3. Useful for visualizing spirochaetes.
4. Useful for visualizing the glycocalyx of certain bacterial species.
5. Cells are mixed with a stain before they are smeared on the slide.
6. Heat fixation of the slide is not recommended.
7. Water is used to remove excess stain from the slide.

D. Multiple Choice Questions

Select the answer that best completes the following statements.

1. For the simple stain procedure, one can use
   a. crystal violet.
   b. eosin.
   c. methylene blue.
   d. nigrosin.
   e. Both (a) and (c) are correct.

2. For the negative stain procedure, one can use
   a. basic fuchsin.
   b. india ink.
   c. nigrosin.
   d. Both (b) and (c) are correct.
   e. All the answers are correct.

3. Before heat fixation, a wet smear of bacterial cells on a slide must first be
   a. air-dried.
   b. blotted dry.
   c. rinsed briefly with water.
   d. stained with a basic dye.
   e. None of the answers are correct.

4. Bacterial capsules can consist of
   a. peptidoglycan.
   b. polysaccharides.
   c. polypeptides with unique amino acids.
   d. Both (b) and (c) are correct.
   e. All the answers are correct.

ANSWERS

Matching Questions

1. _____

2. _____

3. _____

4. _____

5. _____

6. _____

7. _____

ANSWERS

Multiple Choice

1. _____

2. _____

3. _____

4. _____

Gram Staining

In 1884, the Danish physician Christian Gram was trying to develop a staining procedure that would differentiate bacterial cells from eukaryotic nuclei in stained tissue samples. Although Gram was not completely successful in developing a tissue stain, what resulted from his work is the most important stain in bacteriology, the Gram stain. The Gram stain is an example of a differential stain. These staining reactions take advantage of the fact that cells or structures within cells display dissimilar staining reactions that can be distinguished by the use of different dyes. In the Gram stain, the two kinds of cells, gram-positive and gram-negative, can be identified by their respective colors, purple and pink to red after performing the staining method. The procedure is based on the fact that gram-positive bacteria retain a crystal violet-iodine complex through decolorization with alcohol or acetone. Gram-positive bacteria appear as purple when viewed by microscopy. In contrast, alcohol or acetone removes the crystal violet-iodine complex from gram-negative bacteria. These bacteria must, therefore, be counterstained with a red dye, safranin, after the decolorization step in order to be visualized by microscopy. Hence, gram-negative bacteria appear as pink to red cells when viewed by microscopy.

Figure 15.1 illustrates the appearance of cells after each step in the Gram-stain procedure. Note that initially both gram-positive and gram-negative cells are stained by the **primary stain,** crystal violet. In the second step of the procedure, Gram's iodine is added to the smear. Iodine is a **mordant** that complexes with the crystal violet and forms an insoluble complex in gram-positive cells. At this point, both types of cells will still appear

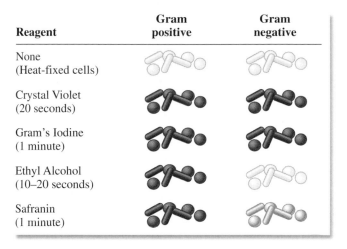

Reagent	Gram positive	Gram negative
None (Heat-fixed cells)		
Crystal Violet (20 seconds)		
Gram's Iodine (1 minute)		
Ethyl Alcohol (10–20 seconds)		
Safranin (1 minute)		

Figure 15.1 Color changes that occur at each step in the Gram-staining process.

as purple. The dye-mordant complex is not removed from gram-positive bacteria but is leached from gram-negative cells by the alcohol or acetone in the **decolorization** step. After decolorization, gram-positive cells are purple but gram-negative cells are colorless. In the final step, a **counterstain,** safranin, is applied, which stains the colorless gram-negative cells. The appearance of the gram-positive cells is unchanged because the crystal violet is a much more intense stain than safranin.

The mechanism for how the Gram stain works is related to chemical differences in the cell walls of gram-positive and gram-negative bacteria (figure 15.2). When viewed by electron microscopy, gram-positive cells have a thick layer of **peptidoglycan** that comprises the

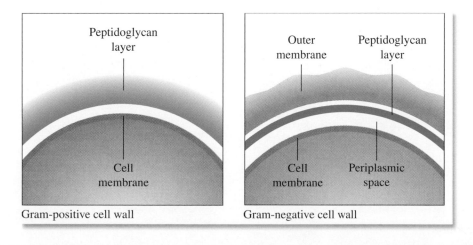

Figure 15.2 Comparison of gram-positive and gram-negative cell walls.

Gram-positive cell wall

Gram-negative cell wall

cell wall of these organisms. In contrast, the cell wall in gram-negative cells consists of an outer membrane that covers a much thinner layer of peptidoglycan. It is these significant differences in structure that probably determines whether the dye-mordant complex is removed from the gram-negative cell or remains associated with the gram-positive cell.

Of all the staining techniques you will use in microbiology, the Gram stain is one of the most important. It will be critical in identifying your unknown bacteria and you will use it routinely in many exercises in this manual. Although this technique seems quite simple, performing it with a high degree of reliability requires some practice and experience. Several factors can affect the outcome of the procedure:

1. It is important to use cultures that are 16–18 hours old. Gram-positive cultures older than this can convert to gram-variable or gram-negative and give erroneous results. (It is important to note that gram-negative bacteria never convert to gram-positive.)
2. It is critical to prepare thin smears. Thin smears allow the observation of individual cells and any arrangement in which the cells occur. Furthermore, the thickness of your smears can affect decolorization. Thick smears can entrap the primary stain, which is not removed by alcohol or acetone. Cells that occur in the entrapped stain can appear gram-positive leading to erroneous results.
3. Decolorization is the most critical step in the Gram-stain procedure. If the destaining reagent is over-applied, the dye-mordant complex can eventually be removed from gram-positive cells, converting them to gram-negative cells.

During this laboratory period, you will be provided an opportunity to stain several different kinds of bacteria to see if you can achieve the degree of success that is required. Remember, if you don't master this technique now, you will have difficulty with your unknowns later.

Staining Procedure

Materials

- slides with heat-fixed smears
- Gram-staining kit and wash bottle
- bibulous paper

1. Cover the smear with **crystal violet** and let stand for *20 seconds* (see figure 15.3).

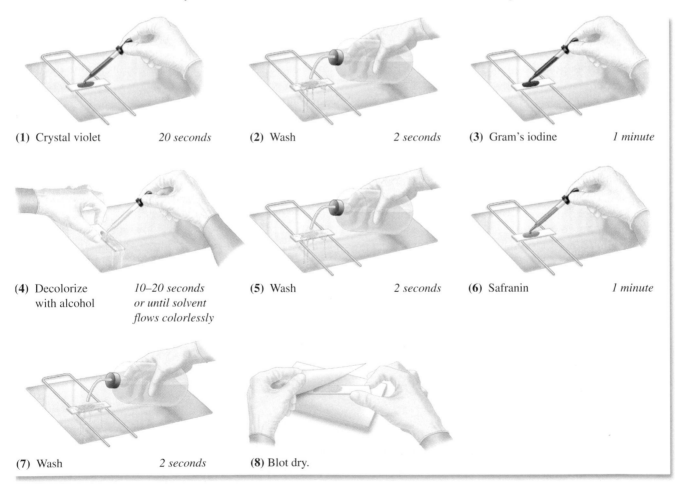

| (1) Crystal violet | *20 seconds* | (2) Wash | *2 seconds* | (3) Gram's iodine | *1 minute* |

| (4) Decolorize with alcohol | *10–20 seconds or until solvent flows colorlessly* | (5) Wash | *2 seconds* | (6) Safranin | *1 minute* |

| (7) Wash | *2 seconds* | (8) Blot dry. |

Figure 15.3 The Gram-staining procedure.

2. Briefly wash off the stain, using a wash bottle of distilled water. Drain off excess water.
3. Cover the smear with **Gram's iodine** solution and let it stand for *one minute*. (Your instructor may prefer only 30 seconds for this step.)
4. Wash off the Gram's iodine. Hold the slide at a 45-degree angle and allow the 95% alcohol to flow down the surface of the slide. Do this until the alcohol is colorless as it flows from the smear down the surface of the slide. *This should take no more than 20 seconds for properly prepared smears.* Note: thick smears can take longer than 20 seconds for decolorization.
5. Stop decolorization by washing the slide with a gentle stream of water.
6. Cover the smear with **safranin** for 1 minute.
7. Wash gently for a few seconds, blot dry with bibulous paper, and air-dry.
8. Examine the slide under oil immersion.

Assignments

The organisms that will be used here for Gram staining represent a diversity of form and staining characteristics. Some of the rods and cocci are gram-positive; others are gram-negative. One rod-shaped organism is a spore-former and another is acid-fast. The challenge here is to make Gram-stained slides of various combinations that reveal their differences.

Materials

- broth cultures of *Staphylococcus aureus*, *Pseudomonas aeruginosa*, and *Moraxella (Branhamella) catarrhalis*
- nutrient agar slant cultures of *Bacillus megaterium* and *Mycobacterium smegmatis*

Mixed Organisms I (Triple Smear Practice Slides) Prepare three slides with three smears on each slide. On the left portion of each slide make a smear of *Staphylococcus aureus*. On the right portion of each slide make a smear of *Pseudomonas aeruginosa*. In the middle of the slide make a smear that is a mixture of both organisms, using two loopfuls of each organism. *Be sure to flame the loop sufficiently to avoid contaminating cultures.*

Gram stain one slide first, saving the other two for later. Examine the center smear. If done properly, you should see purple cocci and pink to red rods as shown in illustration 3, figure 15.4.

Call your instructor over to evaluate your slide. If the slide is improperly stained, the instructor will be able to tell what went wrong by examining all three smears. He or she will inform you how to correct your technique when you stain the next triple smear reserve slide.

Record your results on Laboratory Report 15–17 by drawing a few cells in the appropriate circle.

Mixed Organisms II Make a Gram-stained slide of a mixture of *Bacillus megaterium* and *Moraxella (Branhamella) catarrhalis*.

This mixture differs from the previous slide in that the rods (*B. megaterium*) will be purple and the cocci (*M.B. catarrhalis*) will be large pink to red diplococci. See illustration 4, figure 15.4.

As you examine this slide, look for clear areas in the rods, which represent endospores. Since endospores are refractile and impermeable to crystal violet, they will appear as transparent holes in the cells.

Draw a few cells in the appropriate circle on your Laboratory Report sheet.

Acid-Fast Bacteria To see how acid-fast mycobacteria react to Gram's stain, make a Gram-stained slide of *Mycobacterium smegmatis*. If your staining technique is correct, the organisms should appear gram-positive.

Draw a few cells in the appropriate circle on your Laboratory Report sheet.

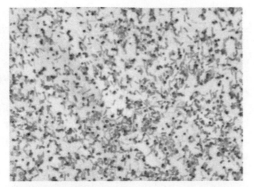

(1) Simple stain
    Bacillus subtilis and Staphylococcus

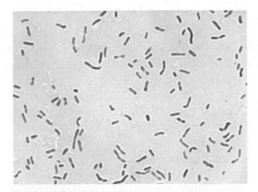

(2) Capsule stain
    Klebsiella pneumoniae

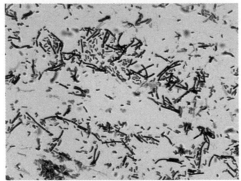

(3) Gram stain
    Bacillus megaterium and Escherichia

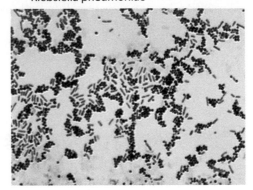

(4) Gram stain
    Escherichia coli and Staphylococcus aureus

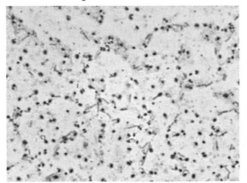

(5) Endospore stain
    Clostridium sporogenes

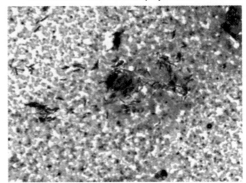

(6) Acid-fast stain
    Mycobacterium smegmatis and
    Staphylococcus aureus

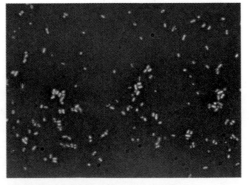

(7) Negative stain of bacterial cells

Figure 15.4   **Photomicrographs of representative stains (630X).**
© The McGraw-Hill Companies, Inc./Auburn University Research Instrumentation Facility/Michael Miller, photographer.

Spore Staining:
Two Methods

When species of bacteria belonging to the genera *Bacillus* and *Clostridia* exhaust essential nutrients, they undergo a complex developmental cycle that produces resting stages called **endospores.** Endospores allow these bacteria to survive environmental conditions that are not favorable for growth. If nutrients once again become available, the endospore can go through the process of germination to form a new vegetative cell and growth will resume. Endospores are very dehydrated structures that are not actively metabolizing. Furthermore, they are resistant to heat, radiation, acids, and many chemicals, such as disinfectants, that normally harm or kill vegetative cells. Their resistance is due in part to the fact that they have a protein coat, or **exosporium,** that forms a protective barrier around the spore. Heat resistance is associated with the water content of endospores. The higher the water content of an endospore, the less heat resistant the endospore will be. During sporulation, the water content of the endospore is reduced to 10–30% of the vegetative cell. This results because calcium ions complex with spore-specific proteins and a chemical, dipicolinic acid. The latter compound is not found in vegetative cells. This complex forms a gel that controls the amount of water that can enter the endospore, thus maintaining its dehydrated state.

Since endospores are not easily destroyed by heat or chemicals, they define the conditions necessary to establish sterility. For example, to destroy endospores by heating, they must be exposed for 15 to 20 minutes to steam under pressure, which generates temperatures of 121° C. Such conditions are produced in an **autoclave.**

The resistant properties of endospores also mean that they are not easily penetrated by stains. For example in Exercise 15, you observed that endospores did not readily Gram stain. If endospore-containing cells are stained by basic stains such as crystal violet, the spores appear as unstained areas in the vegetative cell.

However, if heat is applied while staining with malachite green, the stain penetrates the endospore and becomes entrapped in the endospore. The malachite green is not removed by subsequent washing with decolorizing agents or water. In this instance, heat is acting as mordant to facilitate the uptake of the stain.

Schaeffer-Fulton Method

The Schaeffer-Fulton method, which is depicted in figure 16.1, utilizes malachite green to stain the endospore and safranin to stain the vegetative portion of the cell. Utilizing this technique, a properly stained spore-former will have a green endospore contained in a pink sporangium. Illustration 5, figure 15.4, on page 112 reveals what such a slide looks like under oil immersion.

Prepare a smear of *Bacillus megaterium* and allow the smear to air-dry. Heat fix the dried smear and follow the steps for staining outlined in figure 16.1.

Materials

- 24–36 hour nutrient agar slant culture of *Bacillus megaterium*
- electric hot plate and small beaker (25 ml)
- spore-staining kit consisting of a bottle each of 5% malachite green and safranin

Dorner Method

The Dorner method for staining endospores produces a red spore within a colorless sporangium. Nigrosin is used to provide a dark background for contrast. The six steps involved in this technique are shown in figure 16.2. Although both the sporangium and endospore are stained during boiling in step 3, the sporangium is decolorized by the diffusion of safranin molecules into the nigrosin.

Prepare a slide of *Bacillus megaterium* that utilizes the Dorner method. Follow the steps in figure 16.2.

(1) Cover smear with small piece of paper toweling and saturate it with malachite green. Steam over boiling water for *5 minutes*. Add additional stain if stain boils off.

(2) After the slide has cooled sufficiently, remove the paper toweling and rinse with water for 30 seconds.

(3) Counterstain with safranin for about *20 seconds*.

(4) Rinse briefly with water to remove safranin.

(5) Blot dry with bibulous paper, and examine slide under oil immersion.

Figure 16.1   The Schaeffer-Fulton spore stain method.

Materials

- nigrosin
- electric hot plate and small beaker (25 ml)
- small test tube (10 × 75 mm size)
- test-tube holder
- 24–36 hour nutrient agar slant culture of *Bacillus megaterium*

Quick Spore Stain

A variation on the Schaeffer-Fulton method is a quick method that uses the same stains.

Materials

- *Bacillus megaterium* slant cultures, older than 36 hours
- malachite green stain
- safranin stain
- staining racks
- clothespins

Procedure

1. Prepare a smear of the organism and allow it to air-dry.
2. Grasp the slide with the air-dried smear with a clothespin and pass it through a Bunsen burner flame 10 times. Be careful not to overdo the heating as the slide can break.
3. Immediately flood the smear with malachite green and allow to stand for 5 minutes.
4. Wash the smear with a gentle stream of water.
5. Stain with safranin for 45 seconds. Spores will be green and the vegetative cell will be red.

Laboratory Report

After examining the organisms under oil immersion, draw a few cells in the appropriate circles in Laboratory Report 15–17.

(1) Make a heavy suspension of bacteria by dispersing several loopfuls of bacteria in 5 drops of sterile water.

(2) Add 5 drops of carbolfuchsin to the bacterial suspension.

(3) Heat the carbolfuchsin suspension of bacteria in a beaker of boiling water for *10 minutes*.

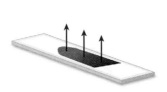

(4) Mix several loopfuls of bacteria in a drop of nigrosin on the slide.

(5) Spread the nigrosin-bacteria mixture on the slide in the same manner as in Exercise 11 (Negative Staining).

(6) Allow the smear to air-dry. Examine the slide under oil immersion.

Figure 16.2 **The Dorner spore stain method.**

Acid-Fast Staining:
Ziehl-Neelsen Method

Bacteria such as *Mycobacterium* and some *Nocardia* have cell walls that contain a high lipid content. One of the cell wall lipids is a waxy material called **mycolic acid.** This material is a complex lipid that is composed of fatty acids and fatty alcohols that have hydrocarbon chains up to 80 carbons in length. It significantly affects the staining properties of these bacteria and prevents them from being stained by many of the stains routinely used in microbiology. The acid-fast stain is an important diagnostic tool in the identification of the *Mycobacterium tuberculosis* the causative agent of tuberculosis, and *Mycobacterium leprae,* the bacterium that causes leprosy in humans.

To facilitate staining of these bacteria, it is necessary to use techniques that make the cells more permeable to stains. In the classic Ziehl-Neelsen staining method, the primary stain, carbolfuchsin, is mixed with phenol and the cells are heated for 5 minutes during the staining procedure. Phenol is necessary for the carbofuchsin to penetrate the waxy cell wall lipids and heating further facilitates the penetration of the stain. Heat acts as a mordant to make the stain complex more permeable to the mycolic acid. Subsequent treatment of the cells with acid-alcohol, a decolorizer, does not remove the entrapped stain. Hence, these cells are termed **acid-fast.** Cells that do not contain mycolic acid in their cell walls are easily decolorized by acid-alcohol and are termed **non-acid-fast.** To be visualized, non-acid-fast bacteria must be counter-stained with methylene blue. In the acid-fast staining method, acid-fast cells appear red to pink in stained preparations, whereas non-acid-fast cells appear blue (figure 15.4, photo 6). The application of heat to cells during staining with carbolfuchsin and phenol is not without concerns. Phenol can vaporize when heated, giving rise to noxious fumes that are toxic to the eyes and mucous membranes. The Kinyoun acid-fast method is a modification in which the concentrations of both carbolfuchsin and phenol are increased but the bacterial cells are not heated during the staining procedure. The increased concentrations of the stain and phenol are sufficient to allow penetration of stain into cells and the carbolfuchsin is not removed by de-staining with acid-alcohol. This procedure has the further advantage that phenol fumes are not generated during the staining of the cells.

In the following exercise, you will prepare an acid-fast stain of a mixture of *Mycobacterium smegmatis* and *Staphylococcus aureus* using the Kinyoun method for acid-fast staining. *M. smegmatis* is a nonpathogenic, acid-fast rod that occurs in soil and on the external genitalia of humans. *S. aureus* is a non-acid-fast coccus that is also part of the normal flora of humans but that can also be a serious opportunistic pathogen.

Materials

- nutrient agar slant culture of *Mycobacterium smegmatis* (48-hour culture)
- nutrient broth culture of *S. aureus*
- electric hot plate and small beaker
- acid-fast staining kit (carbolfuchsin, acid-alcohol, and methylene blue)

Smear Preparation   Prepare a mixed culture smear by placing two loopfuls of *S. aureus* on a slide and transferring a small amount of *M. smegmatis* to the broth on the slide with an inoculating needle. Since the Mycobacteria are waxy and tend to cling to each other in clumps, break up the masses of organisms with the inoculating needle. After air-drying the smear, heat-fix it.

Staining   Follow the staining procedure outlined in figure 17.1.

Examination   Examine under oil immersion and compare your slide with illustration 6, figure 15.4 on page 112.

Laboratory Report   Record your results in Laboratory Report 15–17.

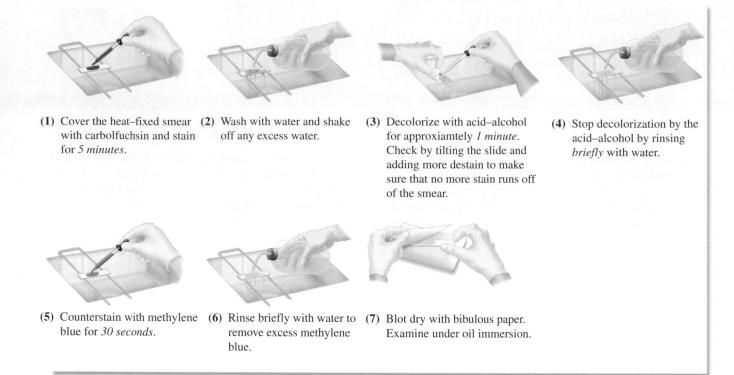

(1) Cover the heat–fixed smear with carbolfuchsin and stain for *5 minutes*.

(2) Wash with water and shake off any excess water.

(3) Decolorize with acid–alcohol for approxiamtely *1 minute*. Check by tilting the slide and adding more destain to make sure that no more stain runs off of the smear.

(4) Stop decolorization by the acid–alcohol by rinsing *briefly* with water.

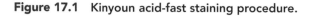

(5) Counterstain with methylene blue for *30 seconds*.

(6) Rinse briefly with water to remove excess methylene blue.

(7) Blot dry with bibulous paper. Examine under oil immersion.

Figure 17.1 Kinyoun acid-fast staining procedure.

Laboratory Report

Student: _____

Date: _____   Section: _____

15–17   Differential Staining: Gram, Spore, and Acid-Fast

A. Results

Gram-Staining (Exercise 15)

1. Draw cells from the Gram-stained slide. Differentiate the bacteria according to the Gram reaction, cell shape, and cell arrangements.

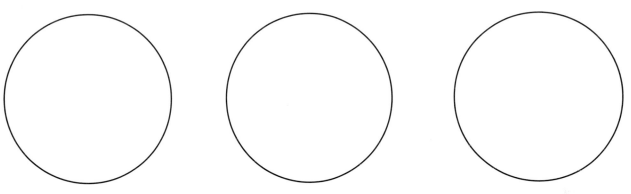

P. aeruginosa & S. aureus
(Gram-stain)

B. megaterium & M. B. catarrhalis
(Gram-stain)

M. smegmatis
(Gram-stain)

Spore Staining (Exercise 16)

2. Draw cells from the spore slide. Differentiate spores from vegetative cells.

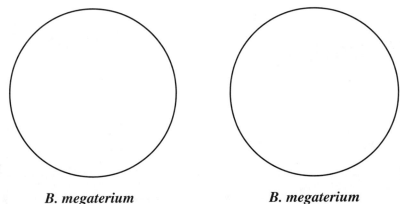

B. megaterium
(Schaeffer-Fulton method)

B. megaterium
(Dorner method)

Acid-Fast Staining (Exercise 17)

3. Draw cells from the acid-fast slide. Differentiate acid-fast from non-acid-fast cells.

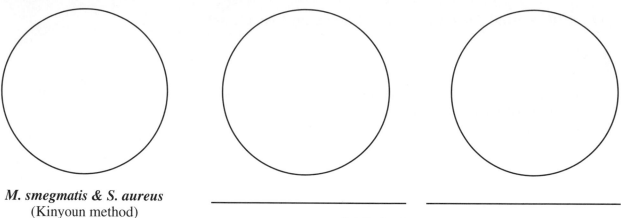

M. smegmatis & S. aureus
(Kinyoun method)

OPTIONAL STAINING

B. Short Answer Questions

1. Which of the three differential stains would likely be the first used when identifying an unknown bacterium? Explain.

2. What is the function of a mordant?

3. For differential staining, how does a counterstain differ from a primary stain?

4. How do gram-positive and gram-negative bacteria differ in cellular structure? How does this contribute to their differential staining properties?

5. Which is the most critical step in the Gram-stain procedure? Why? If this procedure is done incorrectly, how might that affect the final results?

6. How does culture age affect the results of a Gram-stain?

7. How does culture age affect the results of a spore stain?

_____

_____

8. Why must smear thickness be considered before performing a Gram-stain?

_____

_____

9. What color are bacterial endospores after a Gram-stain is performed? What does this tell you about the physical properties of endospores?

_____

_____

10. *Bacillus anthracis,* the causative agent of anthrax, is an endospore-former. Why does this trait enhance its capabilities as a biological weapon?

_____

_____

11. What makes *Mycobacterium* particularly resistant to staining? How are the bacteria in this genus grouped in terms of Gram classification?

_____

_____

12. How do you think the acid-fast nature of *Mycobacterium* contributes to its virulence?

_____

_____

C. Matching Questions

Match the color (pink, purple, no color) to the description of vegetative cells during the Gram-staining procedure. Choices may be used more than once.

1. *Staphylococcus aureus* before primary stain is added.
2. *Pseudomonas aeruginosa* after primary stain is added.
3. *Bacillus megaterium* after the mordant is added.
4. *Staphylococcus aureus* cells after the decolorizer is used.
5. *Moraxella (Branhamella) catarrhalis* after the decolorizer is used.
6. *Bacillus megaterium* after the counterstain is added.
7. *Pseudomonas aeruginosa* after the counterstain is added.

ANSWERS

Matching Questions

1. _____

2. _____

3. _____

4. _____

5. _____

6. _____

7. _____

D. Fill-in-the-Blank Questions

Compare and contrast the three differential staining techniques by completing the table below.

DIFFERENTIAL STAIN

	GRAM-STAIN	SPORE STAIN (Schaeffer-Fulton)	ACID-FAST STAIN (Kinyoun)
Primary stain			
Mordant			
Decolorizer			
Counterstain			
Cell type = color after completion of stain	Gram-positive = Gram-negative =	Sporangium = Vegetative cell =	Acid-fast bacteria = Non-acid-fast bacteria =

E. Multiple Choice

Select the answer that best completes the following statements.

1. Bacterial cell wall is composed of
   a. peptidoglycan.
   b. phospholipids.
   c. proteins.
   d. simple polysaccharides.
   e. Both (b) and (c) are correct.

2. The exosporium, or endospore coat, is composed of
   a. peptidoglycan.
   b. phospholipids.
   c. proteins.
   d. simple polysaccharides.
   e. Both (b) and (c) are correct.

3. Endospores are produced by bacteria in the genus
   a. *Bacillus.*
   b. *Clostridium.*
   c. *Mycobacterium.*
   d. Both (a) and (b) are correct.
   e. Both (a) and (c) are correct.

4. Acid-fast staining is useful for identifying the causative agent of
   a. leprosy.
   b. tetanus.
   c. tuberculosis.
   d. Both (a) and (c) are correct.
   e. All the answers are correct.

ANSWERS

Multiple Choice

1. _____

2. _____

3. _____

4. _____

Motility Determination

The major organelles of motility in bacteria are **fla-gella.** Flagella allow cells to move toward nutrients in the environment or move away from harmful substances, such as acids, in a complicated process called **chemotaxis.** The flagellum is a rigid helical structure that extends as much as 10 microns out from the cell. However, flagella are very thin structures, less than 0.2 microns, and, therefore, they are below the resolution of the light microscope. For flagella to be observed by light microscopy, they must be stained by special techniques. An individual bacterial flagellum is composed of a rigid filament that occurs in the form of a helix. This constitutes the main body of the flagellum structure. The filament is connected to a hook that is attached to a shaft that is inserted into a series of rings whose number differ for gram-positive and gram-negative cells (figure 18.1). Gram-positive cells contain the S and M rings that are situated in the area of the cell membrane. Gram-negative cells also possess the S and M rings and two additional rings, the L and P rings that are associated with the outer membrane and

peptidoglycan of the cell. The shaft, rings, and accessory proteins make up the basal body of the flagellum. The basal body is situated in the cell membrane/cell wall area of the bacterial cell. Rotation of the flagellum is powered by a proton motive force (pmf) that is established when proteins associated with the basal body transport protons across the cell membrane, creating a charge differential across the membrane. The pmf induces the S and M rings to rotate which results in the rotation of the shaft, hook, and filament. Other proteins in the basal body can reverse the direction of rotation of the flagellum. The movement of the rigid and helical filament is analogous to the rotation of a propeller on a boat engine. Hence, the movement of the filament propels the bacterial cell in much the same way that a boat is moved through the water by its engine and propeller. This is in contrast to a eukaryotic flagellum, which causes the cell to move because the flagellum beats like a whip.

Motility and the arrangement of flagella around the cell (figure 18.2) are important taxonomic characteristics that are useful in characterizing bacteria.

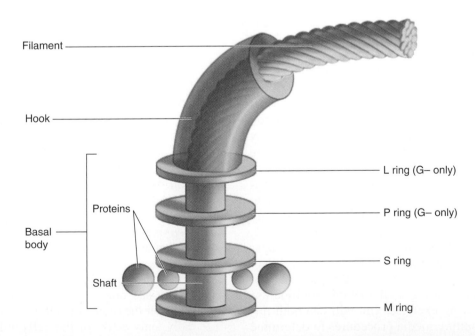

Filament

Hook

Basal body

Proteins

Shaft

L ring (G− only)

P ring (G− only)

S ring

M ring

Figure 18.1 Structure of the Gram-negative bacterial flagellum

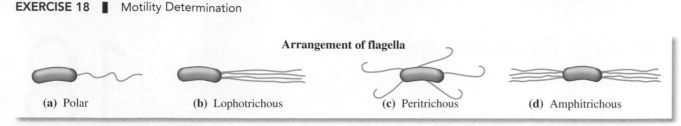

Figure 18.2 **Arrangement of flagella**

Motility can be determined by several methods. It can be determined microscopically by observing cells in a **wet mount.** In this procedure, a drop of viable cells is placed on a microscope slide and covered with a cover glass. The slide is then observed with a phase-contrast microscope. The rapid swimming movement of cells in the microscopic field confirms motility. However, wet mounts can easily dry out by evaporation, which is especially troublesome if observations need to made for prolonged periods of time. Drying can be delayed by using the **hanging drop technique,** shown in figure 18.3. In this procedure, a drop of cells is placed on a cover glass, which is then placed over a special slide that has a concave depression in its center. The coverslip is held in place with petroleum jelly, thus forming an enclosed glass chamber that prevents drying.

For the beginner, true swimming motility under the microscope must be differentiated from **Brownian motion** of cells or movement caused by currents under the cover glass. Brownian motion is movement due to molecular bombardment of cells causing cells to shake or "jiggle about" but not move in any vectorial way. Cells can also appear to move because currents can be created under the cover glass when pressure is exerted by focusing the oil immersion lens or by the wet mount drying out. This causes cells to "sweep" across the field.

Another method for determining motility involves inoculating semisoft agar medium. This medium has an agar concentration of 0.4%, which does not inhibit bacteria from "swimming" through the medium. In this procedure, the organisms are inoculated by stabbing the semisolid medium with an inoculating needle. If the organisms are motile, they will swim away from the line of inoculation into the uninoculated surrounding medium, causing the medium to be turbid. Non-motile bacteria will be found only along the line of inoculation. For pathogenic bacteria, such as the typhoid bacillus, the use of semisoft agar medium to determine motility is often preferred over microscope techniques because of the potential for infection posed by pathogens in making wet mounts.

In the following exercise, you will use both microscopic and culture media procedures to determine motility of bacterial cultures.

First Period

During the first period, you will make wet mounts and hanging drop slides of two organisms: *Proteus vulgaris* and *Micrococcus luteus*. Tube media (semisolid medium or SIM medium) and a soft agar plate will also be inoculated. The media inoculations will have to be incubated to be studied in the next period. Proceed as follows:

Materials

- microscope slides and cover glasses
- depression slide
- 2 tubes of semisolid or SIM medium
- 1 petri plate of soft nutrient agar (20–25 ml of soft agar per plate)
- nutrient broth cultures of *Micrococcus luteus* and *Proteus vulgaris* (young cultures)
- inoculating loop and needle
- Bunsen burner

Wet Mounts Prepare wet mount slides of each of the organisms, using several loopfuls of the organism on the slides. Examine under an oil immersion objective. Observe the following guidelines:

- Use only scratch-free, clean slides and cover glasses. This is particularly important when using phase-contrast optics.
- Label each slide with the name of the organism.
- By manipulating the diaphragm and voltage control, reduce the lighting sufficiently to make the organisms visible. Unstained bacteria are very transparent and difficult to see.
- For proof of true motility, look for directional movement that is several times the long dimension of the bacterium. The movement will also occur in different directions in the same field.
- Ignore Brownian movement. *Brownian movement* is vibrational movement caused by invisible molecules bombarding bacterial cells. If the only movement you see is vibrational and not directional, the organism is non-motile.
- If you see only a few cells exhibiting motility, consider the organism to be motile. Characteristically, only a few of the cells will be motile at a given moment.

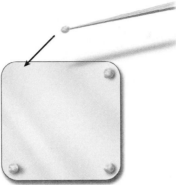

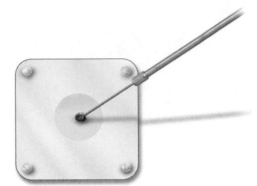

(1) A small amount of Vaseline is placed near each corner of the cover glass with a toothpick.

(2) Two loopfuls of organisms are placed in the cover glass.

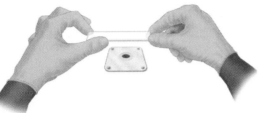

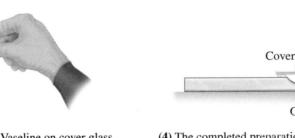

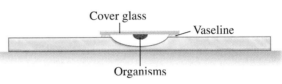

Cover glass

Vaseline

Organisms

(3) Depression slide is pressed against Vaseline on cover glass and quickly inverted.

(4) The completed preparation can be examined under oil immersion.

Figure 18.3   The hanging drop slide.

- Don't confuse water current movements with true motility. Water currents are due to capillary action caused by temperature changes and drying out. All objects move in a straight line in one direction.
- And, finally, always *examine a wet mount immediately,* once it has been prepared, because motility decreases with time after preparation.

Hanging Drop Slides   By referring to figure 18.3, prepare hanging drop slides of each organism. Be sure to use clean cover glasses and label each slide with a china marking pencil. When placing loopfuls of organisms on the cover glass, be sure to flame the loop between applications. Once the slide is placed on the microscope stage, do as follows:

1. Examine the slide first with the low-power objective. If your microscope is equipped with an automatic stop, avoid using the stop; instead, use the coarse adjustment knob for bringing the image into focus. The greater thickness of the depression slide prevents one from being able to focus at the stop point.
2. Once the image is visible under low power, swing the high-dry objective into position and readjust the lighting. Since most bacteria are drawn to the

edge of the drop by surface tension, **focus near the edge of the drop.**
3. If your microscope has phase-contrast optics, switch to high-dry phase. Although a hanging drop does not provide the shallow field desired for phase-contrast, you may find that it works fairly well.
4. If you wish to use oil immersion, simply rotate the high-dry objective out of position, add immersion oil to the cover glass, and swing the oil immersion lens into position.
5. Avoid delay in using this setup. Water condensation may develop to decrease clarity and the organisms become less motile with time.
6. Review all the characteristics of bacterial motility that are stated on page 124 under wet mounts.

Tube Method   Inoculate tubes of semisolid or SIM media with each organism according to the following instructions:

1. Label the tubes of semisolid (or SIM) media with the names of the organisms. Place your initials on the tubes also.
2. Flame and cool the inoculating needle, and insert it into the culture after flaming the neck of the tube.

(1) Wire with organisms is brought into tube without touching walls of tube.

(2) Wire penetrates medium to two-thirds of its depth.

(3) Wire is withdrawn from medium and tube. Neck of tube is flamed and plugged.

Figure 18.4   Stab technique for motility test.

3. Remove the cap from the tube of medium, flame the neck, and stab it two-thirds of the way down to the bottom, as shown in figure 18.4. Flame the neck of the tube again before returning the cap to the tube.
4. Repeat steps 2 and 3 for the other culture.
5. Incubate the tubes at room temperature for 24 to 48 hours.

Plate Method   Mark the bottom of a plate of soft agar with two one-half inch circles about one inch apart. Label one circle ML and the other PV. These circles will be targets for your culture stabs. Put your initials on the plate also.

Using proper aseptic techniques, stab the medium in the center of the ML circle with *M. luteus* and the center of the other circle with *P. vulgaris*. Incubate the plate for 24 to 48 hours at room temperature.

Second Period

Assemble the following materials that were inoculated during the last period and incubated.

Materials

- culture tubes of motility medium that have been incubated
- inoculated petri plate that has been incubated

Compare the two tubes that were inoculated with *M. luteus* and *P. vulgaris*. Look for cloudiness as evidence of motility. *Proteus* should exhibit motility. Does it? Record your results on the Laboratory Report.

Compare the appearance of the two stabs in the soft agar. Describe the differences that exist in the two stabs.

Does the plate method provide any better differentiation of results than the tube method?

Laboratory Report

Complete the Laboratory Report for this exercise.

Laboratory Report

Student: _____

Date: _____ Section: _____

18 Motility Determination

A. Results

1. Which bacterial species exhibited true motility on the slides?

   _____

   _____

2. Draw the appearance of the inoculated SIM tubes. Did the semisolid medium inoculations concur with the slide results?

   _____

   _____

   Micrococcus luteus   *Proteus vulgaris*

B. Short Answer Questions

1. Describe the structure of a flagellum. How do flagella generate cell motility?

   _____

   _____

2. If you compared two motile bacterial species and determined one was considerably more motile than the other, which arrangement of flagella would you expect to be associated with the highly motile species? How would you confirm this supposition?

   _____

   _____

3. Differentiate between the following types of movement observed in a wet mount or hanging drop slide.

   a. true motility.

   _____

   b. Brownian movement.

   _____

   c. water current movement.

   _____

4. Between wet mount and hanging drop slide preparations, which is more resistant to evaporation? Which works best with phase-contrast microscopy?

   _____

   _____

5. What concentration of agar is used in a semisolid medium for motility determination? How does that compare to a typical solid medium (*see Exercise 19*)? Explain.

   _____

   _____

6. Why are semisolid media sometimes preferred over slide techniques for evaluating bacterial motility?

   _____

   _____

7. If SIM medium was used for motility determination for *Proteus vulgaris,* what noticeable change to the medium was observed? (*Hint: see Exercise 43 to find out what the letters "SIM" represent.*)

   _____

   _____

Culture Methods

5

All nutritional types are represented among the protists. This metabolic diversity requires a multiplicity of culture methods. This unit presents those techniques that have proven most successful for the culture of bacteria, molds, and slime molds.

The first four Exercises (19, 20, 21, and 22) pertain to basic techniques applicable to the cultivation of bacteria.

This unit culminates the basic techniques phase of this course. A thorough understanding of microscopy, slide techniques, and culture methods provides a substantial foundation for the remainder of the exercises in this manual. If independent study projects are to be pursued as a part of this course, the completion of this unit will round out the background knowledge and skills for such work.

Culture Media Preparation

The cultivation of microorganisms on an artificial growth medium requires that the medium supply all the nutritional and energy requirements necessary for growth. However, in some cases, we may not know what the specific nutrient requirements are for a certain organism to grow. In order to cultivate such an organism, we construct a medium using rich extracts of meat or plants that would supply all the amino acids, nucleotide bases, vitamins, and other growth factors required by our organism. Such a medium is called a **complex medium** because the exact composition and amounts of the individual amino acids, vitamins, growth factors, and other components that make up the medium are not exactly known. Many of the media used in microbiology are complex, such as nutrient agar, which is used to cultivate a variety of bacteria, especially those used in exercises in this manual. For some organisms we know what the specific nutritional requirements are for growth. For example, for *Escherichia coli,* we can prepare a medium composed of specific components and amounts, such as glucose and various salts. This medium is called a **defined medium** because the specific chemical composition is known and the individual components are weighed out exactly to make up the medium.

Nutritional Requirements of Bacteria

Any medium, be it complex or defined, must supply certain basic nutritional requirements that are necessary for all cells to grow. These include a carbon source, energy source, nitrogen, minerals, vitamins and growth factors, and water.

Carbon Source Organisms can be divided into two groups based on their carbon requirements. **Heterotrophs** obtain their carbon from organic compounds such as polysaccharides, carbohydrates, amino acids, peptides, and proteins. Meat and plant extracts are added to complex media to supply these nutrients. In contrast, **autotrophs** derive their carbon requirements from fixing carbon dioxide. From the latter, they must synthesize all the complex molecules that comprise the bacterial cell.

Energy Source Bacterial cells require energy to carry out biosynthetic processes that lead to growth.

These include synthesizing nucleic acids, proteins, and structural elements such as cell walls. **Chemoorganotrophs** derive their energy needs from the breakdown of organic molecules by fermentation or respiration. Most bacteria belong to this metabolic group. **Chemolithotrophs** oxidize inorganic ions such as nitrate or iron to obtain energy to fix carbon dioxide. Examples of the chemolithotrophs are the nitrifying and iron bacteria. **Photoautotrophs** contain photosynthetic pigments such as chlorophyll or bacteriochlorophyll that convert solar energy into chemical energy by the process of photosynthesis. Energy derived from photosynthesis can then be used by the cell to fix carbon dioxide and synthesize the various cellular materials necessary for growth. For these organisms, no energy source is supplied in the medium but rather energy is supplied in the form of illumination. The cyanobacteria and the green and purple sulfur bacteria are examples of photoautotrophs that carry out photosynthesis. A few photosynthetic bacteria are **photoheterotrophs.** These organisms also derive their energy requirements from photosynthesis but their carbon needs come from growth on organic molecules such as succinate or glutamate. Some of the purple nonsulfur bacteria are found in this category.

Nitrogen Nitrogen is an essential element in biological molecules such as amino acids, nucleotide bases, and vitamins. Some bacteria can synthesize these compounds using carbon intermediates and inorganic forms of nitrogen (e.g., ammonia and nitrate). Others lack this capability and must gain their nitrogen from organic molecules such as proteins, peptides, or amino acids. Beef extract and peptones are incorporated into complex media to provide a source of nitrogen for these bacteria. Some bacteria are even capable of fixing atmospheric nitrogen into inorganic nitrogen, which can then be used for biosynthesis of amino acids. Bacteria such as *Rhizobium* and *Azotobacter* are examples of nitrogen-fixing bacteria.

Minerals Metals are essential in bacterial metabolism because they are cofactors in enzymatic reactions and are integral parts of molecules such as cytochromes, bacteriochlorophyll, and vitamins. Metals required for growth include sodium, potassium, calcium, magnesium, manganese, iron, zinc, copper, cobalt, and phosphorus. Most are required in catalytic or very small amounts.

Vitamins and Growth Factors   Vitamins serve as coenzymes in metabolism. For example, the vitamin niacin is a part of the coenzyme NAD, and flavin is a component of FAD. Some bacteria, like the streptococci and lactobacilli, require vitamins because they are unable to synthesize them. Other bacteria (e.g., *Escherichia coli*) can synthesize vitamins and hence do not require them in media in order to grow. However, sometimes, even in addition to supplying all the normal components, it is necessary to add growth factors for ample growth of certain bacteria. Many pathogens are fastidious and grow better if blood or serum components are incorporated into their media. Blood and serum may provide additional metabolic factors not found in the normal components.

Water   The cell consists of 70% to 80% water. Cells require an aqueous environment because enzymatic reactions and transport only occur in its presence. Furthermore, water maintains the various components of the cytoplasm in solution. When preparing media, it is essential to always use either **dionized** or **distilled water.** Tap water can contain minerals such as calcium, phosphorus, and magnesium ions that could react with peptones and meat extracts to cause unwanted precipitates and cloudiness.

In addition to having the right components, it is important to make sure that the pH of the medium is adjusted to optimal values so that growth is not inhibited. Most bacteria grow best at a neutral pH value around 7. Fungi prefer pH values around 5 for best growth. Most commercial media do not require adjusting the pH, but it would probably be necessary to adjust the pH of a defined synthetic medium. This can be done with acids such as HCl or bases such as sodium hydroxide.

Differential and Selective Media

Media can be made with components that will allow certain bacteria to grow but will inhibit others from growing. Such a medium is a **selective medium.** Antibiotics, dyes, and various inhibitory compounds are often incorporated into media to create selective conditions for growing specific organisms. For example, the dyes eosin and methylene blue when incorporated into EMB media do not affect the growth of gram-negative bacteria but do inhibit the growth of gram-positive bacteria. Incorporation of sodium chloride into mannitol-salt agar selects for *Staphylococcus aureus* but inhibits the growth of other bacteria that cannot tolerate the salt concentration.

A **differential medium** contains substances that cause some bacteria to take on an appearance that distinguishes them from other bacteria. When *Staphylococcus aureus* grows on mannitol-salt agar, it ferments mannitol, changing a pH indicator from red to yellow around colonies. Other staphylococci cannot ferment mannitol and their growth on this medium results in no change in the indicator. EMB is also a differential medium. Gram-negative bacteria that ferment lactose in this medium form colonies with a metallic-green sheen. Non-lactose-fermenting bacteria do not form colonies with the characteristic metallic sheen.

Media can be prepared in liquid or solid form depending on the application of its use. Liquid broth cultures are used to grow large volumes of bacteria. Fermentation studies, indole utilization, and the methyl red and Voges-Proskauer tests are done in broth cultures. Streaking of bacteria and selection of isolated colonies are done on solid media. **Agar,** a complex polysaccharide isolated from seaweed, is added in a concentration of 1.5% to solidify liquid media. The use of agar in bacteriological media was first introduced in Robert Koch's laboratory. Agar has unique properties that make it ideal for use in microbiology. First, it melts at 100° C but does not solidify until it cools to 45° C. Bacteria can be inoculated into agar media at this temperature (e.g., pour plates) without killing the cells. Second, agar is not a nutrient for most bacteria (the exceptions are a few bacteria found in marine environments). Sometimes, agar is added at lower concentrations (e.g., 0.4%) to make semisoft media. This type of media is used in motility studies.

Prior to 1930, it was necessary for laboratory workers to prepare media using various raw materials. This often involved boiling plant material or meat to prepare extracts that would be used in the preparation of media. Today, commercial companies prepare and sell media components, which are used for most routine bacteriological media. It is only necessary to weigh out a measured amount of a specific medium, such as nutrient agar, and dissolve it in water. In some cases, it may be necessary to adjust the pH prior to sterilizing the medium.

Before any medium can be used to grow bacteria or other microorganisms, it must be **sterilized,** that is, any contaminating bacteria introduced during preparation must be killed or removed. Most media can be **autoclaved** to achieve sterilization. This involves heating the media to **121° C** for at least **15 minutes** at **15 psi** of steam pressure. These conditions are sufficient to kill cells and any endospores present. Sometimes a medium may require a component that is heat sensitive and it cannot be subjected to autoclave temperatures. The component can be filter sterilized by passing a solution through a bacteriological filter of 0.45 microns. This filter will retain any cells and endospores that may be present. After filtration, the component can be added to the sterilized medium.

Media Preparation Assignment

In this laboratory period, you will work with your laboratory partner to prepare tubes of media that will be used in future laboratory experiments (figure 19.1). Your instructor will indicate which media you are to prepare. Record in the space below the number of tubes of specific media that have been assigned to you and your partner.

nutrient broth _____
nutrient agar pours _____
nutrient agar slants _____
other _____

Several different sizes of test tubes are used for media, but the two sizes most generally used are either 16 mm or 20 mm diameter by 15 cm long. Select the correct size tubes first, according to these guidelines:

Large tubes (20 mm dia): Use these test tubes for *all pours* (i.e., nutrient agar, Sabouraud's agar, EMB agar, etc.). Pours are used for filling petri plates.

Small tubes (16 mm dia): Use these tubes for all *broths, deeps,* and *slants.*

If the tubes are clean and have been protected from dust or other contamination, they can be used without cleaning. If they need cleaning, scrub out the insides with warm water and detergent, using a test-tube brush. Rinse twice, first with tap water and finally with distilled water to rid them of all traces of detergent. Place them in a wire basket or rack, inverted, so that they can drain. Do not dry with a towel.

Measurement and Mixing

The amount of medium you make for a batch should be determined as precisely as possible to avoid shortage or excess.

Materials

- graduated cylinder, beaker, glass stirring rod
- bottles of dehydrated media
- Bunsen burner and tripod, or hot plate

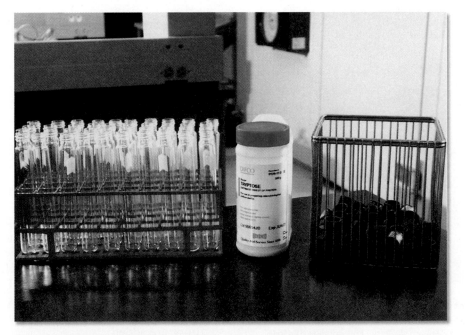

Figure 19.1 Basic supplies for making up a batch of medium.
©The McGraw-Hill Companies/Auburn University Photographic Service

1. Measure the correct amount of water needed to make up your batch. The following volumes required per tube must be taken into consideration:

pours . . . . 12 ml
deeps . . . . 6 ml
slants . . . . 4 ml
broths . . . . 5 ml
broths with fermentation tubes . . . . 5–7 ml

2. Consult the label on the bottle to determine how much powder is needed for 1,000 ml and then determine by proportionate methods how much you need for the amount of water you are using. Weigh this amount on a balance and add it to the beaker of water (figure 19.2). If the medium does not contain agar, the mixture usually goes into solution without heating (figure 19.3).

3. **If the medium contains agar,** heat the mixture on a stirring hot plate (figure 19.4) or on an electric hot plate until it comes to a boil. To safeguard against water loss, *before heating, mark the level of the top of the medium on the side of the beaker with a china marking pencil.* As soon as it "froths up," turn off the heat. If an electric hot plate is used, the medium must be removed from the hot plate or it will boil over the sides of the container.

    Caution: Be sure to keep stirring the medium so that it does not char on the bottom of the beaker.

4. Check the level of the medium with the mark on the beaker to note if any water has been lost. Add sufficient distilled water as indicated. Keep the temperature of the medium at about 60° C to avoid solidification. The medium will solidify at around 40° C.

Adjusting the pH

Although dehydrated media contain buffering agents to keep the pH of the medium in a desired range, the pH of a batch of medium may differ from that stated on the label of the bottle. Before the medium is tubed, therefore, one should check the pH and make any necessary adjustments.

If a pH meter (figure 19.5) is available and already standardized, use it to check the pH of your medium. If the medium needs adjustment, use the bottles of HCl and NaOH to correct the pH. If no meter is available, pH papers will work about as well. Make pH adjustment as follows:

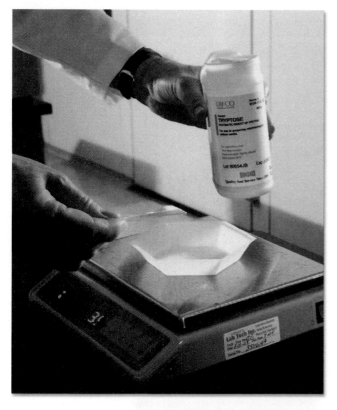

Figure 19.2 Correct amount of dehydrated medium is carefully weighed on a balance.
©The McGraw-Hill Companies/Auburn University Photographic Service

Materials

- beaker of medium
- acid and base kits (dropping bottles of 1N and 0.1N HCl and NaOH)
- glass stirring rod
- pH papers
- pH meter (optional)

1. Dip a piece of pH test paper into the medium to determine the pH of the medium.
2. **If the pH is too high,** add a drop or two of HCl to lower the pH. For large batches use 1N HCl. If the pH difference is slight, use the 0.1N HCl. Use a glass stirring rod to mix the solution as the drops are added.
3. **If the pH is too low,** add NaOH, one drop at a time, to raise the pH. For slight pH differences, use 0.1N NaOH; for large differences use 1N NaOH. Use a glass stirring rod to mix the solution as the drops are added.

Figure 19.3 Dehydrated medium is dissolved in a measured amount of water.
©The McGraw-Hill Companies/Auburn University Photographic Service

Figure 19.4 If the medium contains agar, it must be heated to dissolve the agar.
©The McGraw-Hill Companies/Auburn University Photographic Service

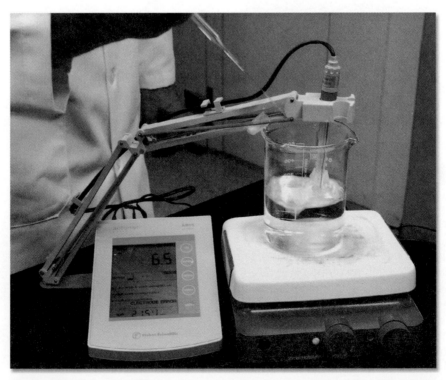

Figure 19.5 The pH of the medium is adjusted by adding acid or base as per recommendations.
©The McGraw-Hill Companies/Auburn University Photographic Service

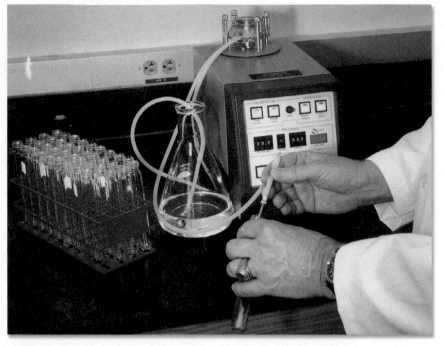

Figure 19.6 An automatic pipetting machine will deliver precise amounts of media to test tubes.
©The McGraw-Hill Companies/Auburn University Photographic Service

Filling the Test Tubes

Once the pH of the medium is adjusted, it must be dispensed into test tubes. If an automatic pipetting machine is to be used, as shown in figure 19.6, it will have to be set up for you by your instructor. These machines can be adjusted to deliver any amount of medium at any desired speed. When large numbers of tubes are to be filled, the automatic pipetting machine should be used.

Materials

- automatic pipetters

1. Follow the instructions provided by your instructor for setting up the automatic pipetter. This will involve adjusting the desired amount of medium to be delivered to each test tube and possibly other settings. If you are using an automatic pipette aid, you will need to repeatedly draw up medium in a pipette and deliver the desired amount by pressing the release button on the pipette aid (figure 19.7).

2. Place the supply tube into the medium and proceed to fill each test tube according to the type of delivery system you are using. Your instructor will help you with this step.

Figure 19.7 Small batches of media can be delivered with hand-held automatic pipetters.
©The McGraw-Hill Companies/Auburn University Photographic Service

3. If you are delivering agar medium, keep the beaker of medium on a stirring hot plate to maintain the agar in solution. A magnetic stirring bar placed in the medium will aid in constantly stirring the solution.

4. If the medium is to be used for fermentation, add a Durham tube to each tube before filling the test tube. This should be placed in the tube *with the open end of the Durham tube down.* When medium is placed in the test tube, the Durham tube may float on the top of the medium but it will submerge during autoclaving.

Capping the Tubes

The last step before sterilization is to provide a closure for each tube. Plastic (polypropylene) caps are suitable in most cases. All caps that slip over the tube end have inside ridges that grip the side of the tube and provide an air gap to allow steam to escape during sterilization (figure 19.8). If you are using tubes with plastic screw-caps, *the caps should not be screwed tightly before sterilization; instead, each one must be left partly unscrewed about a quarter of a turn.*

If no slip-on caps of the correct size are available, it may be necessary to make up some cotton plugs. A properly made cotton plug should hold firmly in the tube so that it is not easily dislodged.

Sterilization

As soon as the tubes of media have been stoppered, they must be sterilized. Organisms on the walls of the tubes, in the distilled water, and in the dehydrated medium will begin to grow within a short period of time at room temperature, destroying the medium.

Prior to sterilization, the tubes of media should be placed in a wire basket with a label taped on the outside of the basket. The label should indicate the type of medium, the date, and your name.

Sterilization must be done in an autoclave (figure 19.9). The following considerations are important in using an autoclave:

- Check with your instructor on the procedure to be used with your particular type of autoclave. Complete sterilization occurs at 250° F (121.6° C). To achieve this temperature the autoclave has to develop 15 pounds per square inch (psi) of steam pressure. To reach the correct temperature, there must be some provision in the chamber for the escape of air. On some of the older units it is necessary to allow the steam to force air out through the door before closing it.

- *Don't overload the chamber.* One should not attempt to see how much media can be packed into it. Provide ample space between baskets of media to allow for circulation of steam.

Figure 19.8 Once the medium has been dispensed, tubes are capped prior to autoclaving.
©The McGraw-Hill Companies/Auburn University Photographic Service

Figure 19.9 Media is sterilized in an autoclave for 15–20 minutes at 15 psi steam pressure.
©The McGraw-Hill Companies/Auburn University Photographic Service

- *Adjust the time of sterilization to the size of load.* Small loads may take only 10 to 15 minutes. An autoclave full of media may require 30 minutes for complete sterilization.

After Sterilization

Slants If you have a basket of tubes that are to be converted to slants, it is necessary to lay the tubes down in a near-horizontal manner as soon as they are removed from the autoclave. The easiest way to do this is to use a piece of rubber tubing (1/2″ dia) to support the capped end of the tube as it rests on the countertop. Solidification should occur in about 30–60 minutes.

Other Media Tubes of broth, agar deeps, nutrient gelatin, etc., should be allowed to cool to room temperature after removal from the autoclave. Once they have cooled down, place them in a refrigerator or cold-storage room.

Storage If tubes of media are not to be used immediately, they should be stored in a cool place. When stored for long periods of time at room temperature media tend to lose moisture. At refrigerated temperatures media will keep for months.

Laboratory Report

Complete the Laboratory Report for Exercise 19.

Laboratory Report

19   Culture Media Preparation

A. Short Answer Questions

1. Differentiate between complex and defined media.

2. Name six basic nutritional requirements supplied in all culture media.

3. What growth factor is often supplied for cultivation of fastidious bacterial pathogens?

4. A powdered complex medium can be stored for months in the laboratory. However, after preparing the medium, it must be sterilized almost immediately. Why?

5. An autoclave is typically used for sterilization of media.

    a. Define sterilization.

    b. Under what conditions are media typically sterilized in an autoclave?

    c. What type of media components cannot be sterilized in an autoclave?

    d. What is an alternative to autoclaving?

6. Mannitol-salt agar (MSA) is a selective and differential medium.

    a. What is a selective medium? What component(s) of MSA make it selective?

b. What is a differential medium? What component(s) of MSA make it differential?

_____

c. This medium is useful for the isolation and characterization of which microorganisms?

_____

7. Agar is a solidifying agent used in media preparation.

a. What is its origin?

_____

b. What makes it ideal for cultivation of microbes?

_____

c. How and why does the agar concentration in semisolid media differ from conventional solid media?

_____

B. Multiple Choice

Select the answer that best completes the following questions.

1. Most bacteria derive their carbon and energy needs from organic molecules and are classified as
   a. chemolithotrophs.
   b. chemoorganotrophs.
   c. photoautotrophs.
   d. photoheterotrophs.

2. The cyanobacteria use solar energy to fix carbon dioxide. They are classified as
   a. chemolithotrophs.
   b. chemoorganotrophs.
   c. photoautotrophs.
   d. photoheterotrophs.

3. *Rhizobium* and *Azotobacter* are examples of nitrogen-fixing soil bacteria that are classified as
   a. chemolithotrophs.
   b. chemoorganotrophs.
   c. photoautotrophs.
   d. photoheterotrophs.

4. Some purple nonsulfur bacteria utilize solar energy but require an organic carbon source. They are classified as
   a. chemolithotrophs.
   b. chemoorganotrophs.
   c. photoautotrophs.
   d. photoheterotrophs.

ANSWERS

Multiple Choice

1. _____

2. _____

3. _____

4. _____

Preparation of Stock Cultures

Once a microorganism has been isolated into a pure culture, it is important to preserve and maintain the organism in a **stock culture.** Establishing a stock culture will insure that the culture is available for future study. Many of the bacterial cultures that you use in exercises in this manual come from stock cultures maintained at your institution. When needed, the culture is grown out and used in an exercise. Some cultures in the exercises were probably purchased from a commercial culture collection such as the American Type Culture Collection (ATCC) located in Maryland. The ATCC is a repository for cultures of bacteria, fungi, yeast, protozoa, algae, and eukaryotic cell lines and hybridomas in the United States. In addition, the ATCC is a source for bacteriophages and viruses that infect eukaryotic cells. Cultures are deposited with the ATCC by scientific investigators who have isolated new or novel microorganisms, or who have genetically modified an existing organism for some purpose. This makes the culture available for study by other scientists or to institutions of higher learning.

Many factors are essential in maintaining a stock culture. First, it important to insure that the culture does not become contaminated. This is not a trivial task as stock cultures must be transferred to a fresh medium routinely to prevent death of the culture. The transfer process is dependent on aseptic technique, which prevents contamination by other microorganisms. When preserving a culture, it is crucial to provide conditions that will minimize physiological or genetic changes to the culture. Last, the method of preservation should result in the greatest survival of cells. When cultures are preserved and maintained, these factors are considered in choosing the type of preservation that will be used. Stock cultures can be maintained in several ways, including freezing, lyophilization, and storage under mineral oil.

Freezing Most stock cultures are maintained at low temperatures. Because micro-ice crystals can form in the cytoplasm during freezing, which can lyse cells, cryoprotectants such as glycerol are often added to cultures to prevent this from occurring. Freezing at very low temperatures, at $-196°$ C in ultracold freezers or in liquid nitrogen, is considered superior to freezing at $-20°$ C in conventional freezers because cultures can be maintained for longer periods of time.

Lyophilization Cultures can be quickly frozen and the water in the cytoplasm removed under vacuum in a process called lyophilization, or freeze-drying. Because water is removed from the cell, metabolic activity cannot take place and, hence, physiological or mutational changes are unlikely to occur. Lyophilization will also allow cultures to be kept for very long periods of time. Lyophilized cells can be revitalized by adding sterile water to the dry cultures and plating them onto fresh media.

Storage under Mineral Oil Some cultures can be stored under a layer of sterile mineral oil and stored in the refrigerator or even at ambient temperature. The layer of oil limits air from reaching the culture and also limits metabolic activity of cells.

The preparation of stock cultures will be particularly important when you are trying to identify an unknown organism because a stock culture will be the source for making numerous inoculations in various test media. Normally, two different stock cultures are established: (1) the **reserve stock culture** and (2) a **working stock culture.** The reserve stock culture is used only to maintain the culture. This stock culture is transferred to fresh media on a routine basis, but it is not used for making transfers for staining or for performing metabolic tests. Limiting the number of times the culture is opened will limit the potential for contamination.

The working stock culture is used for making inoculations into test media. It is the culture from which you will obtain cells for staining reactions and for performing tests to identify your unknown organism. When it becomes old or if it becomes contaminated, it is refreshed by preparing a new culture from the reserve stock culture.

In the following exercise, you will set up cultures to test the difference in viability of cultures stored at $-20°$ C compared to those maintained at ambient temperature. The principles in this exercise will also be applied later when you begin the exercises on the identification of an unknown bacterium, Exercise 39.

When you begin the identification of an unknown, you will be encouraged to refer back to this exercise.

Materials

- cultures of *Escherichia coli* and *Bacillus megaterium* on tryptone agar slants
- tryptone agar slants

First Period

1. Inoculate two tryptone agar slants with *E. coli*. Label each tube with the organism and the date of inoculation. Also, label one tube "freezer" ($-20°$ C) and the other tube "room temperature" (RT).

2. Repeat this same procedure for *B. megaterium.*
3. Incubate the cultures at $37°$ C for 24 hours.
4. Place the freezer tubes at $-20°$ C and the other tubes at room or ambient temperatures. Allow the tubes to remain at these temperatures for 4–6 weeks.

After 4–6 Weeks

1. Culture the respective tubes onto fresh tryptone agar slants and incubate at $37°$ C for 24 hours. Compare the appearance of the cultures and note any difference in the viability of the cultures.
2. Record your results in Laboratory Report 20.

Laboratory Report

Student: _____

Date: _____   Section: _____

20 Preparation of Stock Cultures

A. Results

1. Appearance of slants after growth for 24 hours:

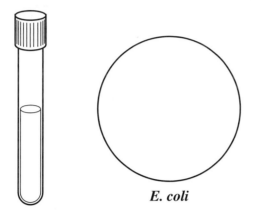

E. coli   B. megaterium

2. Appearance of culture after 4–6 weeks:

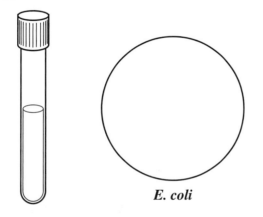

E. coli   B. megaterium

a. Room Temperature

b. Freezer (−20° C)

3. Growth of cultures after 4–6 weeks:

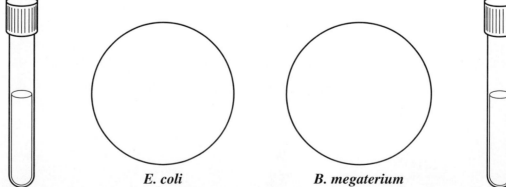

E. coli   B. megaterium

      a. Room Temperature

      b. Freezer ($-20°$ C)

B. Short Answer Questions

1. Why is the practice of aseptic technique essential for preparing a stock culture?

2. What is the importance of having a working stock culture?

3. Why is glycerol often added to freezer stock cultures?

4. What advantages does lyophilization have over freezing?

5. Why might *B. megaterium* cultures be viable when kept at room temperatures whereas the *E. coli* cultures lose viability?

Cultivation of Anaerobes

Bacteria can be classified into groups based upon their metabolic need for oxygen, which comprises about 20% of the atmospheric gases.

Obligate (Strict) Aerobes   These bacteria require oxygen for growth because they carry out respiratory metabolism in which oxygen functions as a terminal electron acceptor. Examples are *Pseudomonas* and *Micrococcus.*

> **Microaerophiles**   These are bacteria that prefer to grow in oxygen concentrations of 5–10% rather than the 20% found in the atmosphere and hence they are more sensitive to oxygen. Their sensitivity to oxygen may be due to the fact that they are limited in their ability to carry out respiration or because they may have oxygen-sensitive proteins and enzymes. *Helicobacter pylori* is a microaerophile that causes stomach ulcers in humans. The oxygen concentration in the stomach is less than the 20% that occurs in the atmosphere.

> **Facultative Aerobes**   These bacteria can grow aerobically or anaerobically depending on the culture conditions. Aerobic growth in the presence of oxygen is by respiration, but if oxygen is absent, they can grow by other modes of metabolism such as fermentation. *Escherichia coli* is a facultative anaerobe.

Anaerobes   The bacteria that belong to this group can be further divided into subgroups depending on their sensitivity to oxygen or their ability to use oxygen under certain conditions:

> **Aerotolerant Anaerobes**   This group of anaerobes can grow in the presence of oxygen and are not usually harmed by its presence in the environment. Their metabolism does not require oxygen but rather involves fermentation in which metabolic intermediates such as pyruvate serve as terminal electron acceptors. An example is *Streptococcus pyogenes,* an important pathogen of humans that causes several diseases such as strep throat and heart and kidney infections.

> **Obligate (Strict) Anaerobes**   Bacteria belonging to this group cannot tolerate oxygen and must be cultivated under conditions in which oxygen is removed, otherwise they are killed. Obligate anaerobes carry out fermentation or employ **anaerobic respiration,** in which inorganic compounds such as sulfate or nitrates replace oxygen as the terminal electron acceptor. Examples of strict anaerobes are *Clostridium* and *Bacteroides.*

The reason for the sensitivity of strict anaerobes to oxygen is not completely understood. Most strict anaerobes have proteins and enzymes that are damaged by oxygen because of the presence of sensitive sulfhydryl groups in proteins. It is also known that toxic forms of oxygen are generated in aerobic environments by photochemistry and even as a result of aerobic metabolism. For example, hydrogen peroxide is a by-product of respiration. Other toxic forms of oxygen include singlet oxygen and the anion superoxide. Hydrogen peroxide is a strong oxidant and superoxide is a highly reactive form of oxygen. Both of these forms of oxygen can interact with biological molecules such as proteins and nucleic acids to seriously damage them and cause cell death. Aerobic bacteria have evolved enzymatic systems that convert some of the toxic forms of oxygen to less harmful compounds. For example, aerobes contain the enzyme **catalase,** which degrades hydrogen peroxide into oxygen and water. Aerobes also contain **superoxide dismutase,** which converts superoxide to oxygen and hydrogen peroxide and the latter can be acted upon by catalase. Another detoxifying enzyme that occurs in aerobes is **peroxidase,** which requires the coenzyme NAD^+ to break down hydrogen peroxide into water. Aerotolerant anaerobes lack catalase but they may possess superoxide dismutase or other mechanisms to deal with superoxide. Strict anaerobes do not possess either catalase or superoxide dismutase but some may have alternative enzymes to deal with superoxide. The sensitivity of strict anaerobes to oxygen may involve a number of factors and is therefore not strictly due to the absence of detoxifying enzymes such as catalase and superoxide dismutase.

The cultivation of obligate anaerobes requires specialized conditions that eliminate oxygen and therefore its toxic forms. This can be achieved in

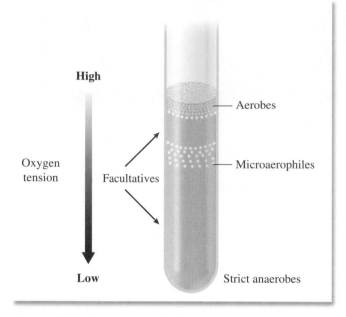

Figure 21.1 Oxygen needs of microorganisms.

anaerobic incubators or in anaerobic jars that employ chemical catalysts to eliminate oxygen. These bacteria can also be cultivated in specialized media that contain chemicals such as thioglycollate, which reacts with oxygen to create anaerobic conditions.

The growth of some bacteria such as the streptococci can be enhanced by cultivation in a candle jar where the concentration of oxygen is less than that in the atmosphere. Cultures are set up in a jar in which a lighted candle is placed. A lid is placed on the jar and tightened, and the candle is extinguished because the oxygen is partially consumed by combustion. The oxygen concentration decreases and the carbon dioxide increases to about 3.5% in the jar.

Figure 21.1 illustrates where the various classes of bacteria grow in a tube in relation to the level of oxygen tension in the medium. In this experiment, you will inoculate various media with several organisms that have different oxygen requirements. The media you will use are fluid thioglycollate (FTM), tryptone glucose yeast extract agar (TGYA), and Brewer's anaerobic agar. Each medium will serve a different purpose. A description for each medium follows:

TGYA Shake   This solid medium will be used to prepare "shake tubes." The medium is not primarily for the cultivation of anaerobes but will be used to determine the oxygen requirements of different bacteria. It will be inoculated in the liquefied state, shaken to mix the organisms throughout the medium, and allowed to solidify. After incubation one determines the oxygen requirements on the basis of where growth occurs in the shake tube: top, middle, or bottom.

FTM   Fluid thioglycollate medium is a rich liquid medium that supports the growth of both aerobic and anaerobic bacteria. It contains glucose, cystine, and sodium thioglycollate to reduce its oxidation-reduction (O/R) potential. It also contains the dye resazurin, which is an indicator for the presence of oxygen. In the presence of oxygen the dye becomes pink. Since the oxygen tension is always higher near the surface of the medium, the medium will be pink at the top and colorless in the middle and bottom. The medium also contains a small amount of agar, which helps to localize the organisms and favors anaerobiasis in the bottom of the tube.

Brewer's Anaerobic Agar   This solid medium is excellent for culturing anaerobic bacteria in petri dishes. It contains thioglycollate, a reducing agent, and resazurin, an oxidation/reduction (O/R) indicator. For strict anaerobic growth, it is essential that plates be incubated in an oxygen-free environment.

To provide an oxygen-free incubation environment for the petri plates of anaerobic agar, we will use the **GasPak anaerobic jar.** Note in figure 21.2

Figure 21.2   The GasPak anaerobic jar.

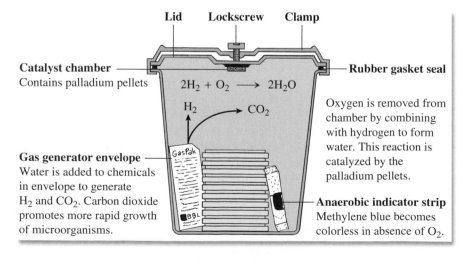

that hydrogen is generated in the jar, which removes the oxygen by forming water. Palladium pellets catalyze the reaction at room temperature. The generation of hydrogen is achieved by adding water to a plastic envelope of chemicals. Note also that CO_2 is produced, which is a requirement for the growth of many fastidious bacteria. To make certain that anaerobic conditions actually exist in the jar, an indicator strip of methylene blue becomes colorless in the total absence of oxygen. If the strip is not reduced (decolorized) within 2 hours, the jar has not been sealed properly, or the chemical reaction has failed to occur.

In addition to doing a study of the oxygen requirements of six organisms in this experiment, you will have an opportunity during the second period to do a microscopic study of the types of endospores formed by three spore-formers used in the inoculations. Proceed as follows:

First Period
(Inoculations and Incubation)

Since six microorganisms and three kinds of media are involved in this experiment, it will be necessary for economy of time and materials to have each student work with only three organisms. The materials list for this period indicates how the organisms will be distributed.

During this period, each student will inoculate three tubes of medium and only one petri plate of Brewer's anaerobic agar. The tubes and all of the plates will be placed in a GasPak jar to be incubated in a 37° C incubator. Students will share results.

Materials

per student:
- 3 tubes of fluid thioglycollate medium
- 3 TGYA shake tubes (liquefied)
- 1 petri plate of Brewer's anaerobic agar

broth cultures for **odd-numbered students:**
- *Staphylococcus aureus, Enterococcus faecalis* and *Clostridium sporogenes*

broth cultures for **even-numbered students:**
- *Bacillus subtilis, Escherichia coli,* and *Clostridium beijerinckii (ATCC 14950)*
- GasPak anaerobic jar, 3 GasPak generator envelopes, 1 GasPak anaerobic generator strip, scissors, one 10-ml pipette, and water baths at student stations (electric hot plate, beaker of water, and thermometer)

1. Set up a 45° C water bath at your station in which you can keep your tubes of TGYA shakes from solidifying. One water bath for you and your laboratory partner will suffice. (Note in the materials list that the agar shakes have been liquefied for you prior to lab time.)

Figure 21.3   Organisms are dispersed in medium by rolling tube gently between palms.

2. Label the six tubes with the organisms assigned to you (one organism per tube), your initials, and assignment number.

   Note: *Handle the tubes gently to avoid taking on any unwanted oxygen into the media. If the tubes of FTM are pink in the upper 30%, they must be boiled a few minutes to drive off the oxygen, then cooled to inoculate.*

3. Heavily inoculate each of the TGYA shake tubes with several loopfuls of the appropriate organism for that tube. To get good dispersion of the organisms in the medium, roll each tube gently between the palms as shown in figure 21.3. To prevent oxygen uptake, do not overly agitate the medium. Allow these tubes to solidify at room temperature.

4. Inoculate each of the FTM tubes with the appropriate organisms.

5. Streak your three organisms on the plate of anaerobic agar in the manner shown in figure 21.4. Note that only three straight-line streaks, well-separated,

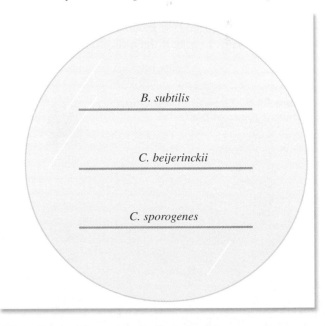

Figure 21.4   Three organisms are streaked on agar plate as straight-line streaks.

are made. Place the petri plate (inverted) in a cannister with the plates of other students that is to go into the GasPak jar.

6. Once all the students' plates are in cannisters, place the cannisters and tubes into the jar.

7. To activate and seal the GasPak jar, proceed as follows:

   a. Peel apart the foil at one end of a GasPak indicator strip and pull it halfway down. The indicator will turn blue on exposure to the air. Place the indicator strip in the jar so that the wick is visible.

   b. Cut off the corner of each of three GasPak gas generator envelopes with a pair of scissors. Place them in the jar in an upright position.

   c. Pipette 10 ml of tap or distilled water into the open corner of each envelope. Avoid forcing the pipette into the envelope.

   d. Place the inner section of the lid on the jar, making certain it is centered on top of the jar. Do not use grease or other sealant on the rim of the jar since the O-ring gasket provides an effective seal when pressed down on a clean surface.

   e. Unscrew the thumbscrew of the outer lid until the exposed end is completely withdrawn into the threaded hole. Unless this is done, it will be impossible to engage the lugs of the jar with the outer lid.

   f. Place the outer lid on the jar directly over the inner lid and rotate the lid slightly to allow it to drop in place. Now rotate the lid firmly to engage the lugs. The lid may be rotated in either direction.

   g. Tighten the thumbscrew by turning clockwise. If the outer lid raises up, the lugs are not properly engaged.

8. Place the jar in a 37° C incubator. After 2 or 3 hours, check the jar to note if the indicator strip has lost its blue color. If decolorization has not occurred, replace the palladium pellets and repeat the entire process.

9. Incubate the tubes and plates for 24 to 48 hours.

Second Period

(Culture Evaluations and Spore Staining)

Remove the lid from the GasPak jar. If vacuum holds the inner lid firmly in place, break the vacuum by sliding the lid to the edge. When transporting the plates and tubes to your desk *take care not to agitate the FTM tubes.* The position of growth in the medium can be easily changed if handled carelessly.

Materials

- tubes of FTM
- shake tubes of TGYA
- 2 Brewer's anaerobic agar plates
- spore-staining kits and slides

1. Compare the six FTM and TGYA shake tubes that you and your laboratory partner share with figure 21.5 to evaluate the oxygen needs of the six organisms.

2. Compare the growths (or lack of growth) on your petri plate and the plate of your laboratory partner.

3. Record your results on Laboratory Report 21.

4. If time permits, make a combined slide with three separate smears of the three spore-formers, using the spore-staining methods in Exercise 16. Draw the organisms in the circles provided in Laboratory Report 21.

Laboratory Report

Complete Laboratory Report 21.

Figure 21.5 Growth patterns for different types of bacteria.

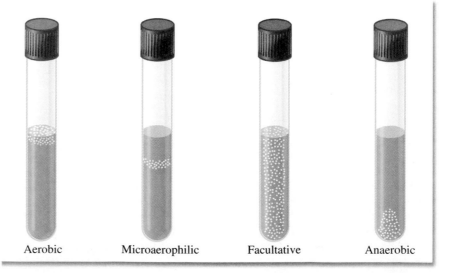

| Aerobic | Microaerophilic | Facultative | Anaerobic |

Laboratory Report

Student: _____

Date: _____ Section: _____

21   Cultivation of Anaerobes

A. Results

1.   Tube Inoculations

After carefully comparing the appearance of the six cultures belonging to you and your laboratory partner, select the best tube for each organism and sketch its appearance in the tubes below. Indicate under
each name the type of medium (FTM or TGYA).

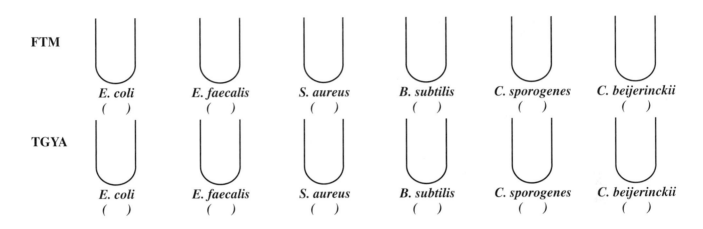

FTM

| E. coli | E. faecalis | S. aureus | B. subtilis | C. sporogenes | C. beijerinckii |
| ( ) | ( ) | ( ) | ( ) | ( ) | ( ) |

TGYA

| E. coli | E. faecalis | S. aureus | B. subtilis | C. sporogenes | C. beijerinckii |
| ( ) | ( ) | ( ) | ( ) | ( ) | ( ) |

2.   Plate Inoculations

After comparing the growths on the two plates of Brewer's anaerobic agar with the growths in the six
tubes, classify each organism as to its oxygen requirements:

Escherichia coli: _____

Bacillus subtilis: _____

Enterococcus faecalis: _____

Clostridium sporogenes: _____

Staphylococcus aureus: _____

Clostridium beijerinckii: _____

3.   Spore Study

If a spore-stained slide is made of the three spore-formers, draw a few cells of each organism in the
spaces provided.

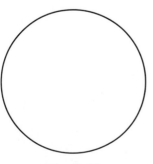

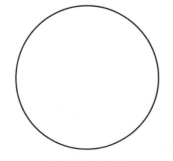

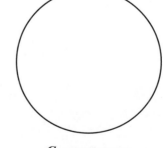

B. subtilis   *C. beijerinckii*   *C. sporogenes*

B. Short Answer Questions

1. What is the role of oxygen for cellular respiration?

   _____

   _____

2. What type of metabolism occurs in the absence of oxygen?

   _____

   _____

3. Name two enzymes that are present in obligate aerobes but lacking in obligate anaerobes. What is the function of each enzyme?

   _____

   _____

4. Differentiate between a microaerophile and an aerotolerant organism.

   _____

   _____

5. Why is resazurin a useful media additive for the study of anaerobes?

   _____

   _____

6. Why is a GasPak anaerobic jar necessary for the culture of anaerobes on plates of Brewer's anaerobic agar but not in tubes of fluid thioglycollate medium (FTM)?

   _____

   _____

Enumeration of Bacteria:
The Standard Plate Count

It is often essential to determine the numbers of bacteria in a sample. For example, the grading of milk is based on the numbers of bacteria present. Whether a patient has a bladder infection is dependent on a certain threshold level of bacteria present in a urine sample. Sometimes it is just important to know how many bacteria are present in food or water to determine purity. Several different methods can be used to determine the number of bacterial cells and each method has its own advantages and disadvantages. The use of one method over another will be dictated by the purpose of the study. The following are some of the methods for determining bacterial numbers:

Microscopic Counts A sample can be diluted and the cells in the sample can be counted with the aid of a microscope. Special slides, such as the Petroff-Hauser chamber, facilitate counting because the slide has a grid pattern and the amount of sample delivered to the grid is known. Milk samples can be counted by microscopic means with a great deal of reliability and confidence.

Most Probable Number (MPN) The number of bacteria in a sample can be determined by the relationship of some growth parameter to statistical probability. The safety of drinking water is dependent on there being no sewage contamination of potable water, and this is tested using the MPN method. Indicator bacteria called **coliforms,** which are found in the intestines of humans and warm-blooded animals, ferment lactose to produce acid and gas. The presence of these bacteria in a water sample suggests the potential for disease. A series of tubes with lactose is inoculated with water samples and the pattern of tubes showing acid and gas is compared to statistical tables that give the probable numbers of coliforms present. You will use this procedure in Exercise 61 to test water for the presence of coliforms. The MPN method is limited to testing where statistical tables have been set up for a particular growth parameter.

Standard Plate Count (SPC) The standard plate, or viable count, is one of the most common methods for determining bacterial numbers in a sample. A sample is diluted in a series of dilution blanks as shown in figure 22.1. Aliquots of the dilutions are

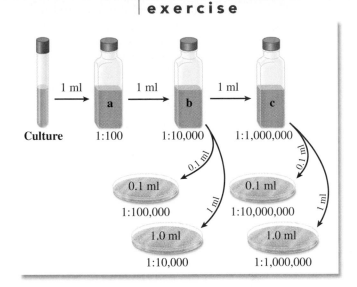

Figure 22.1 Quantitative plating procedure.

then plated onto media and the numbers of colonies are counted after incubation for 24–48 hours. It is assumed that the bacterial cells are diluted to an end point where a single cell divides giving rise to a visible colony on a plate. The number of bacteria in the original sample is determined by multiplying the number of colonies by the dilution factor. However, the assumption that a colony represents a single cell is not always correct because cells in a chain, such as *Streptococcus,* will also give rise to a colony on a plate. Because of the uncertainty in how many actual cells form a colony, counts by the SPC are reported as **colony forming units (cfus).** Only numbers between 30 and 300 cfus are considered statistically valid. If the cfus are greater than 300, there is a probability that overcrowding on the plate could have inhibited some cells from growing. Less than 30 cfus could involve a sampling error and an underestimate of numbers. The SPC method determines only viable cells, whereas a microscopic count determines both living and dead cells. Also, the SPC method is biased because specific conditions and media are used and these factors may exclude certain bacteria in the counts. For example, the SPC would severely underestimate the numbers of bacteria in a soil sample because the conditions and the medium used for the count probably favor heterotrophs that grow aerobically at neutral pH values. These conditions do not allow for the growth of anaerobes, chemolithotrophs, or bacteria that may grow at extremes of pH.

Indirect Methods Sometimes one only wants to know if cells are growing and, therefore, increasing in number. Growth can be related to some parameter that increases with cell division. Growing cells increase their protein, nucleic acid content, and mass because cells are dividing. Thus, measurements of protein, DNA, and dry weight can be used to monitor growth. Likewise, a culture will become more turbid as cells divide, and the **turbidity** of a culture can be determined and related to growth. Cell turbidity can be measured in a spectrophotometer, which measures the **absorbance** or **optical density** of a culture. Oftentimes, a standard curve is constructed that relates optical density to actual numbers of bacteria determined by a SPC. However, one must bear in mind that both living and dead cells will contribute to the culture turbidity, which is also a disadvantage of this method.

In the following exercise, you will use the SPC to determine the numbers of bacteria in a culture. You will also measure the turbidity of a culture and plot the optical density values of diluted samples.

Quantitative Plating Method
(Standard Plate Count)

In determining the number of organisms present in water, milk, and food, the **standard plate count (SPC)** procedure is universally used. It is relatively easy to perform and gives excellent results. We can also use this basic technique to calculate the number of organisms in a bacterial culture. It is in this respect that this assignment is set up.

The procedure consists of diluting the organisms with a series of sterile water blanks as illustrated in figure 22.1. Generally, only three bottles are needed, but more could be used if necessary. By using the dilution procedure indicated here, a final dilution of 1:1,000,000 occurs in blank C. From blanks B and C, measured amounts of the diluted organisms are transferred into empty petri plates. Nutrient agar, cooled to 50° C, is then poured into each plate. After the nutrient agar has solidified, the plates are incubated for 24 to 48 hours and examined. A plate that has between 30 and 300 colonies is selected for counting. From the count it is a simple matter to calculate the number of organisms per milliliter of the original culture. It should be pointed out that greater accuracy can be achieved by pouring two plates for each dilution and averaging the counts. Duplicate plating, however, has been avoided for economic reasons.

Pipette Handling

Success in this experiment depends considerably on proper pipetting techniques. Pipettes may be available to you in metal cannisters or in individual envelopes; they may be disposable or reusable. Because of potential hazards, no mouth pipetting is allowed, and all pipetting is therefore done using pipette aids. Your instructor will indicate the techniques that will prevail in this laboratory. If this is the first time that you have used sterile pipettes, consult figure 22.2, keeping the following points in mind:

- When removing a sterile pipette from a cannister, do so without contaminating the ends of the other pipettes with your fingers. This can be accomplished by *gently* moving the cannister from side to side in an attempt to isolate one pipette from the rest.
- After removing your pipette, replace the cover on the cannister to maintain sterility of the remaining pipettes.
- Don't touch the body of the pipette with your fingers or lay the pipette down on the table before or after you use it. **Keep that pipette sterile** until you have used it, and don't contaminate the table or yourself with it after you have used it.
- Always use a mechanical pipetting device such as the one in illustration 3, figure 22.2.
- Remove and use only one pipette at a time; if you need 3 pipettes for the whole experiment and remove all 3 of them at once, there is no way that you will be able to keep 2 of them sterile while you are using the first one.
- When finished with a pipette, place it in the *discard cannister.* The discard cannister will have a disinfectant in it. At the end of the period, reusable pipettes will be washed and sterilized by the laboratory assistant. Disposable pipettes will be discarded. Students have been known to absent-mindedly return used pipettes to the original sterile cannister, and, occasionally, even toss them into the wastebasket. We are certain that no one in this laboratory would *ever* do that!

Diluting and Plating Procedure

Proceed as follows to dilute out a culture of *E. coli* and pour four plates, as illustrated in figure 22.1.

Materials

per 4 students:
- 1 bottle (40 ml) broth culture of *E. coli*

per student:
- 1 bottle (80 ml) nutrient agar
- 4 petri plates

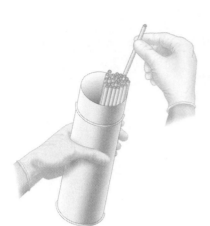

(1) Reusable pipettes may be available in disposable envelopes or metal cannisters. When using pipettes from cannisters be sure to cap them after removing a pipette.

(2) Never touch the tip or barrel of a pipette with your fingers. Contaminating the pipette will contaminate your work.

(3) Use a pipette aid for all pipetting in this laboratory. Pipetting by mouth is too hazardous.

(4) After using a pipette place it in the discard cannister. Even "disposable" pipettes must be placed here.

Figure 22.2   Pipette-handling techniques.

- 1.1 ml pipettes
- 3 sterile 99 ml water blanks
- cannister for discarded pipettes

1. Liquefy a bottle of nutrient agar. While it is being heated, label three 99 ml sterile water blanks **A, B,** and **C.** Also, label the four petri plates **1:10,000, 1:100,000, 1:1,000,000,** and **1:10,000,000.** In addition, indicate with labels the amount to be pipetted into each plate (**0.1 ml** or **1.0 ml).**
2. Shake the culture of *E. coli* and transfer 1 ml of the organisms to blank A, using a sterile 1.1 ml pipette. After using the pipette, place it in the discard cannister.
3. Shake blank A 25 times in an arc of 1 foot for 7 seconds with your elbow on the table as shown in figure 22.3. Forceful shaking not only brings about good distribution, but it also breaks up clumps of bacteria.

4. With a different 1.1 ml pipette, transfer 1 ml from blank A to blank B.
5. Shake water blank B 25 times in same manner.
6. With another sterile pipette, transfer 0.1 ml from blank B to the 1:100,000 plate and 1.0 ml to the 1:10,000 plate. With the same pipette, transfer 1.0 ml to blank C.
7. Shake blank C 25 times.
8. With another sterile pipette, transfer from blank C 0.1 ml to the 1:10,000,000 plate and 1.0 ml to the 1:1,000,000 plate.
9. After the bottle of nutrient agar has boiled for 8 minutes, cool it down in a water bath at 50° C for **at least 10 minutes.**
10. Pour one-fourth of the nutrient agar (20 ml) into each of 4 plates. Rotate the plates **gently** to get

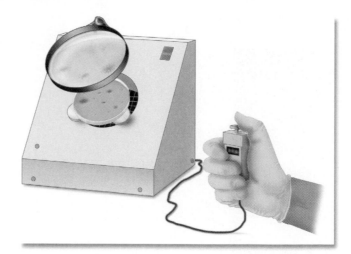

Figure 22.4 Colony counts are made on a Quebec counter, using a mechanical hand tally.

Figure 22.3 Standard procedure for shaking water blanks.

adequate mixing of medium and organisms. **This step is critical!** Too little action will result in poor dispersion and too much action may slop inoculated medium over the edge.

11. After the medium has cooled completely, incubate at 35° C for 48 hours, inverted.

Counting and Calculations

Materials

- 4 culture plates
- Quebec colony counter
- mechanical hand counter
- felt-tip pen (optional)

1. Lay out the plates on the table in order of dilution and compare them. *Select the plates that have no fewer than 30 nor more than 300 colonies for your count.* Plates with less than 30 or more than 300 colonies are statistically unreliable.

2. Place the plate on the Quebec colony counter with the lid removed. See figure 22.4. Start counting at the top of the plate, using the grid lines to prevent counting the same colony twice. Use a mechanical hand counter. Count every colony, regardless of how small or insignificant. Record counts on the table in section A of Laboratory Report 22.

   Alternative Counting Method: Another way to do the count is to remove the lid and place the plate upside down on the colony counter. Instead of using the grid to keep track, use a felt-tip pen to mark off each colony as you do the count.

3. Calculate the number of bacteria per ml of undiluted culture using the data recorded in section A of Laboratory Report 22. Multiply the number of colonies counted by the dilution factor (the reciprocal of the dilution).

   Example: If you counted 220 colonies on the plate that received 1.0 ml of the 1:1,000,000 dilution: $220 \times 1,000,000$ (or 2.2×10^8) bacteria per ml. If 220 colonies were counted on the plate that received 0.1 ml of the 1:1,000,000 dilution, then the above results would be multiplied by 10 to convert from number of bacteria per 0.1 ml to number of bacteria per 1.0 ml (2,200,000,000, or 2.2×10^9).

   Use only two significant figures. If the number of bacteria per ml was calculated to be 227,000,000, it should be recorded as 230,000,000, or 2.3×10^8.

Determination of Growth by Optical Density

Turbidity can give a quick indication that a culture is growing but it does not give actual cell numbers. For an actively growing culture, turbidity will increase with time. The turbidity of a culture can be measured in a spectrophotometer because a bacterial culture acts like a colloidal suspension. As light passes through the culture, it will be absorbed by the bacterial cells and the light emerging from the culture will be proportionally decreased by the number of

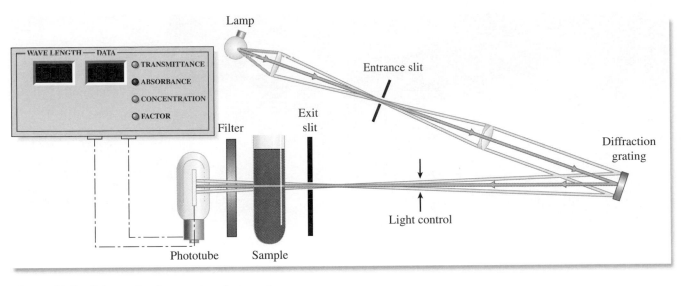

Figure 22.5 Schematic of a spectrophotometer.

cells present. Therefore, within certain defined limits, the amount of light absorbed is proportional to the number of cells present.

Figure 22.5 illustrates the path of light through a spectrophotometer. A beam of white light passes through a series of lenses and a slit and onto a diffraction grading where the light is separated into different wavelengths of the visible spectrum. A specific wavelength of monochromatic light can be selected from the diffraction grading by an exit slit. Adjusting the wavelength control on the instrument will reorient the diffraction grading so that a different wavelength can be selected by the exit slit. The monochromatic light then passes through a sample and activates a photo-mutiplier tube that measures the **absorbance (optical density, O.D.)** on a galvanometer. The higher the absorbance, the greater the concentration of bacterial cells. In the following exercise, you will demonstrate the relationship between O.D. and cell turbidity by measuring the optical density values for various dilutions of a culture.

There should be a direct proportional relationship between the concentration of bacterial cells and the absorbance (optical density, O.D.) of the culture. To demonstrate this principle, you will measure the O.D. of various dilutions of the culture provided to you. These values will be plotted on a graph as a function of culture dilution. You may find that there is a linear relationship between concentration of cells and O.D. only up to a certain O.D. At higher O.D. values, the relationship may not be linear. That is, for a doubling in cell concentration, there may be less than a doubling in O.D.

Materials

- broth culture of *E. coli* (same one as used for plate count)
- spectrophotometer cuvettes (2 per student)
- 4 small test tubes and test-tube rack
- 5 ml pipettes
- bottle of sterile nutrient broth (20 ml per student)

1. Calibrate the spectrophotometer using the procedure described in figure 22.6. These instructions apply specifically to the Bausch and Lomb Spectronic 20 digital spectrophotometer. It is important to blank the spectrophotometer by adjusting the instrument to an absorbance of 0 (zero) using uninoculated nutrient broth. The medium contains components that cause it to be slightly colored and, hence, it will absorb some light, adding to the light absorbance of the bacterial culture tubes. Blanking the instrument using the uninoculated medium will subtract the absorbance resulting from the medium. In handling the cuvettes, keep the following in mind:
   a. Rinse the cuvette with distilled or deionized water to clean it before using.
   b. Keep the lower part of the cuvette free of liquids, smudges, and fingerprints by carefully wiping the surface only with Kim wipes or lint-free tissue provided. Do not use paper towels or handkerchiefs for this purpose. If smudges, liquids, or fingerprints occur on the cuvette surface, they can contribute to light absorbance and erroneous readings.

(a) Turn on the instrument with the on/off switch (left front knob). Allow the spectrophotometer to warm up for 15 minutes. Set the wavelength to 550 nm and position the filter to correspond to this wavelength. Select the Transmittance mode and make sure the cuvette holder cover is closed. Adjust the digital readout to zero % transmittance using the left front (on/off) knob (figure 22.6a).

(b) Set the mode to Absorbance. Insert a cuvette containing sterile nutrient broth into the sample holder and close the cover (figure 22.6b). Adjust the Absorbance to 0 (zero) using the right front knob. This may require several turns of the knob (figure 22.6c).

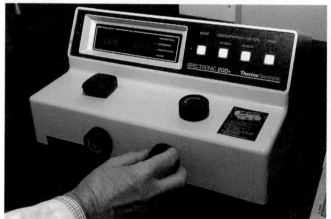

(c) Remove the blank and insert a cuvette with one of the bacterial cell samples. Close the cover and read the absorbance /O.D.

(d) Occasionally blank the instrument to zero with the sterile nutrient broth during the course of reading the cell samples.

Figure 22.6 Calibration procedure for the B & L Spectronic 20 digital spectrophotometer.
© The McGraw-Hill Companies/Auburn University Photographic Services.

    c. Insert the cuvette into the sample holder with the index line aligned with the index line on the cuvette holder. Properly seat the cuvette by exactly aligning the lines on the cuvette and holder.

    d. Handle cuvettes with care as they are of optical quality and expensive.

2. Label a cuvette 1:1 (near top of tube) and four test tubes 1:2, 1:4, 1:8, and 1:16. These tubes will be used for the serial dilutions shown in figure 22.7.

3. With a 5 ml pipette, dispense 4 ml of sterile nutrient broth into tubes 1:2, 1:4, 1:8, and 1:16.

4. Shake the culture of *E. coli* vigorously to suspend the organisms, and with the same 5 ml pipette, transfer 4 ml to the 1:1 cuvette and 4 ml to the 1:2 test tube.

5. Mix the contents in the 1:2 tube by drawing the mixture up into the pipette and discharging it into the tube three times.

6. Transfer 4 ml from the 1:2 tube to the 1:4 tube, mix three times, and go on to the other tubes in a similar manner. Tube 1:16 will have 8 ml of diluted organisms.

7. Measure the optical density of each of the five tubes, starting with the 1:16 tube first. The contents of each of the test tubes must be transferred to a cuvette for measurement. Be sure to close the lid on the sample holder when making measurements. A single cuvette can be used for all the measurements.

8. Record the O.D. values in the table of Laboratory Report 22.

9. Plot the O.D. values on the graph of Laboratory Report 22.

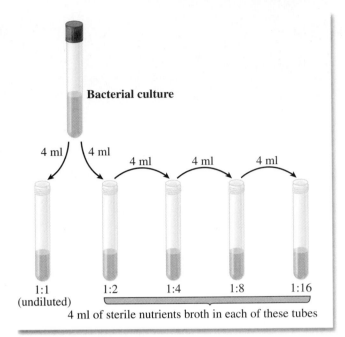

Figure 22.7   Dilution procedure for cuvettes.

Laboratory Report

Student: _____

Date: _____ Section: _____

22 Enumeration of Bacteria: The Standard Plate Count

A. Results

1. Quantitative Plating Method

   a. Record your plate counts in this table:

DILUTION PLATED	ml PLATED	NUMBER OF COLONIES
1:10,000	1.0	
1:100,000	0.1	
1:1,000,000	1.0	
1:10,000,000	0.1	

   b. How many cells per ml were there in the undiluted culture?_____

2. Optical Density Determination

   a. Record the optical density values for your dilutions in the following table.

DILUTION	OPTICAL DENSITY
1:1	
1:2	
1:4	
1:8	
1:16	

b. Plot the optical densities versus the concentration of organisms. Complete the graph by drawing a line between plot points.

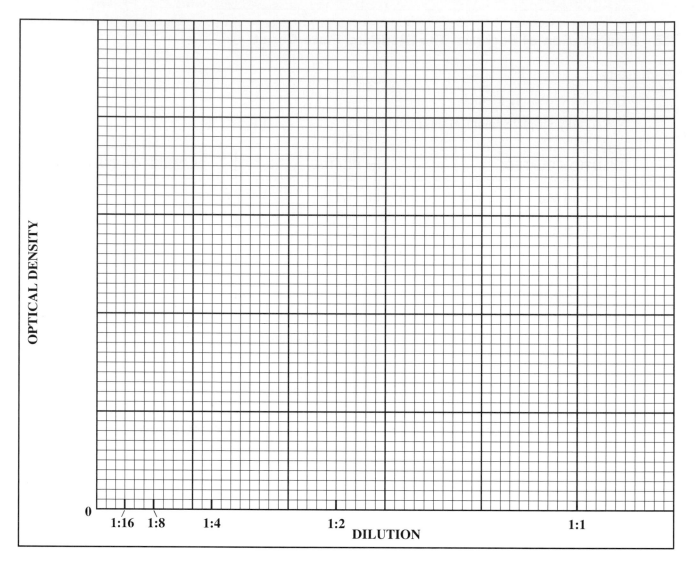

c. What is the maximum O.D. that is within the linear portion of the curve? _____

d. What is the corrected or true O.D. of the undiluted culture? (Hint: If the O.D. for the 1:2 dilution but not the 1:1 dilution is within the linear portion of the curve, then the O.D. of the 1:1 dilution should not be considered correct. The correct or true O.D. of the undiluted culture in this example could be estimated by multiplying the O.D. of the 1:2 dilution by 2.) _____

e. What is the correlation between corrected O.D. and cell number for your culture? _____

_____

_____

B. Short Answer Questions

1. When performing a standard plate count, why are the counts reported as colony forming units (CFUs) rather than cells?

   _____

   _____

2. How would you inoculate a plate to get a 1:10 dilution? A 1:100 dilution?

   _____

   _____

3. Why is it necessary to perform a plate count in conjunction with the turbidimetry procedure?

   _____

   _____

Slime Mold Culture

The slime molds are eukaryotic, heterotrophic microorganisms that exist in cool, shady, and moist places in the woods where they are found on decaying logs, dead leaves, and other organic material. Unlike bacteria and other microorganisms, they ingest their food by phagocytosis in a manner similar to the amoebas. The vegetative cells are unique in that they lack cell walls. However, when fruiting bodies are formed, cell walls are present in these structures.

The slime molds are taxonomically placed in the Division Gymnomycota in the Kingdom Myccteae. Their classification as protozoans or fungi is a result of their having intermediate characteristics of both organisms.

Figure 23.1 illustrates the life cycle of one type of slime mold, the plasmodial type. The genus *Physarum* is the one to be studied in this experiment. The assimilative stage of this organism is the **plasmodium.** This multinucleate structure is slimy in appearance and moves slowly by flowing its cytoplasm in amoeboid fashion over surfaces on which it feeds. Most species feed on bacteria and possibly on other small organisms that they encounter.

Plasmodial growth continues as long as adequate nutrients and moisture are available. However, because environmental conditions can change, the organism will form a **sclerotia** or a **sporangium.** A **sclerotium** is a hardened mass of irregular shape that forms from the plasmodium when moisture and temperature conditions become less than ideal. When conditions improve, the sclerotium reverts back to a plasmodium. Figure 23.2 is a photograph of two sclerotia that formed on a laboratory culture. **Sporangia** are fruiting structures that form under conditions similar to those that induce the formation of sclerotia. Exactly why sporangia form instead of sclerotia is still not clearly understood. Sporangia form by the separation of the plasmodium into many rounded mounds of protoplasm that extend upward on stalks. The nuclei within the sporangia undergo meiosis to become haploid spores with tough cell walls. The sclerotia and sporangia of figures 23.2 and 23.3 were photographed on the same culture of laboratory-grown *Physarum.*

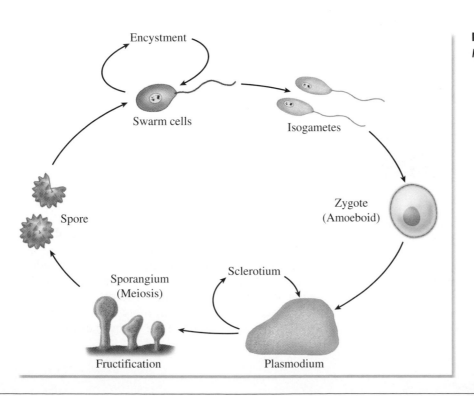

Figure 23.1 Life cycle of *Physarum polycephalum.*

Figure 23.2 Sclerotia of *Physarum polycephalum* (3X).

Figure 23.3 Sporangium of *Physarum polycephalum* (actual size is 3mm tall).

Image originally published in Henry C. Aldrich and John W. Daniel, *Cell Biology of Physarum and Didymium*. New York: Academic Press, 1982. Used with permission

Both sclerotia and spores may survive adverse environmental conditions for long periods of time. Once environmental conditions improve, the spores germinate to produce flagellated pear-shaped **swarm cells.** These swarm cells may do one of three things: (1) they may encyst if conditions suddenly become adverse, (2) they may divide one or more times to form isogametes, or (3) they may act as isogametes and unite directly to form a **zygote.** Once a zygote is formed, it takes on an amoeboid form and undergoes a series of mitotic divisions to produce a plasmodium. This completes the life cycle.

Three procedures will be described here for the study of *Physarum polycephalum:* (1) moist chamber culture, (2) agar culture method, and (3) spore germination technique. The techniques used will be determined by the availability of time and materials.

Moist Chamber Culture

To grow large numbers of plasmodia, sclerotia, and sporangia that can be used for an entire class, one needs to create a rather large moisture chamber. Any covered glass or plastic container that is 10 to 12 inches square or round is suitable.

Materials

- sclerotia of *Physarum polycephalum*
- container for culture ($10\frac{1}{2}$" dia Pyrex casserole dish with cover or 10–12" square plastic box with cover)
- glass petri dish cover
- sharp scalpel
- rolled oat flakes (long-cooking type)
- 10" dia filter paper or paper toweling

1. In the center of the container place a petri dish cover, open end down. Lay a large piece of filter paper or paper toweling over the petri dish and saturate with distilled water. The petri dish provides a raised area above any excess water that may make the paper too wet.
2. With a sharp scalpel transfer a small fragment of sclerotium from the *Physarum* culture to the filter paper. A sclerotium may vary from dark orange to brown in color. See figure 23.2. Moisten the sclerotium with a drop of distilled water.
3. After a few hours the organism will begin to seek food. At this point, place a flake of rolled oats near the edge of the spreading growth for it to feed on.
4. Incubate the moist chamber in a dark place at room temperature. Add distilled water and oat flakes periodically as needed. It is better to add a few fresh flakes daily than to overfeed by applying all flakes at once. Such a culture should keep for several weeks. To promote the formation of sclerotia, allow some of the water to evaporate by

leaving the lid partially open for a while. To bring about sporangia formation, withhold food while keeping the culture moist.

Agar Culture Method

(Plasmodial Study)

An actively metabolizing plasmodium is dark yellow and streaked with vessels. The streaming of protoplasm in these vessels is best observed under the microscope. To be able to study this unique structure, it is best to culture the organism on nonnutrient agar. Make such a culture as follows:

Materials

- rolled oat flakes
- scalpel
- petri plate with 15 ml of nonsterile, nonnutrient agar

1. Lift some occupied oat flakes from the filter paper in the moist chamber and transfer to a plate of nonsterile, nonnutrient 1.5% agar. Maintain this culture by adding fresh oat flakes periodically, but don't add water.
2. After a well-developed plasmodium has formed, study the streaming protoplasm under low power of the microscope. Observation is made by transmitted light through the agar on the microscope stage. Look for periodical reversal of direction of flow.
3. Cut one of the vessels through in which the flow is active and observe the effect.
4. Transfer a piece of plasmodium to another part of the medium and watch it reconstitute itself.
5. Leave the cover slightly open on the petri dish for several days and note any changes that might occur as time goes by.

Spore Germination

The observation of spore germination can be achieved with a hanging drop slide. Once sporangia are in abundance, one can make such a slide as follows:

Materials

- depression slides (sterile)
- plain microscope slides (sterile)
- cover glasses (sterile)
- Vaseline
- toothpicks
- sporangia of *Physarum polycephalum*
- Bunsen burner
- 70% alcohol

1. With a toothpick, place a small amount of Vaseline near each corner of the cover glass. (See figure 18.2, page 124.)
2. Saturate a sporangium with a drop of 70% alcohol on the center of a sterile plain microscope slide.
3. As soon as the alcohol has evaporated, add a drop of distilled water and place another sterile slide over the wet sporangium.
4. Crush the sporangium with thumb pressure on the upper slide. Separate the two slides to expose the crushed sporangium.
5. Transfer a few loopfuls of crushed sporangial material to a drop of distilled water on a sterile cover glass.
6. Place the depression slide over the cover glass, make contact, and quickly invert to produce a completed hanging drop slide.
7. Examine under low and high power.

Laboratory Report

Complete Laboratory Report 23.

Laboratory Report

Student: _____

Date: _____ Section: _____

23   Slime Mold Culture

A. Results

1. What happened when the flow of protoplasm on a plasmodium was interrupted by cutting with a scalpel?

_____

_____

2. Describe your observations of the crushed spores on the hanging drop slide.

_____

_____

_____

B. Short Answer Questions

1. Why has the classification of slime molds been difficult?

_____

_____

2. Define the function(s) of the following structures in the life cycle of *Physarum:*

a. plasmodium.

_____

b. sclerotium.

_____

c. sporangium.

_____

d. swarm cell.

_____

e. zygote.

_____

Slide Culture:
Fungi

The isolation, culture, and microscopic examination of fungi require the use of suitable selective media and special microscopic slide techniques. Simple wet mounts prepared from fungal cultures usually do not reveal the arrangement of spores on fruiting bodies because the manipulation of the culture disrupts the fruiting structures and the hyphae of the culture. The type of fruiting structure and spore arrangement and morphology are important in the identification and taxonomy of these microorganisms. One way to preserve the integrity of the fruiting structure is to prepare a slide culture that can then be stained. This allows the observation of the fruiting structure *in situ* and does not disrupt the arrangement of the spores. In this exercise, a slide culture method will be used to prepare stained slides of molds. The method is superior to wet mounts in that the hyphae, sporangiophores, and spores remain more or less intact when stained.

When fungi are collected from the environment, as in Exercise 8, Sabouraud's agar is most frequently used. It is a simple medium consisting of 1% peptone, 4% glucose, and 2% agar-agar. The pH of the medium is adjusted to 5.6, which favors the growth of fungi but inhibits most bacterial growth.

Unfortunately, for some fungi the pH of Sabouraud's agar is too low and the glucose content is too high. A better medium for these organisms is one suggested by C. W. Emmons that contains only 2% glucose, with 1% neopeptone, and an adjusted pH of 6.8–7.0. To inhibit bacterial growth, 40 mg of chloramphenicol is added to one liter of the medium.

In addition to the above two media, cornmeal agar, Czapek solution agar, and others are available for special applications in culturing molds.

Figure 24.1 illustrates the procedure that will be used to produce a fungal culture on a slide that can be stained directly on the slide. Note that a sterile cube of Sabouraud's agar is inoculated on two sides with spores from a mold colony. Figure 24.2 illustrates how the cube is held with a scalpel blade as inoculation takes place. The cube is placed in the center of a microscope slide with one of the inoculated surfaces placed against the slide. On the other inoculated surface of the cube is placed a cover glass. The assembled slide is incubated at room temperature for 48 hours in a moist chamber (petri dish with a small amount of water). After incuba-

tion, the cube of medium is carefully separated from the slide and discarded.

During incubation the fungal culture will grow over the glass surfaces of the slide and cover glass. By adding a little stain to the slide, a semipermanent slide can be made by placing a cover glass over it. The cover glass can also be used to make another slide by placing it on another clean slide with a drop of stain on it. Before the stain (lactophenol cotton blue) is used, it is desirable to add to the hyphae a drop of alcohol, which acts as a wetting agent.

First Period

(Slide Culture Preparation)

Proceed as follows to make slide cultures of one or more mold colonies.

Materials

- petri dishes, glass, sterile
- filter paper (9 cm dia, sterile)
- glass U-shaped rods
- fungal culture plate (mixture)
- 1 petri plate of Sabouraud's agar or Emmons' medium per 4 students
- scalpels
- inoculating loop
- sterile water
- microscope slides and cover glasses (sterile)
- forceps

1. Aseptically, with a pair of forceps, place a sheet of sterile filter paper in a petri dish.
2. Place a sterile U-shaped glass rod on the filter paper. (Rod can be sterilized by flaming, if held by forceps.)
3. Pour enough sterile water (about 4 ml) on filter paper to completely moisten it.
4. With forceps, place a sterile slide on the U-shaped rod.
5. *Gently* flame a scalpel to sterilize, and cut a 5 mm square block of the medium from the plate of Sabouraud's agar or Emmons' medium.
6. Pick up the block of agar by inserting the scalpel into one side as illustrated in figure 24.2. Inoculate both top and bottom surfaces of the

Figure 24.1 Procedure for making two stained slides from slide culture.

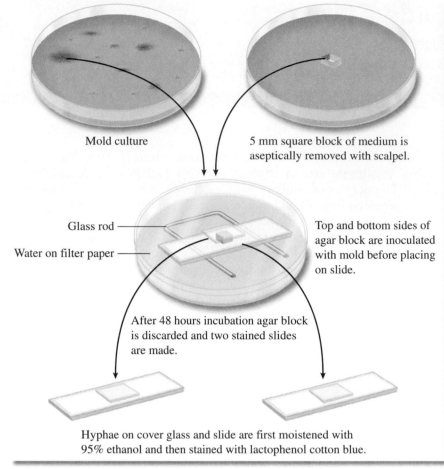

Mold culture

5 mm square block of medium is aseptically removed with scalpel.

Glass rod

Water on filter paper

Top and bottom sides of agar block are inoculated with mold before placing on slide.

After 48 hours incubation agar block is discarded and two stained slides are made.

Hyphae on cover glass and slide are first moistened with 95% ethanol and then stained with lactophenol cotton blue.

cube with spores from the mold colony. Be sure to flame and cool the loop prior to picking up spores.

7. Place the inoculated block of agar in the center of a microscope slide. Be sure to place one of the inoculated surfaces down.

8. Aseptically, place a sterile cover glass on the upper inoculated surface of the agar cube.

9. Place the cover on the petri dish and incubate at room temperature for 48 hours.

10. After 48 hours, examine the slide under low power. If growth has occurred, you should see hyphae and

Figure 24.2 Inoculation technique.

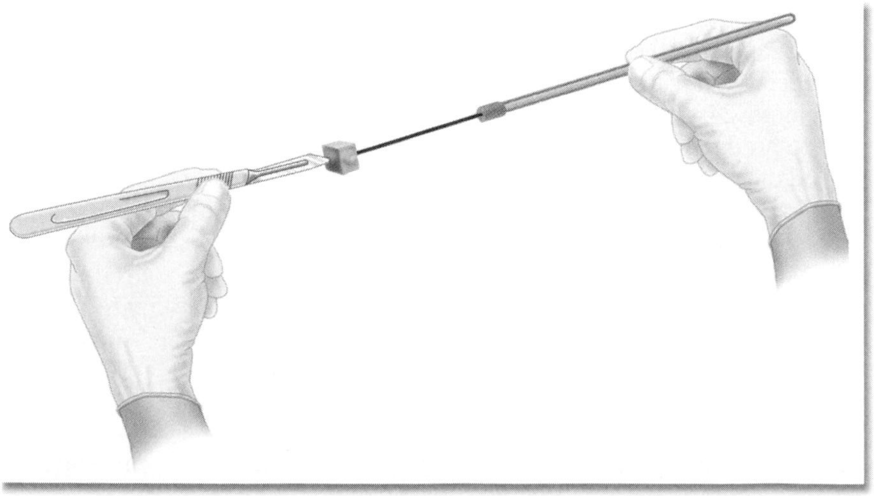

spores. If growth is inadequate and spores are not evident, allow the fungus to grow another 24–48 hours before making the stained slides.

Second Period

(Application of Stain)

As soon as there is evidence of spores on the slide, prepare two stained slides from the slide culture, using the following procedure:

Materials

- microscope slides and cover glasses
- 95% ethanol
- lactophenol cotton blue stain
- forceps

1. Place a drop of lactophenol cotton blue stain on a clean microscope slide.

2. Remove the cover glass from the slide culture and discard the block of agar.
3. Add a drop of 95% ethanol to the hyphae on the cover glass. As soon as most of the alcohol has evaporated, place the cover glass, mold side down, on the drop of lactophenol cotton blue stain on the slide. This slide is ready for examination.
4. Remove the slide from the petri dish, add a drop of 95% ethanol to the hyphae, and follow this up with a drop of lactophenol cotton blue stain. Cover the entire preparation with a clean cover glass.
5. Compare both stained slides under the microscope; one slide may be better than the other one.

Laboratory Report

There is no Laboratory Report for this exercise.

Bacterial Viruses

Viruses differ from bacteria in being much smaller and therefore below the resolution of the light microscope. The smallest virus is one million times smaller than a typical eukaryotic cell. Viruses are obligate intracellular parasites that require a host cell in order to replicate and reproduce, and hence they cannot be grown on laboratory media. Despite these obstacles, we can detect their presence by the effects that they have on their host cells.

Viruses infect all types of cells, eukaryotic and prokaryotic. They are composed of RNA or DNA but never both, and a protein coat, or capsid, that surrounds the nucleic acid. Their dependence on cells is due to their lack of metabolic machinery necessary for the synthesis of viral components. By invading a host cell, they can utilize the metabolic systems of the host cells to achieve their replication.

The study of viruses that parasitize plant and animal cells is time-consuming and requires special tissue culture techniques. Bacterial viruses are relatively simple to study, utilizing ordinary bacteriological techniques. It is for this reason that bacterial viruses will be studied here. However, the principles learned from studying the viruses that infect bacteria apply to viruses of eukaryotic cells.

Viruses that infect bacterial cells are called bacteriophages, or phages. They are diverse in their morphology and size. Some of the simplest ones have single-stranded DNA. Most phages are tadpole-shaped, with "heads" and "tails" as seen in figure 25.1. The capsid (head) may be round, oval, or polyhedral and is composed of individual protein subunits called capsomeres. It forms a protective covering around the viral genome. The tail structure or sheath is composed of a contractile protein that surrounds a hollow core, which is a conduit for the delivery of viral nucleic acid into the host cell. At the end of the tail is a base plate with tail fibers and spikes attached to it. The tail fibers bind to chemical groups on the surface of the bacterial cell and are responsible for recognition. Lysozyme associated with the tail portion of the virus erodes and weakens the cell wall of the host cell. This facilitates the injection of the viral nucleic acid by the sheath contracting and forcing the hollow core through the weakened area in the cell wall.

Infections by viruses can have two outcomes. The lytic cycle involves virulent phages that cause lysis and death of the host cell. The lysogenic cycle involves temperate phages, which can integrate their DNA into host cell DNA and alter the genetics of the host cell.

173

In the lytic cycle, the virus assumes control of cell metabolism and uses the cell's metabolic machinery to manufacture phage components (i.e., nucleic acid, capsid, sheath, tail fibers, spikes, and base plates). Mature phage particles are assembled and released from the cell where they can in turn invade new host cells. The result of a lytic infection for the host cell is almost always death. (fig. VI.1)

In the lysogenic cycle, the viral DNA of the temperate virus is integrated into host DNA and no mature phages are made. Cells grow normally and are immune to further infections by the same phage. There is no visible evidence to indicate that a virus is even present in the cell. In some cases, the virus can carry genes that confer new genetic capabilities on the virally infected cell, or lysogen. For example, when *Corynebacterium diphtheriae* is infected with a certain lysogenic phage, because the phage carries a toxin gene in its genome, the host cells begin to produce a potent toxin responsible for the symptoms of diphtheria. This phenomenon is known as lysogenic conversion and is responsible for some of the toxins produced by various pathogens. Periodically, the lysogenic phage DNA can excise from the host DNA and initiate the lytic cycle and the production of mature phages. This results in the lysis of the host cell.

Visual evidence for lysis can be demonstrated by mixing phages with host cells and plating them onto media. The bacteria form a confluent lawn of growth, and where the phages cause lysis of the bacterial cells, there will be seen clear areas called plaques.

Some of the most studied bacteriophages are those that infect *Escherichia coli*, such as the T-even phages and lambda phage. They are known as the *coliphages*. Because *E. coli* is an intestinal bacterium, the coliphages can readily be isolated from raw sewage and coprophagous (dung-eating) insects such as flies. The exercises in this section will demonstrate some of the techniques for isolating, assaying, and determining the burst size of bacteriophages. It is recommended that you thoroughly understand the various stages in the lytic cycle before you begin the experiments in this section.

174

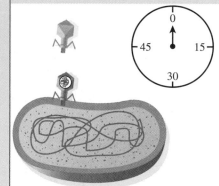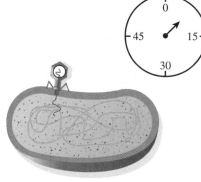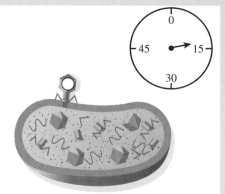

(1) **Adsorption:** Phage is adsorbed to specific receptor site on the bacterial cell surface. This is **Time Zero**.

(2) Phage DNA enters cell to initiate **Eclipse Stage**. Bacterial DNA is degraded within minutes.

(3) Phage capsids, tails, and phage DNA begin to appear within 12 minutes as phage reorients cell metabolism to its own fabrication processes.

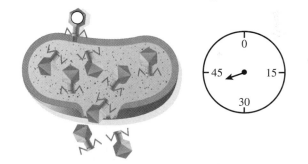

(4) Components of phage are assembled into mature infective virions. The eclipse period ends with first appearance of infective phage in cell.

(5) Cell wall opens up due to enzymatic action to release mature virions. **Burst size** is the number of units released by cell. Total time: 40 minutes.

Figure VI.1   The lytic cycle of a virulent bacteriophage.

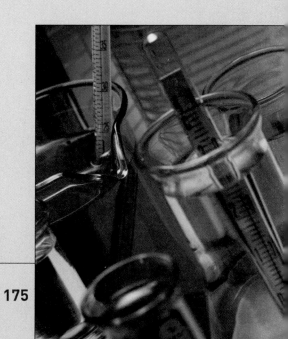

Determination of a Bacteriophage Titer

Bacteriophages are viruses that infect bacterial cells. They were first described by Twort and d'Herelle in 1915 when they both noted that bacterial cultures spontaneously cleared and the bacteria-free liquid that remained could cause new cultures of bacteria to also clear. Because it appeared that the cultures were being "eaten" by some unknown agent, d'Herelle coined the term *bacteriophage,* which means "bacterial eater." Like all viruses, bacteriophages, or phages, for short, are **obligate intracellular parasites,** that is they must invade a host cell in order to replicate and reproduce. This is due to the fact that viruses are composed primarily of only a single kind of nucleic acid molecule encased in a protein coat, or **capsid,** that protects the nucleic acid. All viruses lack metabolic machinery, such as energy systems, and protein synthesis components necessary for independent replication. In order to replicate and reproduce, they must use the host cell's metabolic machinery to synthesize their various component parts.

Viruses also exhibit specificity for their hosts. For example, a certain bacteriophage may only infect a specific strain of a bacterium. Examples are the T-even bacteriophages that infect *Escherichia coli* B, whereas other phages infect *E. coli* K12, a different strain of the organism. A phage that infects *Staphylococcus aureus* does not infect *E. coli* and vice versa. This specificity can be used in phage typing of pathogens (Exercise 28).

The structure of a T_4 bacteriophage is shown in figure 25.1, A phage consists of a **nucleocapsid,** which is the nucleic acid and protein capsid. The nucleocapsid is attached to a protein **sheath** that is contractile and contains a hollow tube in its center. The sheath sits on a **base plate** to which **tail fibers** and **spikes** are attached. Most of the phage structure is necessary for delivery of the phage nucleic acid into its host. A single virus or phage particle is also called a **virion.**

The steps in a lytic phage infection of a bacterial cell are basically as follows:

Recognition The bacteriophage recognizes its host by the tail fibers binding to chemical groups on the surface of the host cell. For example, these chemi-

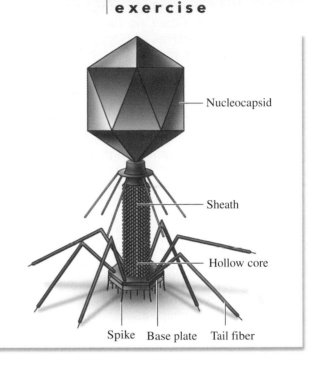

Figure 25.1 Bacteriophage.

cal groups can be carbohydrate groups that are part of the lipopolysaccharide molecule in gram-negative cells. Part of the specificity in phage infections resides in the differences in cell surface structures of bacteria. If the groups that are recognized by tail fibers are not present on a bacterial cell, a phage cannot bind to the cell and cause an infection.

Penetration The phage particle settles onto the surface of the host cell, and **lysozyme** that is associated with the phage tail begins to erode a localized area of the cell wall, thus weakening it. The sheath contracts, forcing a hollow core through the weakened area of the cell wall. As contraction occurs, the viral nucleic acid is injected through the core into the bacterial cytoplasm.

Replication Only the phage nucleic acid enters the host cell, while the capsid and remainder of the phage structure remains on the outside of the bacterial cell. Some phages also inject a nuclease along with the viral nucleic acid. This nuclease specifically degrades host DNA. As a result, the host cell is killed and does not carry out metabolic functions that would compete with viral replication. However, the virus leaves intact the host cell metabolic machinery for producing energy and for synthesizing nucleic acids and proteins, all of which are needed for viral replication. The various components of the virus come together and are assembled into mature phage particles by the process of **self-assembly.** Genes on the viral genome also encode for the synthesis of lysozyme, which begins to degrade the cell wall and weaken it from inside the cell.

Release The combination of the weakened cell wall brought about by the action of lysozyme plus the pressure exerted by virus particles in the cell causes the cell to burst, releasing the phages into the environment where they can then infect other susceptible cells. One phage particle infecting one host cell can produce as many as 200 viral progeny. This number is called the **burst size.**

If bacterial cells are mixed with a bacteriophage in soft agar, the bacteria will grow to form a **conflent lawn** of cells. Phage will infect the cells, causing them to undergo lysis and form clear areas in the confluent lawn called **plaques.** Each plaque is formed by the progeny of a single phage that has replicated and lysed the bacterial cells in that area. Like colony-forming units, **plaque-forming units** can be counted to determine the number of viral particles in a suspension.

In the following exercise, you will work in pairs and determine the number of phage particles or plaque-forming units in a suspension of T_4 bacteriophages. You will use *E. coli* B as the host cell for this experiment.

Materials

- 18–24 hour broth culture of *Escherichia coli* B
- 2 ml suspension of T_4 bacteriophages with a titer of at least 10,000 phages/ml

- 5 trypticase soy agar (TSA) plates. These should be warmed to 37° C before use.
- 5 tubes of soft agar (0.7% agar). Prior to use, melt and hold at 50° C in a water bath.
- 5 tubes of 9.0 ml trypticase soy (TS) broth
- 1-ml sterile pipettes
- pipette aids

1. Label the 5 TSA plates with your name and dilutions from 1:10 to 1:100,000.
2. Label 5 TS broth tubes with the dilutions 1:10 to 1:100,000 (figure 25.2).
3. Prepare serial 10-fold dilutions of the phage stock suspensions by transferring 1 ml of the phage suspension to the first dilution blank. Mix well and transfer 1 ml of the first dilution to the second dilution blank (1:100). Repeat this same procedure until the original phage stock has been diluted 1:100,000 (figure 25.2).
4. Aseptically transfer 2 drops of *E. coli* B broth culture to each of the 5 soft agar overlay tubes.
5. Transfer 1 ml of the first (1:10) phage dilution tube to a soft agar overlay and mix thoroughly but gently. After mixing, pour the contents of the soft agar tube onto the respective TSA plate. Make sure that the soft agar completely covers the surface of the TSA plate. This can be accomplished by gently swirling the plate several times after pouring and while the soft agar is still liquid.
6. Repeat this procedure for each dilution of the phage suspension.
7. Incubate the plates at 37° C for 24 hours. If the exercise cannot be completed at this time, refrigerate the plates until the next laboratory period.
8. Observe the plates. Plaques will appear as clear areas in the bacterial lawn. Count the plaques on the plates. Only include counts between 25 to 250 plaques. This can be facilitated with a bacterial colony counter. Multiply the number of plaques times the dilution factor to determine the number of phage particles in the original suspension of phages.
9. Record the phage titer in Laboratory Report 25.

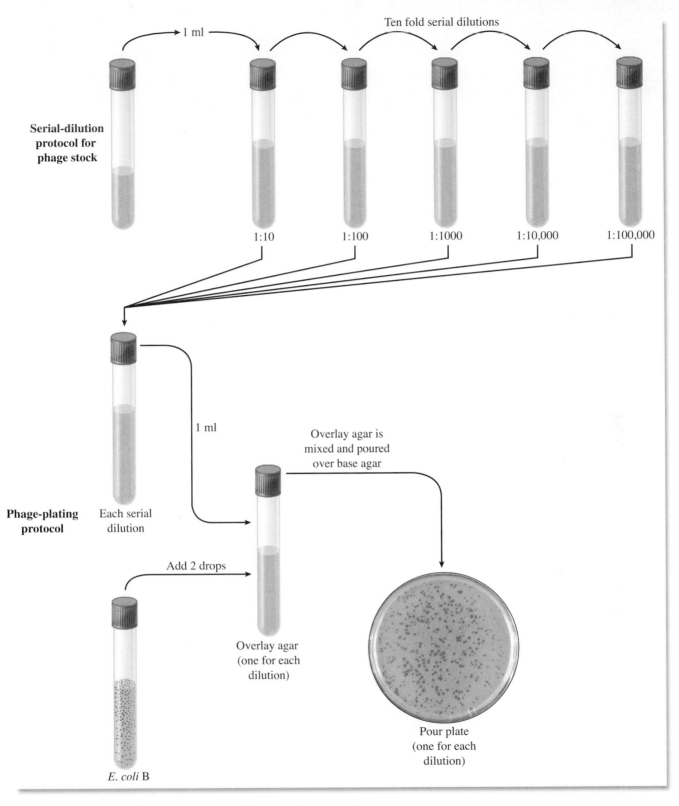

Ten fold serial dilutions

1 ml

Serial-dilution protocol for phage stock

1:10   1:100   1:1000   1:10,000   1:100,000

Phage-plating protocol

Each serial dilution

1 ml

Overlay agar is mixed and poured over base agar

Add 2 drops

Overlay agar (one for each dilution)

E. coli B

Pour plate (one for each dilution)

Figure 25.2 Procedure for determining the titer of bacteriophage.

Laboratory Report

Student: _____

Date: _____ Section: _____

25 Determination of a Bacteriophage Titer

A. Results

1. Dilution giving countable plaques between 25 and 250

2. Number of plaques counted on this plate

3. Number of phage particles in the original stock

B. Short Answer Questions

1. On gram-negative bacteria, to what chemical groups might bacteriophage attach?

    _____

    _____

2. Why are viruses called obligate intracellular parasites?

    _____

    _____

3. What type of viral infection can lead to altered host cell genetics?

    _____

    _____

4. What is lysogenic conversion?

    _____

    _____

5. Why could you isolate bacteriophage specific for *E. coli* from flies?

    _____

    _____

6. Compare and contrast bacteriophage plaques with bacterial colonies.

    _____

    _____

Burst Size Determination:
A One-Step Growth Curve

The average number of mature phage particles (virions) released by the lysis of a single cell is called the **burst size.** This number can vary between 20 and 200. The burst size can be determined by adding a small amount of phage to a known quantity of bacterial host cells and lysing the cells at 5-minute intervals by adding chloroform to the culture. The lysed cell material, containing virions, is then mixed with new bacterial host cells, plated out and incubated. By counting the number of plaques in the various time samples, it is possible to determine the burst size of coliphage T_4 when it infects *E. coli* strain B.

Adsorption

Figure 26.1 illustrates the procedure of this experiment. The first step is to add the phage to the susceptible bacteria. As soon as the two are mixed, adsorption begins. The phages collide in random fashion with the bacterial cells and attach their tails to specific receptor sites on the surfaces of host cells. The adsorption process can be stopped at any time by dilution. **Time zero** of adsorption is the time of mixture of phage and bacteria.

Note in figure 26.1 that 0.1 ml of coliphage T_4 (2×10^8/ml) and 2 ml of *E. coli* (5×10^8/ml) are mixed in the first tube, which is labeled ADS. The ratio of phage to bacteria in this case is 0.02, which calculates out in this manner:

$$\frac{0.1 \times 2 \times 10^8}{2 \times 5 \times 10^8} = \frac{0.2 \times 10^8}{10 \times 10^8} = 0.02, \quad \text{or} \quad 1/50$$

This ratio is called the **multiplicity of infection,** or **m.o.i.**

By referring back to figure VI.1 on page 000, we can see what is occurring in this experiment. Note that during the adsorption stage, DNA in the capsid passes down through the tail into the host through a hole produced in the cell wall by enzymatic action at the tip of the phage tail.

Eclipse Stage

As soon as the phage DNA is injected into the host cell, the phage enters the **eclipse period.** During this period, which lasts approximately 12 minutes for T_4, the entire physiology of the host cell is reoriented toward the production of phage components: capsids, tail fibers, base plates sheaths, and viral nucleic acid. During this period only phage components are present with no mature virions in the lysate. If the lysate were assayed for phage, no plaques would be formed.

Maturation Stage

Late in the eclipse period, the phage components come together by **self-assembly** to form mature infective virions. The phage now enters the maturation stage of replication. The lysis of cultures with chloroform beyond 12 minutes after time zero will reveal the presence of mature virions that can produce plaques on assay plates.

Lysis by the phage of the population of infected cells does not occur all at once, but rather the number of phage particles produced is characterized by a normal distribution curve with a **rise period.** The rise period lasts for several minutes and represents the increase in the numbers of mature phage produced. The peak of the curve is the burst size for the phage infection. In this exercise, you will determine the burst size for T_4.

Two Methods

To accommodate the availability of time and materials, there are two options for performing this experiment. The first option is for students to work in pairs to perform the entire experiment. Figure 26.1 illustrates the procedure for this method. The other option, which requires much less media and time, utilizes a team approach in which students, working in pairs, do just a portion of the experiment; in this case, data are pooled to complete the experiment. Figure 26.3 illustrates the procedure for this method. Your instructor will indicate which method will be used.

The Entire Experiment

To perform the experiment in its entirety, follow the procedures that are shown in figure 26.1.

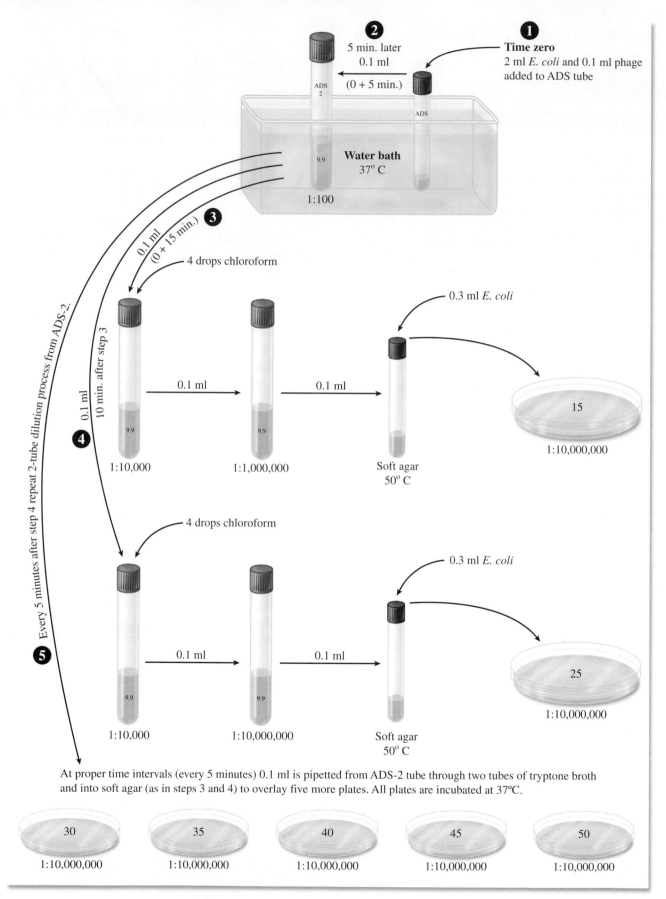

Figure 26.1 Procedure for entire experiment.

Materials

- 1 sterile serological tube (for ADS tube)
- 15 tubes tryptone broth (9.9 ml in each one)
- 8 tubes of nutrient soft agar (3 ml per tube)
- 8 petri plates of tryptone agar
- 16 pipettes (1 ml size)
- 1 dropping bottle of chloroform
- 1 small wire basket to hold 7 tubes of soft agar in water bath
- 1 wire test-tube rack
- 2 water baths (37° C and 50° C)
- 1 culture of *E. coli*, strain B (5 ml) with concentration of 5×10^8 per ml
- 1 tube of T_4 phage (2×10^8 per ml)

Preliminaries

1. Liquefy 8 tubes of soft nutrient agar by boiling in a beaker of water. Cool to 50° C and place in a wire basket or rack in 50° C water bath.
2. Label a sterile serological tube "ADS" to signify the adsorption tube.
3. Label 1 tube of tryptone broth "ADS-2."
4. Label 8 tryptone agar plates: control, 15, 25, 30, 35, 40, 45, and 50.
5. Arrange the ADS, ADS-2, and the 14 tryptone broth tubes in a rack as shown in figure 26.2. Place the rack in a 37° C water bath.
6. Dispense 3 to 4 drops of chloroform in each of the 7 tubes of tryptone broth that are in the front row.

Inoculations and Dilutions

1. Pipette 0.1 ml of *E. coli*. strain B, into a tube of liquefied soft nutrient agar and pour into the control plate. Swirl the plate gently to spread evenly. This plate will indicate whether any phage was in the original bacterial culture. Set this plate aside to harden.
2. With the same pipette as above, transfer 0.3 ml of *E. coli* into each of the tubes of soft nutrient agar. Keep the tubes in the 50° C water bath.

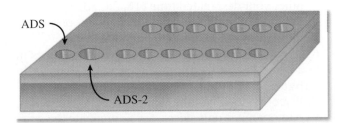

Figure 26.2 Tube arrangement.

3. Still using the same pipette, transfer 2.0 ml of *E. coli* into the ADS tube.
4. With a fresh pipette, deliver 0.1 ml of T_4 phage into the ADS tube and *immediately* record the time (*time zero*) of this mixing with *E. coli* in the following table. Mix gently and allow to remain in the 37° C water bath for 5 minutes.
5. While the mixture is incubating, fill in the table, recording all the projected times so that you will know when each step is to begin.

	TIME	PLATE
Time zero	_____	none
Step 6 time (5 min later)	_____	none
Step 7 time (10 min later)	_____	15
Step 8 time (10 min later)	_____	25
Five minutes later	_____	30
Five minutes later	_____	35
Five minutes later	_____	40
Five minutes later	_____	45
Five minutes later	_____	50

6. After the 5-minute incubation time, transfer 0.1 ml of the mixture to the ADS-2 tube, gently mix, and incubate at 37° C for another 10 minutes.
7. After 10 minutes, transfer 0.1 ml from the ADS-2 tube to the first front row tube of tryptone broth. Keep the ADS-2 tube in the water bath. Mix this dilution tube gently and transfer 0.1 ml to the adjacent tryptone broth tube in the second row. Mix this tube gently, also.
8. Transfer 0.1 ml from the second tube of tryptone broth to a tube of soft nutrient agar, mix gently, flame the tube neck, and pour the soft agar over the tryptone agar plate that is labeled "15."

   Swirl the plate carefully to disperse the soft agar mixture evenly.
9. Follow the above procedure 10 minutes later to produce a soft agar overlay plate on the plate labeled "25."
10. Repeat at the allotted times for 30-, 35-, 40-, 45-, and 50-minute plates.
11. Invert and incubate all plates for 24–48 hours at 37° C.

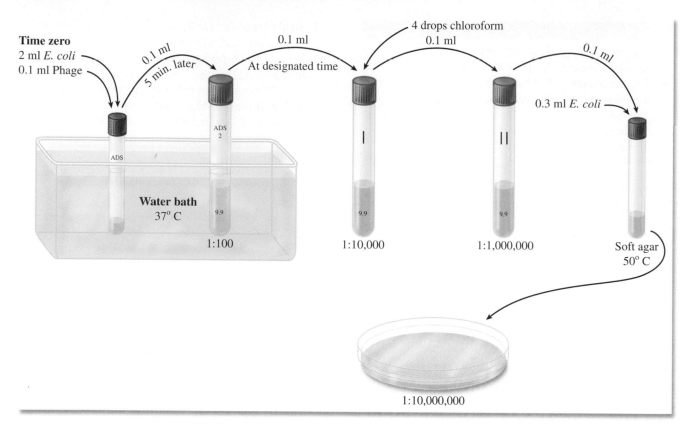

Figure 26.3 Abbreviated procedure (team) method.

Examination of the Plates

Once the plates have been incubated, count the plaques on all the plates, using a Quebec colony counter and hand tally counter. Record all counts on Laboratory Report 26 and determine burst size.

Abbreviated Procedure

(Team Method)

Performance of this experiment in teams will require a minimum of seven pairs of students. Each pair of students (team) will follow the procedure shown in figure 26.3 to produce one soft agar overlay plate for a designated time.

Materials

per team:
- 1 sterile serological tube
- 3 tubes of tryptone broth (9.9 ml per tube)
- 2 tubes of soft nutrient agar (3 ml per tube)
- 2 petri plates of tryptone agar
- 4 1 ml pipettes
- 1 dropping bottle of chloroform
- 1 wire test-tube rack (small)

- 1 small beaker (150 ml)
- 1 tube of T_4 phage (2×10^8 per ml)
- 1 culture of *E. coli*, strain B (5×10^8 per ml)
- water bath at 37° C (a small pan that will hold a test-tube rack)

Preliminaries

1. Liquefy 2 tubes of soft nutrient agar in boiling water. Use a small beaker. Cool the water to 50° C and keep the tubes of media at this temperature.
2. Label a sterile serological tube "ADS" to signify the adsorption tube.
3. Label 1 tube of tryptone broth "ADS-2."
4. Label the other tryptone tubes "I" and "II."
5. Label 1 tryptone agar plate "control" and the other your designated time (15, 25, 30, 35, 40, 45, or 50). *Your instructor will assign you a specific time.* Put your names on both plates.
6. Arrange the ADS, ADS-2, and 2 tryptone tubes in a small test-tube rack in same order as shown in figure 26.2.
7. Place the rack of tubes in a pan of 37° C water. Although it is only necessary to incubate the ADS and ADS-2 tubes, it will be more convenient if they are all together.

Inoculations and Dilutions

1. Pipette 0.1 ml of *E. coli,* strain B, into a tube of liquefied soft nutrient agar and pour it into the control plate. Swirl the plate gently to spread evenly.

    This plate will indicate whether any phage was in the original bacterial culture. Set this plate aside to harden.

2. With the same pipette, transfer 0.3 ml of *E. coli* to the other tube of soft nutrient agar. Keep this tube in the beaker of water at 50° C.

3. Still using the same pipette, transfer 2.0 ml of *E. coli* into the ADS tube.

4. With a *fresh pipette,* deliver 0.1 ml of T_4 phage into the ADS tube.

    Record this time (time zero): _____

5. After 5 minutes, transfer 0.1 ml of the *E. coli*–phage mixture from the ADS tube to ADS-2 tube. Mix the ADS-2 tube gently.

6. After the designated time (time zero plus designated time), transfer 0.1 ml from ADS-2 tube to tryptone broth tube I. Mix gently.

7. Add 3 or 4 drops of chloroform to tube I.

8. With a *fresh pipette,* transfer 0.1 ml from tube I to tube II. Mix tube II gently.

9. With *another fresh pipette,* transfer 0.1 ml from tube II to the tube of soft agar.

10. After mixing the soft agar tube, pour it over the tryptone agar plate. Swirl the plate carefully to disperse the soft agar. Set aside to cool for a few minutes.

11. Incubate both plates at 37° C for 24–48 hours.

Examination of Plates

Once the plates have been incubated, examine both of them on a Quebec colony counter. The control plate should be free of plaques. Count the plaques on the other plate, using a hand tally counter if the number is great. Record your count in Laboratory Report 26 and on the chalkboard.

Laboratory Report

Student: _____

Date: _____ Section: _____

26 Burst Size Determination: A One-Step Growth Curve

A. Results

1. Plaque Counts
    Record the counts of plaques on each of the plates in the following table. Record the peak number of plaques as the *burst size*. The drop in plaque numbers after a peak results from adsorption of mature phage virions on other bacterial cells and cell debris.

15	25	30	35	40	45	50

2. Dilution Interpretation
    Answer the following questions to clarify your understanding of the dilutions that occur in this experiment.

    a. How many cells were present in each milliliter of the original bacterial culture? _____

    _____

    _____

    b. How many bacterial cells (total) were dispensed into the ADS tube? _____

    _____

    _____

    c. If the bacterial dilution per plate is 1:10,000,000, how many bacterial cells were distributed to each plate?

    _____

    _____

    d. How many phage virions were present in 1 ml of the original phage suspension?

    _____

    _____

    e. How many phage virions were present in the 0.1 ml of phage suspension that was added to the ADS tube?

    _____

    _____

    f. What was the numerical ratio of phage virions to bacterial cells in the ADS tube? _____

    _____

    _____

g. How many bacterial cells were placed in the ADS-2 tube?_____

_____

_____

B. Short Answer Questions

1. In terms of phage infection, define burst size.

   _____

   _____

2. During an experimental infection, what is the ratio of phage to bacteria called?

   _____

   _____

3. How was phage adsorption to bacterial cells stopped in this experiment?

   _____

   _____

4. What was the purpose of plating a sample of *E. coli* strain B to which no phage was added?

   _____

   _____

Isolation of Phage from Flies

As stated earlier, coprophagous insects (insects that feed on fecal material and dung, as well as raw sewage) contain various kinds of bacterial viruses. House flies are coprophagous because they deposit their eggs in fecal material where the young larva feed, grow, pupate, and emerge as adult flies. This type of environment is heavily populated by *E. coli* and the various bacteriophages that infect this bacterium.

Fly Collection

To increase the probability of success in isolating phage, it is desirable that one use 20 to 24 houseflies. A smaller number might be sufficient; the larger number, however, increases the probability of initial success. Houseflies should not be confused with the smaller blackfly or the larger blowfly. An ideal spot for collecting these insects is a barnyard or riding stable. One should not use a cyanide killing bottle or any other chemical means. Flies should be kept alive until just prior to crushing and placing them in the growth medium. There are many ways that one might use to capture them—use your ingenuity!

Enrichment

Within the flies' digestive tracts are several different strains of *E. coli* and bacteriophage. Our first concern is to enhance the growth of both organisms to ensure an adequate supply of phage. To accomplish this the flies must be ground up with a mortar and pestle and then incubated in a special growth medium for a total of 48 hours. During the last 6 hours of incubation, a lysing agent, sodium cyanide, is included in the growth medium to augment the lysing properties of the phage.

Figures 27.1 and 27.2 illustrate the procedure.

Materials

- bottle of phage growth medium* (50 ml)
- bottle of phage lysing medium* (50 ml)
- Erlenmeyer flask (125 ml capacity) with cotton plug
- mortar and pestle (glass)
- *see Appendix C for composition

1. Into a clean nonsterile mortar place 24 freshly killed houseflies. Pour half of the growth medium into the mortar and grind the flies to a fine pulp with the pestle.
2. Transfer this fly-broth mixture to an empty flask. Use the remainder of the growth medium to rinse out the mortar and pestle, pouring all the medium into the flask.
3. Wash the mortar and pestle with soap and hot water before returning them to the cabinet.
4. Incubate the fly-broth mixture for 42 hours at 37° C.
5. At the end of the 42-hour incubation period, add 50 ml of lysing medium to the fly-broth mixture. Incubate this mixture for another 6 hours.

Centrifugation

Before attempting filtration, you will find it necessary to separate the fly fragments and miscellaneous bacteria from the culture medium. If centrifugation is incomplete, the membrane filter will clog quickly and filtration will progress slowly. To minimize filter clogging, a triple centrifugation procedure will be used. To save time in the event filter clogging does occur, an extra filter assembly and an adequate supply of membrane filters should be available. These filters have a maximum pore size of 0.45 μm, which holds back all bacteria, allowing only the phage virions to pass through.

Materials

- centrifuge
- 6–12 centrifuge tubes
- 2 sterile membrane filter assemblies (funnel, glass base, clamp, and vacuum flask)
- package of sterile membrane filters (0.45 μm)
- sterile Erlenmeyer flask with cotton plug (125 ml size)
- vacuum pump and rubber hose

1. Into 6 or 8 centrifuge tubes, dispense the enrichment mixture, filling each tube to within 1/2" of the top. Place the tubes in the centrifuge so that the load is balanced. Centrifuge the tubes at 2,500 rpm for 10 minutes.
2. Without disturbing the material in the bottom of the tubes, decant all material from the tubes to within 1" of the bottom into another set of tubes.
3. Centrifuge this second set of tubes at 2,500 rpm for another 10 minutes. While centrifugation is taking place, rinse out the first set of tubes.

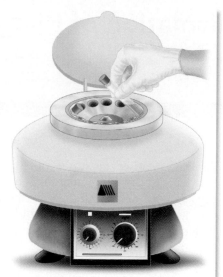

(1) Twenty to twenty-four flies are ground up in phage growth medium with a mortar and pestle.

(2) Crushed flies are incubated in growth medium for 42 hours at 37° C. After adding lysing medium it is incubated for another 6 hours.

(3) Fly-broth culture is triple-centrifuged at 2,500 rpm.

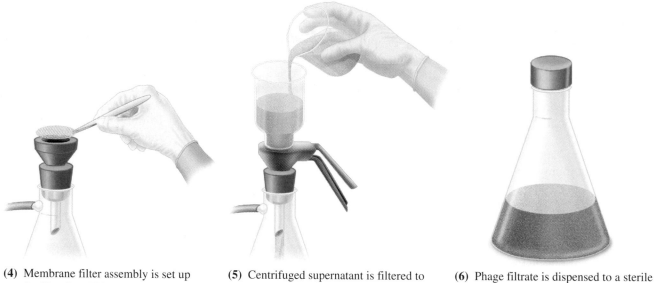

(4) Membrane filter assembly is set up for filtration. This step must be done aseptically.

(5) Centrifuged supernatant is filtered to produce bacteria-free phage filtrate.

(6) Phage filtrate is dispensed to a sterile Erlenmeyer flask from which layered plates will be made. (Fig. 29.2)

Figure 27.1 Procedure for preparation of bacteriophage filtrate from houseflies.

4. When the second centrifugation is complete, pour off the top two-thirds of each tube into the clean set of tubes and centrifuge again in the same manner.

Filtration

While the third centrifugation is taking place, aseptically place a membrane filter on the glass base of a sterile filter assembly (illustration 4, figure 27.1). Use flamed forceps. Note that the filter is a thin sheet with grid lines on it. Place the glass funnel over the filter and fix the clamp in place. Hook up a rubber hose between the vacuum flask and pump.

Now, carefully decant the top three-fourths of each tube into the filter funnel. Take care not to disturb the material in the bottom of the tube. Turn on the vacuum pump. If centrifugation and decanting have been performed properly, filtration will occur almost instantly. If the filter clogs before you have enough

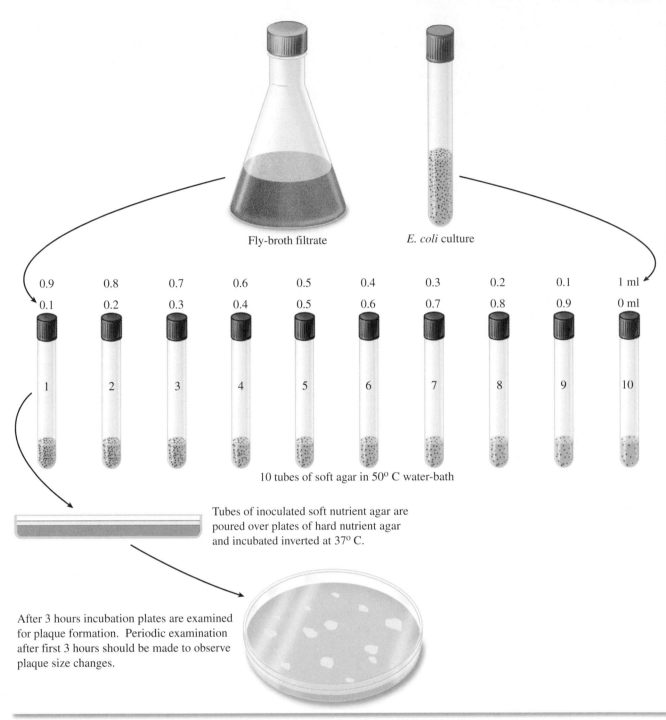

Fly-broth filtrate

E. coli culture

| 0.9 | 0.8 | 0.7 | 0.6 | 0.5 | 0.4 | 0.3 | 0.2 | 0.1 | 1 ml |
| 0.1 | 0.2 | 0.3 | 0.4 | 0.5 | 0.6 | 0.7 | 0.8 | 0.9 | 0 ml |

1   2   3   4   5   6   7   8   9   10

10 tubes of soft agar in 50° C water-bath

Tubes of inoculated soft nutrient agar are poured over plates of hard nutrient agar and incubated inverted at 37° C.

After 3 hours incubation plates are examined for plaque formation. Periodic examination after first 3 hours should be made to observe plaque size changes.

Figure 27.2 Inoculation of *Escherichia coli* with bacteriophage from fly-broth filtrate.

filtrate, recentrifuge all material and pass it through the spare filter assembly.

Aseptically, transfer the final filtrate from the vacuum flask to a sterile 125 ml Erlenmeyer flask that has a sterile closure. Putting the filtrate in a small flask is necessary to facilitate pipetting. Be sure to flame the necks of both flasks while pouring from one to the other.

Inoculation and Incubation

To demonstrate the presence of bacteriophage in the fly-broth filtrate, a strain of phage-susceptible *E. coli* will be used. To achieve an ideal proportion of phage to bacteria, a proportional dilution method will be used. The phage and bacteria will be added to tubes of soft nutrient agar that will be layered over plates of hard nutrient agar. Soft nutrient agar contains only half as much agar as ordinary nutrient agar. [This medium and *E. coli* provide an ideal "lawn" for phage growth.] Its jellylike consistency allows for better diffusion of phage particles; thus, more even development of plaques occurs.

Figure 27.2 illustrates the overall procedure. It is best to perform this inoculation procedure in the morning so that the plates can be examined in late afternoon. As plaques develop, one can watch them increase in size with the multiplication of phage and simultaneous destruction of *E. coli.*

Materials

- nutrient broth cultures of *Escherichia coli* (ATCC #8677 phage host)
- flask of fly-broth filtrate
- 10 tubes of soft nutrient agar (5 ml per tube) with metal caps
- 10 plates of nutrient agar (15 ml per plate, and prewarmed at 37° C)
- 1 ml serological pipettes, sterile

1. Liquefy 10 tubes of soft nutrient agar and cool to 50° C. Keep tubes in water bath to prevent solidification.
2. With a marking pen, number the tubes of soft nutrient agar 1 through 10. Keep the tubes sequentially arranged in the test-tube rack.

3. Label 10 plates of prewarmed nutrient agar 1 through 10. Also, label plate 10 negative control. Prewarming these plates will allow the soft agar to solidify more evenly.
4. With a 1 ml serological pipette, deliver 0.1 ml of fly-broth filtrate to tube 1, 0.2 ml to tube 2, etc., until 0.9 ml has been delivered to tube 9. Refer to figure 27.2 for sequence. **Note that no fly-broth filtrate is added to tube 10.** This tube will be your negative control.
5. With a fresh 1 ml pipette, deliver 0.9 ml of *E. coli* to tube 1, 0.8 ml to tube 2, etc., as shown in figure 27.2. **Note that tube 10 receives 1.0 ml of *E. coli.*** Make sure to gently but throughly mix all the tubes.
6. After flaming the necks of each of the tubes, pour them into similarly numbered plates.
7. When the agar has cooled completely, put the plates, inverted, into a 37° C incubator.
8. **After about 3 hours** incubation, examine the plates, looking for plaques. If some are visible, measure them and record their diameters on Laboratory Report 27.
9. If no plaques are visible, check the plates again in another **2 hours.**
10. Check the plaque size again at **12 hours,** if possible, recording your results. Incubate a total of 24 hours.
11. Complete Laboratory Report 27.

Laboratory Report

Student: _____

Date: _____ Section: _____

27 Isolation of Phage from Flies

A. Results

1. Plaque Size Increase
    With a china marking pencil, circle and label three plaques on one of the plates and record their sizes in millimeters at 1-hour intervals.

TIME	PLAQUE SIZE (millimeters)		
	Plaque No. 1	Plaque No. 2	Plaque No. 3
When first seen			
1 hour later			
2 hours later			
3 hours later			

   a. Were any plaques seen on the negative control plate? _____

   b. Do plates 1, 2, and 3 show a progressive increase in number of plaques with increased amount of

       fly-broth filtrate? _____

   c. Did the phage completely "wipe out" all bacterial growth on any of the plates? _____

       If so, which plates? _____

2. Observations
    Count all the plaques on each plate and record the counts in the following table. If the plaques are very numerous, use a Quebec colony counter and hand counting device. If this exercise was performed as a class project with individual students doing only one or two plates from a common fly-broth filtrate, record all counts on the chalkboard on a table similar to the one below.

Plate Number	1	2	3	4	5	6	7	8	9	10
E. coli (ml)	0.9	0.8	0.7	0.6	0.5	0.4	0.3	0.2	0.1	1.0
Filtrate (ml)	0.1	0.2	0.3	0.4	0.5	0.6	0.7	0.8	0.9	0
Number of plaques										

B. Short Answer Questions

1. How does the life cycle of houseflies contribute to the presence of *E. coli* bacteriophage in their guts?

    _____

    _____

2. From what other environments might *E. coli* bacteriophage be readily isolated?

    _____

    _____

3. Differentiate between:

    a. capsid and sheath.

    _____

    b. lysis and lysogeny.

    _____

    c. virulent phage and temperate phage.

    _____

4. What enzyme is used by phage to penetrate the bacterial cell wall?

    _____

    _____

Phage Typing

The host specificity of bacteriophage is such that it is possible to differentiate strains of individual species of bacteria based on their susceptibility to various kinds of bacteriophage. In epidemiological studies, where it is important to discover the source of a specific infection, determining the phage type of the causative organism can be an important tool in solving the riddle. For example, if it can be shown that the phage type of *S. typhi* in a patient with typhoid fever is the same as the phage type of an isolate from a suspected carrier, chances are excellent that the two cases are epidemiologically related. Since most bacteria are probably infected by bacteriophage, it is theoretically possible to classify each species into strains based on their phage susceptibility. Such phage-typing groups have been determined for *Staphylococcus aureus, Salmonella typhi,* and several other pathogens. The following table illustrates the lytic phage groups for *S. aureus.*

LYTIC GROUP	PHAGES IN GROUP
I	29 52 52A 79 80
II	3A 3B 3C 55 71
III	6 7 42E 47 53 54 75 77 83A
IV	42D
not allotted	81 187

In bacteriophage typing, a suspension of the organism to be typed is uniformly swabbed over an agar surface. The bottom of the plate is marked off into squares and each square labeled to indicate which phage type is applied to the square. A small drop of each bacteriophage type is added to its respective square. After incubation, the plate is examined to determine which phage caused lysis of the test organism. In this exercise, you will determine which phage cause the lysis of *S. aureus* (see figure 28.1).

Materials

- 1 petri plate of tryptone yeast extract agar
- bacteriophage cultures (available types)
- nutrient broth cultures of *S. aureus* with cotton swabs

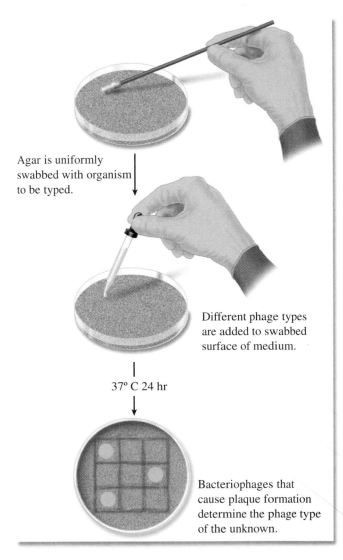

Agar is uniformly swabbed with organism to be typed.

Different phage types are added to swabbed surface of medium.

37° C 24 hr

Bacteriophages that cause plaque formation determine the phage type of the unknown.

Figure 28.1   Bacteriophage typing.

1. Mark the bottom of a plate of tryptone yeast extract agar with as many squares as there are phage types to be used. Label each square with the phage type numbers.
2. Uniformly swab the entire surface of the agar with the organisms.
3. Deposit 1 drop of each phage in its respective square.
4. Incubate the plate at 37° C for 24 hours and record the lytic group and phage type of the culture.
5. Record your results in Laboratory Report 28.

Laboratory Report

Student: _____

Date: _____ Section: _____

28 Phage Typing

A. Results

1. To which phage types was this strain of *S. aureus* susceptible?

    _____

2. To what lytic group does this strain of staphylococcus belong?

    _____

B. Short Answer Questions

1. What factors are responsible for bacteriophage host specificity?

    _____

    _____

2. Why is phage typing an important clinical tool?

    _____

    _____

3. How do viruses contribute to bacterial evolution? Why might this be of concern to health officials?

    _____

    _____

Environmental Influences and Control of Microbial Growth

The exercises in the following section are concerned with the effects that factors such as temperature, pH, water activity, UV light, antibiotics, disinfectants, and antiseptics have on the growth of bacteria. The microbiologist is concerned with providing the optimum conditions for the growth of microorganisms. In contrast, the medical practitioner is concerned with limiting microbial growth to prevent disease. Understanding both of these will enhance the job of each of these individuals.

In Part 4, the primary concern was in formulating a medium that contained all the essential nutrients to support the growth of a microorganism. However, very little emphasis was placed on other limiting factors such as temperature, oxygen, or pH. Even though all its nutritional needs are provided, an organism may fail to grow if these other factors are not considered. The total environment must be sustained to achieve the desired growth of microorganisms.

Temperature:
Effects on Growth

Microorganisms grow over a broad temperature range that extends from below 0°C to greater than 100° C. Based on their temperature requirements, they can be divided into four groups that define their optimal growth:

psychrophiles: optimal growth between –5° C and 20° C; these bacteria can be found in the supercooled waters of the arctic and antarctic.

mesophiles: optimal growth between 20° C and 50° C; most bacteria fall into this class, for example most pathogens grow between 35° C and 40° C.

thermophiles: optimal growth between 50° C and 80° C; bacteria in this group occur in soils where the midday temperature can reach greater than 50°C or in compost piles where fermentation activity can cause temperatures to exceed 60–65° C.

hyperthermophiles: growth optimum above 80° C; many of the Archaea occupy environments that are heated by volcanic activity where water is superheated to above 100° C. These organisms have been isolated from thermal vents deep within the ocean floor and from volcanic heated hot springs.

It should be noted that one single organism is not capable of growth over the entire range but would be restricted to one of the temperature classes. However, some bacteria within the classes are capable of growth at temperatures lower or higher than their optima. For example, some mesophilic bacteria such as *Proteus, Pseudomonas, Campylobacter,* and *Leuconostoc* can grow at 4° C, refrigerator temperatures, and cause food spoilage. These bacteria are referred to as **psychrotrophs.**

Temperature can affect several metabolic factors in the cell. Enzymes are directly affected by temperature and any one enzyme will have a minimum, optimum, and maximum temperature for activity. Maximal enzyme activity will occur at the optimum temperature. At temperatures above the maximum, enzymes will begin to denature and lose activity. Below the minimum temperature, chemical activity slows down and some denaturation can also occur. In addition to the effects on enzyme activity, temperature can also greatly affect cell membranes and transport. As temperature decreases, transport of nutrients into the cell also decreases due to fluidity changes in the membrane. If the temperature increases above the maximum of an organism, membrane lipids can be destroyed resulting in serious damage to the membrane and death of the organism. Last, ribosomes can be directly affected by temperature, and if extremes of temperature occur, they will cease to function adequately.

In this experiment, we will attempt to measure the effects of various temperatures on two physiological reactions: pigment production and growth rate. Nutrient broth and nutrient agar slants will be inoculated with three different organisms that have different optimum growth temperatures. One organism, *Serratia marcescens,* produces a red pigment called *prodigiosin,* which is produced only in a certain temperature range. It is our goal here to determine the optimum temperature for prodigiosin production and the approximate optimum growth temperatures for all three microorganisms. To determine optimum growth temperatures, we will be incubating cultures at five different temperatures. A spectrophotometer will be used to measure turbidity densities in the broth cultures after incubation.

First Period

(Inoculations)

To economize on time and media, it will be necessary for each student to work with only two organisms and seven tubes of media. Refer to table 29.1 to determine your assignment. Figure 29.1 illustrates the procedure.

Materials

- nutrient broth cultures of *Serratia marcescens, Geobacillus stearothermophilus,* and *Escherichia coli*

per student:
- 2 nutrient agar slants
- 5 tubes of nutrient broth

1. Label the tubes as follows:
   Slants: Label both of them *S. marcescens;* label one tube 25° C and the other tube 38° C.

Table 29.1 Inoculation Assignments

STUDENT NUMBER	*S. marcescens*	*G. stearothermophilus*	*E. coli*
1, 4, 7, 10, 13, 16, 19, 22, 25	2 slants and 5 broths		
2, 5, 8, 11, 14, 17, 20, 23, 26	2 slants	5 broths	
3, 6, 9, 12, 15, 18, 21, 24, 27	2 slants		5 broths

Broths: Label each tube of nutrient broth with your other organism and one of the following five temperatures: 5° C, 25° C, 38° C, 42° C, or 55° C.

2. Using a wire loop, inoculate each of the tubes with the appropriate organisms.

3. Place each tube in one of the five baskets that is labeled according to incubation temperature.
   Note: The instructor will see that the 5° C basket is placed in the refrigerator and the other four are placed in incubators that are set at the proper temperatures.

Second Period
(Tabulation of Results)

Materials

- slants and broth cultures that have been incubated at various temperatures
- spectrophotometer and cuvettes
- tube of sterile nutrient broth

1. Compare the nutrient agar slants of *S. marcescens*. Using colored pencils, draw the appearance of the growths on Laboratory Report 29.

2. Shake the broth cultures and compare them, noting the differences in turbidity. Those tubes that appear to have no growth should be compared with a tube of sterile nutrient broth.

3. If a spectrophotometer is available, determine the turbidity of each tube following the instructions in Exercise 22.

4. If no spectrophotometer is available, record turbidity by visual observation. The Laboratory Report indicates how to do this.

5. Exchange results with other students to complete data collection for the experiment.

Laboratory Report

After recording all data, answer the questions in Laboratory Report 29.

Two nutrient agar slants are streaked with *S. marcescens* and incubated at different temperatures for pigment production.

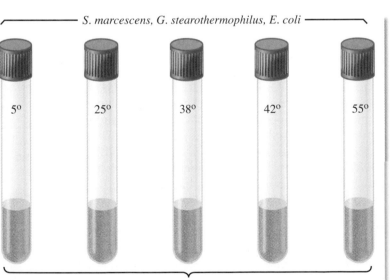

Five nutrient broths are inoculated with one of three organisms and incubated at five different temperatures to determine optimum growth temperatures for each organism.

Figure 29.1 Inoculation procedure.

Laboratory Report

Student: _____

Date: _____ Section: _____

29 Temperature: Effects on Growth

A. Results

1. Pigment Formation and Temperature

    a. Draw the appearance of the growth of *Serratia marcescens* on the nutrient agar slants using colored pencils.

    b. Which temperature seems to be closest to the optimum temperature for pigment formation?

    _____

    c. How is pigment production in *S. marcescens* controlled by temperature?

    _____

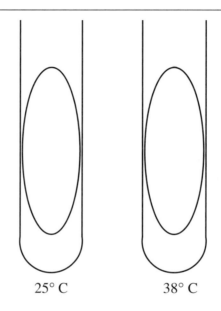

    25° C   38° C

2. Growth Rate and Temperature

    a. If a spectrophotometer is available, dispense the cultures into labeled cuvettes and determine the optical density of each culture.
        If no spectrophotometer is available, record only the visual readings as + , + +, + + +, and none.

Temp. °C	NAME OF ORGANISM			NAME OF ORGANISM		
	Visual Reading	Spectrophotometer		Visual Reading	Spectrophotometer	
		O.D.			O.D.	
5						
25						
38						
42						
55						

b. Growth curves of *Serratia marcescens* and *Escherichia coli* as related to temperature.

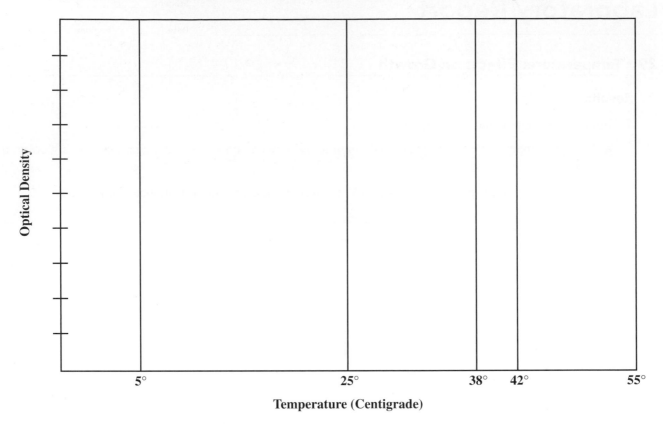

c. On the basis of the above graph, estimate the optimum growth temperature of the two organisms.

Your organism: _____

Your organism: _____

d. To get more precise results for the above graph, what would you do?

_____

_____

B. Short Answer Questions

1. Define the optimal growth temperature for each of the following classes of bacteria.

a. hyperthermophile.

_____

b. thermophile.

_____

c. mesophile.

_____

d. psychrophile.

_____

2. Differentiate between psychrophile and psychrotroph.

_____

_____

3. Why are psychrotrophic bacteria of concern to those in the food-service industry?

_____

_____

4. What is the optimum growth temperature for most human pathogens? Explain.

_____

_____

5. Name three cellular components involved in metabolism that are influenced by temperature changes.

_____

_____

Temperature:
Lethal Effects

Microorganisms are killed by elevated temperatures primarily because of the susceptibility of their macromolecules to heat. Elevated temperature cause proteins to denature and unfold, resulting in the loss of their tertiary structure and biological activity. Because most proteins are enzymes, the metabolic capabilities of an organism are irreversibly damaged by heat. Nucleic acids can also be denatured by heat, resulting in the loss of DNA and RNA structure and death to cells. However, small molecules in the cell such as NAD^+ and other coenzymes are also damaged or destroyed by elevated temperatures and loss of these essential factors contributes to cell death. Some organisms however, are more resistant to heat than others. The endospores of bacteria such as *Bacillus* and *Clostridium* are more resistant to heat than are vegetative cells because endospores, unlike vegetative cells, contain compounds such a calcium dipicolinate that protect the endospore from heat. Also, endospore-specific proteins can bind to nucleic acids and prevent denaturation.

In attempting to compare the susceptibility of different organisms to elevated temperatures, it is necessary to use some yardstick of measure. Two methods of comparison are used: the **thermal death point** and the **thermal death time.** The thermal death point (TDP) is the temperature at which an organism is killed in 10 minutes. The thermal death time (TDT) is the time required to kill a suspension of cells or spores at a given temperature. Because various factors such as pH, moisture, medium composition, and age of the cells can greatly influence results, these variables must be clearly stated.

The thermal death point and thermal death time are important in the food industry because canned foods must be heated to temperatures that will kill the endospores of *Clostridium botulinum* and *Clostridium perfringens,* two bacteria involved in food poisoning.

In this exercise, you will subject cultures of three different organisms to temperatures of 60°, 70°, 80°, 90°, and 100° C. At intervals of 10 minutes, organisms will be removed and plated out to test their viability. The spore-former *Bacillus megaterium* will be compared with the non-spore-formers *Staphylococcus aureus* and *Escherichia coli.* The overall procedure is illustrated in figure 30.1.

Note in figure 30.1 that *before* the culture is heated, a **control plate** is inoculated with 0.1 ml of the organism. When the culture is placed in the wa-

ter bath, a tube of nutrient broth with a thermometer inserted into it is placed in the bath at the same time. Timing of the experiment starts when the thermometer reaches the test temperature.

Due to the large number of plates that have to be inoculated to perform the entire experiment, it will be necessary for each member of the class to be assigned a specific temperature and organism to work with. Table 30.1 provides assignments by student number. After the plates have been incubated, each student's results will be tabulated on a Laboratory Report chart at the demonstration table. The instructor will have copies made of it to give each student so that everyone will have all the pertinent data needed to draw the essential conclusions.

Although this experiment is not difficult, it often fails to turn out the way it should because of student error. Common errors are (1) omission of the control plate inoculation, (2) putting the thermometer in the culture tube instead of in a tube of sterile broth, and (3) not using fresh sterile pipettes when instructed to do so.

Materials

per student:
- 5 petri plates
- 5 pipettes (1 ml size)
- 1 tube of nutrient broth
- 1 bottle of nutrient agar (60 ml)
- 1 culture of organisms

class equipment:
- water baths set up at 60°, 70°, 80°, 90°, and 100° C

broth cultures:
- *Staphylococcus aureus, Escherichia coli,* and *Bacillus megaterium* (minimum of 5 cultures of each species per lab section)

1. Consult table 30.1 to determine what organism and temperature has been assigned to you. If several thermostatically controlled water baths have been provided in the lab, locate the one that you will use. If a bath is not available for your temperature, set up a bath on an electric hot plate or over a tripod and Bunsen burner.

    If your temperature is 100° C, a hot plate and beaker of water are the only way to go. When setting up a water bath use hot tap water to start with to save heating time.

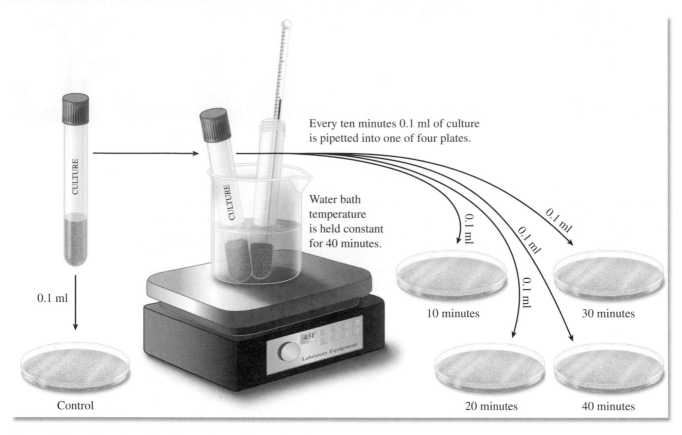

Figure 30.1 Procedure for determining thermal endurance.

2. Liquefy a bottle of 60 ml of nutrient agar and cool to 50° C. This can be done while the rest of the experiment is in progress.
3. Label five petri plates: **control, 10 min, 20 min, 30 min,** and **40 min.**
4. Shake the culture of organisms and transfer 0.1 ml of organisms with a 1 ml pipette to the control plate.
5. Place the culture and a tube of sterile nutrient broth into the water bath. Remove the cap from the tube of nutrient broth and insert a thermometer into the tube. *Don't make the mistake of inserting the thermometer into the culture of organisms!*
6. As soon as the temperature of the nutrient broth reaches the desired temperature, record the time here: _____.
   Watch the temperature carefully to make sure it does not vary appreciably.

7. After 10 minutes have elapsed, transfer 0.1 ml from the culture to the 10-minute plate with a fresh 1 ml pipette. Repeat this operation at 10-minute intervals until all the plates have been inoculated. *Use fresh pipettes each time and be sure to shake the culture before each delivery.*
8. Pour liquefied nutrient agar (50° C) into each plate, rotate, and cool.
9. Incubate at 37° C for 24 to 48 hours. After evaluating your plates, record your results on the chart in Laboratory Report 30 and on the chart on the demonstration table.

Laboratory Report

Complete Laboratory Report 30 once you have a copy of the class results.

Table 30.1 Inoculation Assignments

ORGANISM	Student Number				
	60° C	70° C	80° C	90° C	100° C
Staphylococcus aureus	1, 16	4, 19	7, 22	10, 25	13, 28
Escherichia coli	2, 17	5, 20	8, 23	11, 26	14, 29
Bacillus megaterium	3, 18	6, 21	9, 24	12, 27	15, 30f

Laboratory Report

Student: _____

Date: _____ Section: _____

Temperature: Lethal Effects

A. Results

Examine your five petri plates, looking for evidence of growth. Record on the chalkboard, using a chart similar to the one below, the presence or absence of growth as (+) or (−). When all members of the class have recorded their results, complete this chart.

ORGANISM	60° C					70° C					80° C					90° C					100° C				
	C*	10	20	30	40	C*	10	20	30	40	C*	10	20	30	40	C*	10	20	30	40	C*	10	20	30	40
S. aureus																									
E. coli																									
B. megaterium																									

* Control tubes

1. If they can be determined from the above information, record the **thermal death point** for each of the organisms.

   S. aureus: _____ E. coli: _____ B. megaterium: _____

2. From the following table, determine the **thermal death time** for each organism at the tabulated temperatures.

ORGANISM	THERMAL DEATH TIME				
	60° C	70° C	80° C	90° C	100° C
S. aureus					
E. coli					
B. megaterium					

B. Short Answer Questions

1. What is the importance of inoculating a control plate in this experiment?

   _____

   _____

2. To measure the culture temperature, why is the thermometer placed in a tube separate from the culture?

   _____

   _____

3. *Bacillus megaterium* has a high thermal death point and long thermal death time, but it is not classified as a thermophile. Explain.

_____

_____

4. Give three reasons why endospores are much more resistant to heat than are vegetative cells.

    a. _____

    b. _____

    c. _____

5. List four diseases caused by spore-forming bacteria.

    a. _____

    b. _____

    c. _____

    d. _____

6. Endospores are extremely resistant to heat but can be destroyed by heat if the proper conditions are applied. Describe the heating conditions necessary to kill endospores when the following are sterilized:

    a. Glassware treated in an oven _____

    b. Media in an autoclave _____

    c. Aseptic transfer of bacteria _____

pH and Microbial Growth

Another factor that exerts a strong influence on growth is the hydrogen ion concentration designated by the term **pH** ($-\log_{10}[H^+]$). The hydrogen ion concentration affects proteins and other charged molecules in the cell. Each organism will have an optimal pH at which it grows best. If pH values exceed the optimum for an organism, the solubility of charged molecules can be adversely affected and molecules can precipitate out of solution. For example, the pH can directly affect the charge on amino acids in proteins and result in denaturation and loss of enzyme activity.

Microorganisms can be subdivided into groups based on their ability to grow at different pH values. Bacteria that grow at or near neutral pH are termed **neutrophiles.** Most bacteria are neutrophiles although many can grow over a range of 2–3 pH units. Bacteria that grow at acidic pH values are **acidophiles.** For example, the chemolithotroph, *Thiobacillus thiooxidans,* is an **acidophile** that grows at pH 1; it derives its energy needs from the oxidation of sulfide, producing sulfuric acid that lowers the pH of its environment to a value of 1. *Thiobacillus* is often found in mining effluents that are enriched in metal sulfide ores. Most fungi and yeast are acidophiles that prefer to grow at pH values between 4–6. Media such as potato dextrose agar and Sabouraud's agar for cultivating fungi are adjusted to pH values around 5, which selects for and promotes the growth of these microorganisms. Bacteria that grow at alkaline pH values are termed **alkaliphiles.** True alkalinophilic bacteria are found growing in environments such as soda lakes and high carbonate soils where the pH can reach 10 or above. Many of these organisms belong to the genus *Bacillus.* Some bacteria are alkaline tolerant such as the opportunistic pathogen *Alcaligenes faecalis* that can cause urinary tract infections. This bacterium degrades urea to produce ammonia that increases the pH. It is interesting to note that even though the bacteria described above grow at extremes of pH, they maintain their cytoplasm at or near neutral pH to prevent damage and destruction to charged species and macromolecules.

Because pH can influence or inhibit the growth of microorganisms, it has been used in food preservation. Fermentation of foods can yield acids such as lactic acid and acetic acid, which lower the pH of the fermented food, thus preventing the growth of many microorganisms and the spoilage of the food. Examples are pickles, yogurt, and some cheeses. However, fungi can grow at acidic pH values, and they can often spoil fermented foods such as cheese.

In this exercise, we will test the degree of inhibition of microorganisms that results from media containing different pH concentrations. Note in the materials list that tubes of six different hydrogen concentrations are listed. Your instructor will indicate which ones, if not all, will be tested.

First Period
Materials

per student:
- 1 tube of nutrient broth of pH 3.0
- 1 tube of nutrient broth of pH 5.0
- 1 tube of nutrient broth of pH 7.0
- 1 tube of nutrient broth of pH 8.0
- 1 tube of nutrient broth of pH 9.0
- 1 tube of nutrient broth of pH 10.0

class materials:
- broth cultures of *Escherichia coli*
- broth cultures of *Staphylococcus aureus*
- broth cultures of *Alcaligenes faecalis**
- broth cultures of *Saccharomyces cerevisiae***

1. Inoculate a tube of each of these broths with one organism. Use the organism following your assigned number from the table below:

STUDENT NUMBER	ORGANISM
1, 5, 9, 13, 17, 21, 25	*Escherichia coli*
2, 6, 10, 14, 18, 22, 26	*Staphylococcus aureus*
3, 7, 11, 15, 19, 23, 27	*Alcaligenes faecalis**
4, 8, 12, 16, 20, 24, 28	*Saccharomyces cerevisiae***

---

*Sporosarcina ureae can be used as a substitute for *Alcaligenes faecalis.*
**Candida glabrata is a good substitute for *Saccharomyces cerevisiae.*

2. Incubate the tubes of *E. coli, S. aureus,* and *A. faecalis* at 37° C for 48 hours. Incubate the tubes of *S. ureae, C. glabrata,* and *S. cerevisiae* at 20° C for 48 to 72 hours.

Second Period

Materials

- spectrophotometer
- 1 tube of sterile nutrient broth
- tubes of incubated cultures at various pHs

1. Use the tube of sterile broth to calibrate the spectrophotometer and measure the O.D. of each culture (page 000, Exercise 22). Record your results in the tables on Laboratory Report 31.
2. Plot the O.D. values in the graph on Laboratory Report 31 and answer all the questions.

Laboratory Report

Student: _____

Date: _____ Section: _____

31   pH and Microbial Growth

A. Results

1. If a spectrophotometer is available, dispense the cultures into labeled cuvettes and determine the O.D. values of each culture. To complete the tables, get the results of the other three organisms from other members of the class, and delete the substitution organisms in the tables that were not used.

    If no spectrophotometer is available, record only the visual reading as +, + +, + + +, and none.

pH	*Escherichia coli*			*Staphylococcus aureus*		
	Visual Reading	Spectrophotometer		Visual Reading	Spectrophotometer	
		O.D.			O.D.	
3						
5						
7						
8						
9						
10						

pH	*Alcaligenes faecalis* or *Sporosarcina ureae*			*Saccharomyces cerevisiae* or *Candida glabrata*		
	Visual Reading	Spectrophotometer		Visual Reading	Spectrophotometer	
		O.D.			O.D.	
3						
5						
7						
8						
9						
10						

2. Growth Curves

Once you have recorded all the O.D. values in the two tables, plot them on the following graph. Use different colored lines for each species.

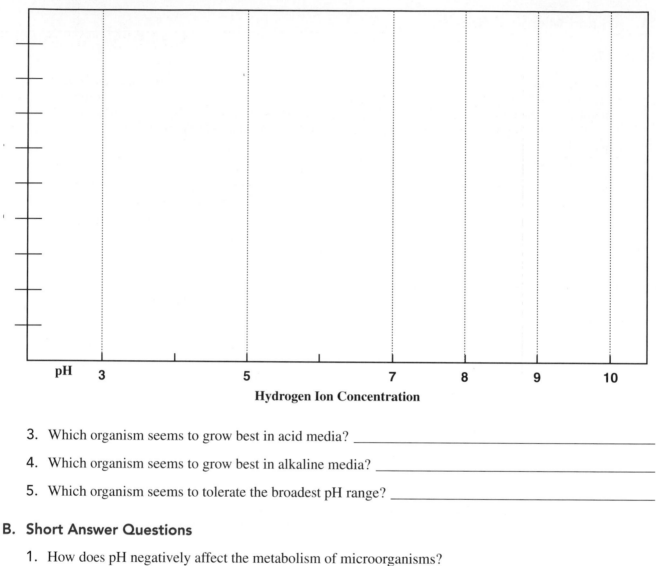

pH   3   5   7   8   9   10

Hydrogen Ion Concentration

3. Which organism seems to grow best in acid media? _____

4. Which organism seems to grow best in alkaline media? _____

5. Which organism seems to tolerate the broadest pH range? _____

B. Short Answer Questions

1. How does pH negatively affect the metabolism of microorganisms?

_____

_____

2. Define three groups of microorganisms in regard to their optimum pH for growth.

_____

_____

3. How would the pH of the culture medium be influenced by sugar fermentation? By urea hydrolysis?

_____

_____

The growth of bacteria can be profoundly affected by the availability of water in an environment. The availability of water is defined by a physical parameter called the water activity, A_w. It is determined by measuring the ratio of the water vapor pressure of a solution to the water vapor pressure of pure water. The values for water activity vary between 0 and 1.0, and the closer the value is to 1.0, the more water is available to a cell for metabolic purposes. Water activity and hence its availability decrease with increases in the concentration of solutes such as salts. This results because water becomes involved in breaking ionic bonds and forming solvation shells around charged species to maintain them in solution.

In the process of **osmosis,** water diffuses from areas of low solute concentration where water is more plentiful to areas of high solute concentration where water is less available. Because there is normally a high concentration of nutrients in the cytoplasm relative to the outside of the cell, water will naturally diffuse into a cell. A medium where solute concentrations on the outside of the cell are lower than the cytoplasm is designated as **hypotonic** (figure 32.1). In general, bacteria are not harmed by hypotonic solutions because the rigid cell wall protects the membrane from being damaged by the osmotic pressure exerted against it. It also prevents the membrane from disrupting when water diffuses across the cell membrane into the cytoplasm.

Environments where the solute concentration is the same inside and outside the cell are termed **isotonic.** Animal cells require isotonic environments or else cells will undergo lysis because only the frag-ile cell membrane surrounds the cell. Tissue culture media for growing animal cells provides an isotonic environment to prevent cell lysis.

Hypertonic environments exist when the solute concentration is greater on the outside of the cell relative to the cytoplasm, and this causes water to diffuse out of the cytoplasm. When this develops, the cell undergoes **plasmolysis** resulting in a loss of water, dehydration of the cytoplasm, and shrinkage of the cell membrane away from the cell wall. In these situations, considerable and often irreversible damage can occur to the metabolic machinery of the cell. Low water activity and hypertonic environments have been used by humans for centuries to preserve food. Salted meat and fish, and jams and jellies with high sugar content resist contamination because very little water is available for cells to grow. Most bacteria that might contaminate these foods would undergo immediate plasmolysis.

Microorganisms can be grouped based on their ability to cope with low water activity and high osmotic pressure. Most bacteria grow best when the water activity is around 0.9 to 1.0. In contrast, **halophiles** require high concentrations of sodium chloride to grow. Examples are the halophilic bacteria that require 15–30% sodium chloride to grow and maintain integrity of their cell walls. These bacteria, which belong to the Archaea, are found in salt lakes and brine solutions, and occasionally growing on salted fish. Some microorganisms are **halotolerant** and are capable of growth in moderate concentrations of salt. For example, *Staphylococcus aureus* can tolerate sodium chloride concentrations that approach 3 *M* or 11%.

Another group of organisms is the **osmophiles,** which are able to grow in environments where sugar concentrations are excessive. An example is *Xeromyces,* a yeast that can contaminate and spoil jams and jellies.

In this exercise, we will test the degree of inhibition of organisms that results with media containing different concentrations of sodium chloride. To accomplish this, you will streak three different organisms on four plates of media. The specific organisms used differ in their tolerance of salt concentrations. The salt concentrations will be 0.5, 5, 10, and 15%. After incubation for 48 hours and several more days, comparisons will be made of growth differences to determine their degrees of salt tolerances.

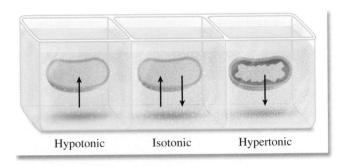

Hypotonic   Isotonic   Hypertonic

Figure 32.1 Osmotic variabilities.

Materials

per student:
- 1 petri plate of nutrient agar (0.5% NaCl)
- 1 petri plate of nutrient agar (5% NaCl)
- 1 petri plate of nutrient agar (10% NaCl)
- 1 petri plate of milk salt agar (15% NaCl)

cultures:
- *Escherichia coli* (nutrient broth)
- *Staphylococcus aureus* (nutrient broth)
- *Halobacterium salinarium* (slant culture)

1. Mark the bottoms of the four petri plates as indicated in figure 32.2.
2. Streak each organism in a straight line on the agar, using a wire loop.
3. Incubate all the plates for 48 hours at room temperature. Record your results on Laboratory Report 32.
4. Continue the incubation of the milk salt agar plate for several more days in the same manner, and record your results again in Laboratory Report 32.

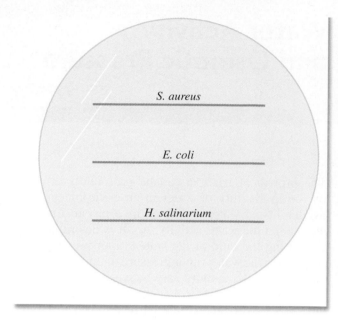

Figure 32.2 Streak pattern.

Laboratory Report

Student: _____

Date: _____ Section: _____

32 Water Activity and Osmotic Pressure

A. Results

1. Record the amount of growth of each organism at the different salt concentrations, using +, + +, + + +, and none to indicate degree of growth.

ORGANISM	SODIUM CHLORIDE CONCENTRATION							
	0.5%		5%		10%		15%	
	48 hr	96 hr	48 hr	96 hr	48 hr	96 hr	48 hr	96 hr
Escherichia coli								
Staphylococcus aureus								
Halobacterium salinarium								

2. Evaluate the salt tolerance of the above organisms.

a. Tolerates very little salt: _____

b. Tolerates a broad range of salt concentration: _____

c. Grows only in the presence of high salt concentration: _____

3. How would you classify *Halobacterium salinarium* as to salt needs? Check one.

Obligate halophile _____   Facultative halophile _____

B. Short Answer Questions

1. Why are bacteria generally resistant to hypotonic environments whereas animal cells are not?

_____

_____

2. How do hypertonic environments negatively affect most bacteria cells?

_____

_____

3. Why are staphylococci well-suited for the colonization of skin?

_____

_____

4. Differentiate between halophiles and osmophiles. Which type would most likely cause spoilage of jams and jellies?

_____

_____

Ultraviolet (UV) light is nonionizing short wavelength radiation that falls between 4 nm and 400 nm in the visible spectrum (figure 33.1). In general, for electromagnetic radiation, the shorter the wavelength, the more damaging it is to cells, thus UV light is much more germicidal than either visible light or infrared radiation. Most bacteria are killed by the effects of ultraviolet light and UV light is routinely used to sterilize surfaces, such as the work areas of transfer hoods used for the inoculation of cultures. The primary lethal effects of ultraviolet light are due to its mutagenic properties. UV radiation at **260 nm** is the most germicidal because this wavelength is the specific wavelength at which DNA maximally absorbs UV light. When DNA absorbs UV light, it causes the formation of **pyrimidine dimers.** These form when a covalent bond is formed between two adjacent thymine or cytosine molecules in a DNA strand (figure 33.2). Dimers essentially cause the DNA molecule to become deformed so that the DNA polymerase cannot replicate DNA strands past the site of dimer formation, nor can genes past this point be transcribed.

Cells have evolved various repair mechanisms to deal with mutational changes in DNA in order to insure that fidelity of replication occurs. One system is the **SOS system,** which enzymatically removes the dimers and inserts in their place new pyrimidine molecules. Unlike DNA polymerase, enzymes of the SOS system can move beyond the point where dimers occur in the molecule. However, if the exposure of UV light is sufficient to cause massive numbers of dimers to form in the DNA of a cell, the SOS system is unable to effectively cope with this situation and it begins to make errors by inserting incorrect bases for the damaged bases, eventually resulting in cell death.

The killing properties of UV light depend on several factors. Time of exposure is important as well as the presence of materials that will block the radiation from reaching cells. For example, plastic can block UV radiation and plastic lenses are an effective means to protect the eyes from UV damage. Endospores are more resistant to UV light than are vegetative cells for several reasons. First, the DNA of endospores is protected by small, acid-soluble proteins that bind to DNA and alter its conformation, thereby protecting it from photochemical damage. Second, a unique spore photo-product is generated by UV light in endospores that functions in enzymatic repair of damaged DNA during endospore germination.

In Exercise 33, you will examine the germicidal effects for UV light on *Bacillus megaterium,* an endospore former, and *Staphylococcus aureus,* a non-endospore former. One-half of each plate will be shielded from the radiation to provide a control comparison. *Bacillus megaterium* and *Staphylococcus aureus* will be used to provide a comparison of the relative resistance of vegetative cells and endospores.

Exposure to ultraviolet light may be accomplished with a lamp as shown in figure 33.3 or with a UV box that has built-in ultraviolet lamps. The UV exposure effectiveness varies with the type of setup used. The

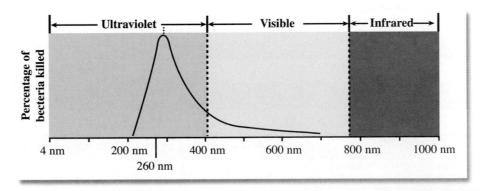

Figure 33.1   Lethal effects of light at various wavelengths.

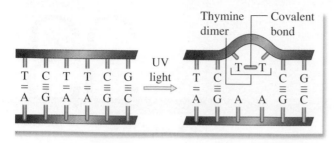

Figure 33.2 Formation of thymine dimers by UV light.

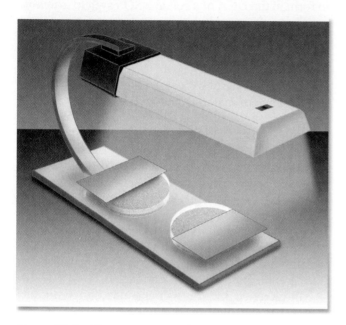

Figure 33.3 Plates are exposed to UV light with 50% coverage.

exposure times given in table 33.1 work well for a specific type of mercury arc lamp. Note in the results that space is provided under the times for adding in different timing. Your instructor will inform you as to whether you should write in new times that will be more suited to the equipment in your lab. Proceed as follows to do this experiment.

Materials

- petri plates of nutrient agar (one or more per student)
- ultraviolet lamp or UV exposure box
- timers
- cards (3″ × 5″)
- nutrient broth cultures of *S. aureus* with swabs
- saline suspensions of *B. megaterium* with swabs

1. Refer to table 33.1 to determine which organism you will work with. You may be assigned more than one plate to inoculate. If different times are to be used, your instructor will inform you what times to write in. Since there are only 16 assignment numbers in the table, more student assignment numbers can be written in as designated by your instructor.
2. Label the bottoms of the plates with your assignment number and your initials.
3. Using a cotton-tipped swab that is in the culture tube, swab the entire surface of the agar in each plate. Before swabbing, express the excess culture from the swab against the inner wall of the tube.
4. Put on the protective goggles provided. Place the plates under the ultraviolet lamp *with the lids removed.* Cover one-half of each plate with a 3″ × 5″ card as shown in figure 33.3. Note that if your number is 8 or 16, you will not remove the lid from your plate. The purpose of this exposure is to determine to what extent plastic protects cells from the effects of UV light.

> **CAUTION:** Before exposing the plates to UV light, put on protective goggles. Avoid looking directly into the UV light source. These rays can cause cataracts and eye injury.

5. After exposing the plates for the correct time durations, re-cover them with their lids, and incubate them inverted at 37° C for 48 hours.

Laboratory Report

Record your observations in Laboratory Report 33 and answer all the questions.

Table 33.1 Student Inoculation Assignments

	EXPOSURE TIMES (STUDENT ASSIGNMENTS)							
S. aureus	1	2	3	4	5	6	7	8
	10 sec	20 sec	40 sec	80 sec	2.5 min	5 min	10 min	20 min*
B. megaterium	9	10	11	12	13	14	15	16
	1 min	2 min	4 min	8 min	15 min	30 min	60 min	60 min*

*These petri plates will be covered with dish covers during exposure.

Laboratory Report

Student: _____

Date: _____ Section: _____

33 Ultraviolet Light: Lethal Effects

A. Results

1. Your instructor will construct a table similar to the one below on the chalkboard for you to record your results. If substantial growth is present in the exposed area, record your results as + + +. If three or fewer colonies survived, record +. Moderate survival should be indicated as + +. No growth should be recorded as −. Record all information in the table.

ORGANISMS	EXPOSURE TIMES							
S. aureus	10 sec	20 sec	40 sec	80 sec	2.5 min	5 min	10 min	10*min
Survival								
B. megaterium	1 min	2 min	4 min	8 min	15 min	30 min	60 min	60*min
Survival								

* plates covered during exposure

2. How many times more resistant were *B. megaterium* spores than *S. aureus* vegetative cells?

_____

_____

3. Why was half of each plate covered with an index card?

_____

_____

4. What was the purpose of leaving the cover on one set of petri dishes?

_____

_____

B. Short Answer Questions

1. Describe the damaging effects of UV radiation on living cells.

_____

_____

2. Why does exposure to UV radiation cause death in vegetative cells but not endospores?

_____

_____

3. At which wavelength is UV radiation most germicidal? Explain.

4. What limited protection do cells have against the damaging effects of UV radiation?

5. What types of damage to human tissues can result from prolonged exposures to UV radiation? What protective measures can be taken to limit these types of damage, both during the experiment as well as in everyday life?

6. Which *B. megaterium* culture, logarithmic or stationary phase, would show the best survival following exposure to UV radiation? Explain.

The Effects of Lysozyme on Bacterial Cells

34

Most bacterial cells are surrounded by a cell wall that contains a common and important component called **peptidoglycan,** which is only found in pro-karyotes. Peptidoglycan is a polymer of alternating N-acetyl-glucosamine (NAG) and N-acetyl-muramic acid (NAM), which are linked covalently by a β (1–4) bond (figure 34.1a).

Each N-acetyl-muramic acid contains a short chain of four amino acids that serves to cross-link the adja-cent polymers of the amino sugars, thus forming a lat-tice network or sac around the exterior of the bacterial cell that provides rigidity and also defines the shape of the cell (i.e., coccus, rod, or spiral) (figure 34.1b). There is some diversity in the structure of the peptide

cross-link. For example, *Staphylococcus aureus* has an inter bridge composed of five glycine residues that connects the tetra peptides of the N-acetyl muramic acid molecules. Other species of *Staphylococcus* may have both serine and glycine making up the inter bridges. *Escherichia coli,* a gram-negative bacterium lacks an inter bridge and in this case the tetrapeptides are directly linked through a peptide bond between an amino acid in one tetrapeptide and another amino acid in the adjacent chain.

The cell wall of gram-positive bacteria is com-posed of 90% peptidoglycan. In addition, most gram-positive bacteria have **teichoic acids** in their cell walls. Teichoic acids are polymers of ribitol-phosphate

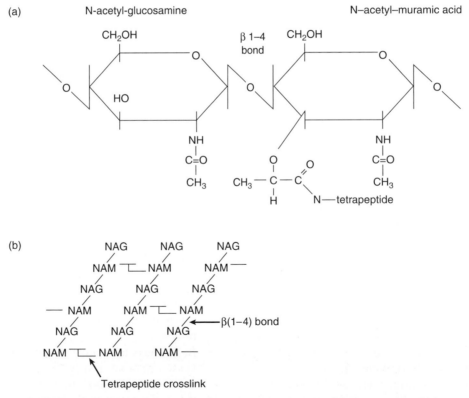

Figure 34.1 (a) Structure of peptidoglycan showing the β (1–4) bond between N-acetyl-glucosamine (NAG) and N-acetyl-muramic acid (NAM). (b) Structure of peptidoglycan showing the alternating units of NAG and NAM bonded by the β (1–4) bond and cross-linked via the tetrapeptide component.

or glycerol-phosphate that can bind covalently to peptidoglycan or to the cytoplasmic membrane. They are responsible for the net negative charge on gram-positive cells. They may function in the expansion of peptidoglycan during cell growth. Bacterial cell enzymes called **autolysins** partially open up the peptidoglycan polymer by breaking β 1–4 bonds between the N-acetyl-muramic acid and N-acetyl-glucosamine molecules so that new monomeric subunits of peptidoglycan can be inserted into the existing cell wall and allow for expansion. This process must be limited because if large gaps in peptidoglycan are produced by the autolysins, the cell membrane can rupture through the gaps. Teichoic acids may play a role in limiting the amount of degradation of peptidoglycan by the autolysins. It is interesting to note that teichoic acids are found in *Staphylococcus* but not in *Micrococcus*.

The cell wall of gram-negative bacteria in contrast to that of gram-positive bacteria is composed of a thin layer of peptidoglycan and an outer membrane that encloses the peptidoglycan. Thus, the peptidoglycan of gram-negative bacteria is not accessible from the external environment. The outer membrane of gram-negative bacteria forms an additional permeability barrier in these organisms. Gram-negative bacteria also lack teichoic acids in their cell walls.

Humans are born with intrinsic non-immune factors that protect them from infection by bacteria. One of these factors is **lysozyme,** which is found in most body fluids such as tears and saliva. Lysozyme, like the autolysins, degrades the β 1–4 bond between the amino sugar molecules in peptidoglycan, thus causing breaks in the lattice and weakening the cell wall. As a result, the solute pressure of the cytoplasm can cause the cell membrane to rupture through these breaks, resulting in cell lysis and death.

The presence of lysozyme in tears, for example, can protect our eyes from infection because bacteria that try to extablish an infection undergo lysis as a result of the lysozyme in tears that washes the surface of the eye. The egg white of hen's eggs also contains lysozyme where it also may play a protective function for the developing embryo.

Gram-negative bacteria are usually more resistant to lysozyme than are gram-positive bacteria because the gram-negative outer membrane prevents lysozyme from reaching the peptidoglycan layer. However, some gram-positive bacteria, such as *Staphylococcus aureus,* are resistant to lysozyme, presumably because of the presence of teichoic acids in their cell walls, which limits the action of autolysins and

lysozyme. In the following exercise you will study the effects of lysozyme on gram-positive and gram-negative bacteria. The source of lysozyme will be human saliva and the egg white of hen's eggs.

Materials

per class of 30–40
- 2–3 hen's eggs

per pair of students:
- 3 plates of nutrient agar
- small vessels for collecting saliva
- sterile cotton swabs
- small sterile test tubes
- sterile Pasteur pipettes
- pipette bulbs

broth cultures of:
- *Escherichia coli*
- *Micrococcus luteus*
- *Staphylococcus aureus*

1. Using a marking pen, divide each plate into three sectors and label them 1, 2 and 3.
2. Using a sterile cotton swab, uniformly inoculate one plate with *E. coli.* Do this by first swabbing the plate in one direction with a cotton swab dipped in the culture. Then turn the plate 90 degrees and swab at a right angle to the first direction.
3. Repeat this procedure for the second and third plates using *M. luteus* and *S. aureus.*
4. The instructor will break an egg and separate the white portion from the yolk. Using a sterile Pasteur pipette, collect a small amount of egg white and transfer it to a small test tube.
5. Collect saliva by expelling some into the small vessel provided.
6. Using a sterile Pasteur pipette, tranfer one drop of egg white to sector 1 on the plate inoculated with *E. coli.* To sector 2, deliver 1 drop of saliva on this same plate. Sector 3 will serve as a growth control for the organism (figure 34.2).
7. Repeat this same procedure for the remaining plates inoculated with *M. luteus* and *S. aureus.*
8. Allow the liquids to absorb into the agar medium and then incubate the plates at 37° C for 24 hours.
9. In the next laboratory period, observe the plates. If the organism is affected by lysozyme, the cells exposed to the enzyme will undergo lysis, forming a clear area in the sector. Which culture was affected by the enzyme? Were all bacteria affected in the same way? Explain.

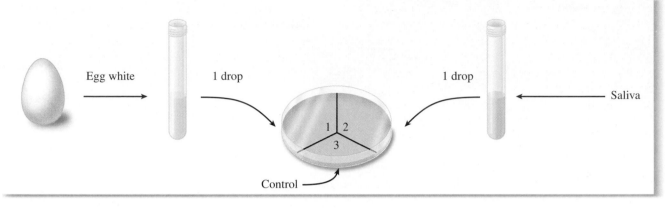

Figure 34.2 Procedure for setting up the lysozyme plates.

Laboratory Report

Student: _____

Date: _____ Section: _____

34 The Effects of Lysozyme on Bacterial Cells

A. Results

1. Record whether lysis occurred to the cells on the plate using a + or −.

	Saliva SECTOR				Egg White SECTOR		
	1	2	3		1	2	3
Staphylococcus aureus							
Escherichia coli							
Micrococcus luteus							

2. Why did lysozyme affect the test organisms differently?

_____

_____

3. What was the purpose of including a quadrant to which no lysozyme was added?

_____

_____

B. Short Answer Questions

1. What is the function of peptidoglycan in bacterial cells?

_____

_____

2. How does lysozyme specifically affect peptidoglycan?

_____

_____

3. Where in nature can lysozyme be found? Why is it produced in these environments?

_____

_____

4. Based on the results, what might one conclude about the types of bacteria that are involved in eye infections?

_____

_____

Evaluation of Alcohol:
Its Effectiveness as an Antiseptic

As an antiseptic, alcohol is widely used to treat the skin before inoculations or the drawing of blood from veins. Prepackaged alcohol swabs used by nurses and laboratory technologists are indispensable for treating the skin prior to these procedures. The question that often arises is: How really effective is alcohol in routine use? When the skin is swabbed prior to penetration, are all, or mostly all, of the surface bacteria killed? To determine alcohol effectiveness, as it might be used in routine skin disinfection, we are going to perform a very simple experiment that utilizes four thumbprints and a plate of enriched agar. Class results will be pooled to arrive at a statistical analysis.

Figure 35.1 illustrates the various steps in this test. Note that the petri plate is divided into four quadrants. On the left side of the plate an unwashed left thumb is first pressed down on the agar in the lower left quadrant of the plate. Next, the left thumb is pressed down on the upper left quadrant. With the left thumb, we are trying to establish the percentage of bacteria that are removed by simple contact with the agar.

On the right side of the plate, an unwashed right thumb is pressed down on the lower right quadrant of the plate. The next step is to either dip the right thumb into alcohol or to scrub it with an alcohol swab and dry it. Half of the class will use the dipping method and the other half will use alcohol swabs. Your instructor will indicate what your assignment will be.

The last step is to press the dried right thumb on the upper right quadrant of the plate.

After the plate is inoculated, it is incubated at 37° C for 24–48 hours. Colony counts will establish the effectiveness of the alcohol.

Materials

- 1 petri plate of veal infusion agar
- small beaker
- 70% ethanol
- alcohol swab

1. Perform this experiment with unwashed hands.
2. With a marking pencil, mark the bottom of the petri plate with two perpendicular lines that divide it into four quadrants. Label the left quadrants **A** and **B** and the right quadrants **C** and **D** as shown in figure 35.1. (*Keep in mind that when you turn the plates over to label them, the A and B quadrants will be on the right and C and D will be on the left.*)
3. Press the pad of your left thumb against the agar surface in the A quadrant.
4. Without touching any other surface, press the left thumb into the B quadrant.
5. Press the pad of your right thumb against the agar surface of the C quadrant.
6. Disinfect the right thumb by one of the two following methods:

2

Without touching any other surface the left thumb is pressed against the agar in quadrant B.

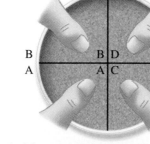

6

The pad of the treated right thumb is pressed against the agar in the D quadrant.

5

The alcohol-treated right thumb is allowed to completely air-dry.

4

The pad of the right thumb is immersed in 70% alcohol or scrubbed with an alcohol swab for 10 seconds.

1

The pad of the unwashed left thumb is momentarily pressed against the agar in quadrant A.

3

The pad of the unwashed right thumb is momentarily pressed against the agar in quadrant C.

Figure 35.1   Procedure for testing the effectiveness of alcohol on the skin.

- dip the thumb into a beaker of 70% ethanol for 5 seconds, or
- scrub the entire pad surface of the right thumb with an alcohol swab.

7. Allow the alcohol to completely evaporate from the skin.

8. Press the right thumb against the agar in the D quadrant.

9. Incubate the plate at 37° C for 24–48 hours.

10. Follow the instructions in Laboratory Report 35 for evaluating the plate and answer all of the questions.

Laboratory Report

Student: _____

Date: _____ Section: _____

35 Evaluation of Alcohol: Its Effectiveness as an Antiseptic

A. Results

1. Count the number of colonies that appear on each of the thumbprints and record them in the following table. If the number of colonies has increased in the second press, record a 0 in percent reduction. Calculate the percentages of reduction and record these data in the appropriate column. Use this formula:

$$\text{Percent reduction} = \frac{(\text{Colony count 1st press}) - (\text{Colony count 2nd press})}{(\text{Colony count 1st press})} \times 100$$

LEFT THUMB (Control)			RIGHT THUMB (Dipped)			RIGHT THUMB (Swabbed)		
Colony Count 1st Press	Colony Count 2nd Press	Percent Reduction	Colony Count 1st Press	Colony Count 2nd Press	Percent Reduction	Colony Count 1st Press	Colony Count 2nd Press	Percent Reduction
Av. % Reduction, Left (C)			Av. % Reduction, Right (D)			Av. % Reduction, Right (S)		

2. In general, what effect does alcohol have on the level of skin contaminants? _____ _____

3. Is there any difference between the effects of dipping versus swabbing? _____

    Which method appears to be more effective? _____

4. There is definitely survival of some microorganisms even after alcohol treatment. Without staining or microscopic scrutiny, predict what types of microbes are growing on the medium where you made the right thumb impression after treatment. _____

    _____

    _____

    _____

    _____

B. Short Answer Questions

1. For what purposes is alcohol a useful antiseptic?

    _____

    _____

2. What advantages does alcohol have over hand soap for antisepsis of the skin?

    _____

    _____

3. Why does treatment of human skin with alcohol not create a completely sterile environment?

    _____

    _____

Antimicrobic Sensitivity Testing:
The Kirby-Bauer Method

Once the causative bacterial agent of a specific disease in a patient has been isolated and identified, it is up to the attending physician to prescribe an antimicrobial agent that will inhibit or kill the pathogen without causing serious harm to the patient. Antimicrobial agents such antibiotics vary in their effectiveness against various pathogenic bacteria. For example, some antibiotics are more effective against gram-positive or gram-negative bacteria whereas some are broad spectrum and can be used against both kinds of bacteria. Oftentimes, antibiotics and antimicrobials must be tested against a pathogen to determine if the organism is sensitive or resistant to the agents. This requires a relatively simple testing method that is reliable and yields results in as short a time as possible. The Kirby-Bauer procedure of sensitivity testing is such a method for testing antibiotics and antimicrobials. **Antimicrobials** are compounds that kill or inhibit microorganisms. **Antibiotics** are antimicrobials, usually of low molecular weight, produced by microorganisms that inhibit or kill other microorganisms. Two common examples are penicillin and streptomycin. However, many times antibiotics are chemically altered to make them more effective in their mode of action. These are referred to as **semi-synthetics.** Some antimicrobials such as the sulfa drugs are chemically synthesized in the laboratory and are not the result of microbial biosynthesis. The sulfa drugs are **synthetics** and were used clinically to treat bacterial infections before the discovery of penicillin. In this exercise, both types of agents will be tested, as shown in figure 36.1.

The test is performed by uniformly streaking a medium with the test organism. Paper disks containing specific concentrations of an antibiotic or antimicroibial are deposited on the agar surface. A chemotherapeutic agent in a disk diffuses out, forming a concentration gradient. If the agent inhibits or kills the test organism, there will be a zone around the disk where no growth occurs called the **zone of inhibition.** The zone of inhibition, however, varies with the diffusibility of the agent, the size of the inoculum, the type of medium, and many other factors. Only by taking all these variables into consideration can a reliable method be worked out.

The **Kirby-Bauer method** is a standardized system that takes all variables into consideration. It is sanctioned by the U.S. FDA and the Subcommittee on Antimicrobial Susceptibility Testing of the National Committee for Clinical Laboratory Standards. Although time is insufficient here to consider all facets of this test, its basic procedure will be followed.

The recommended medium in this test is Mueller-Hinton II agar. Its pH should be between 7.2 and 7.4, and it should be poured to a uniform thickness of 4 mm in the petri plate. This requires 60 ml in a 150 mm plate and 25 ml in a 100 mm plate. For certain fastidious microorganisms, 5% defibrinated sheep's blood is added to the medium.

Inoculation of the surface of the medium is made with a cotton swab from a broth culture. In clinical applications, the broth turbidity has to match a defined standard. Care must also be taken to express excess broth from the swab prior to inoculation.

High-potency disks are used that may be placed on the agar with a mechanical dispenser or sterile forceps. To secure the disks to the medium, it is necessary to press them down onto the agar.

After 16 to 18 hours incubation, the plates are examined and the diameters of the zones are measured to the nearest millimeter. [To determine the significance of the zone diameters, consult table 36.1.]

In this exercise, we will work with four microorganisms: *Staphylococcus aureus, Escherichia coli, Proteus vulgaris,* and *Pseudomonas aeruginosa.* Each student will inoculate one plate with one of the four organisms and place the disks on the medium by whichever method is available. Since each student will be doing only a portion of the total experiment, student assignments will be made. Proceed as follows:

First Period

(Plate Preparation)

Materials

- 1 petri plate of Mueller-Hinton II agar
- nutrient broth cultures (with swabs) of *S. aureus, E. coli, P. vulgaris,* and *P. aeruginosa*
- disk dispenser (BBL or Difco)
- cartridges of disks (BBL or Difco)
- forceps and Bunsen burner
- zone interpretation charts (Difco or BBL)

(1) The entire surface of a plate of nutrient medium is swabbed with organism to be tested.

(2) Handle of dispenser is pushed down to place 12 disks on the medium. In addition to dispensing disks, this dispenser also tamps disks onto medium.

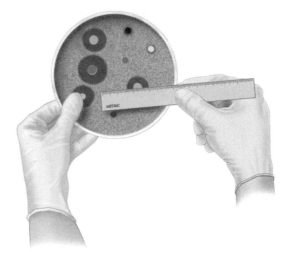

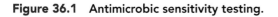

(3) Cartridges (Difco) can be used to dispense individual disks. Only 4 or 5 disks should be placed on small (100 mm) plates.

(4) After 18 hours incubation, the zones of inhibition (diameters) are measured in millimeters. Significance of zones is determined from Kirby-Bauer chart.

Figure 36.1 Antimicrobic sensitivity testing.

1. Select the organisms you are going to work with from the following table.

ORGANISM	STUDENT NUMBER
S. aureus	1, 5, 9, 13, 17, 21, 25
E. coli	2, 6, 10, 14, 18, 22, 26
P. vulgaris	3, 7, 11, 15, 19, 23, 27
P. aeruginosa	4, 8, 12, 16, 20, 24, 28

2. Label your plate with the name of your organism.
3. Inoculate the surface of the medium with the swab after expressing excess fluid from the swab by pressing and rotating the swab against the inside walls of the tube above the fluid level. Cover the surface of the agar evenly by swabbing in three directions. A final sweep should be made of the agar rim with the swab.
4. Allow **3 to 5 minutes** for the agar surface to dry before applying disks.

5. Dispense disks as follows:
   a. If an automatic dispenser is used, remove the lid from the plate, place the dispenser over the plate, and push down firmly on the plunger. With the sterile tip of forceps, tap each disk lightly to secure it to medium.

   b. If forceps are used, sterilize them first by flaming before picking up the disks. Keep each disk at least 15 mm from the edge of the plate. Place no more than 13 on a 150 mm plate, no more than 5 on a 100 mm plate. Apply light pressure to each disk on the agar with the tip of

Table 36.1 Zones of Inhibition in the Kirby-Bauer Method of Antimicrobic Sensitivity Testing

ANTIBIOTIC	CODE	POTENCY	Zone of Inhibition (mm)		
			RESISTANT	INTERMEDIATE	SENSITIVE
Amikacin	AN-30	30 μg			
Enterobacteriaceae					
P. aeruginosa, Acinetobactor					
staphylococci			≤14	15–16	≥17
Amoxicillin/Clavulinic acid	AmC-30	20/10 μg			
Enterobacteriaceae			≤13	14–17	≥18
Staphylococcus spp.			≤19	—	≥20
Haemophilus spp.			≤19	—	≥20
Ampicillin	AM-10	10 μg			
Enterobacteriaceae			≤13	14–16	≥17
Staphylococcus spp.			≤28	—	≥29
Enterococcus spp.			≤16	—	≥17
Listeria monocytogenes			≤19	—	≥20
Haemophilus spp.			≤18	19–21	≥22
β-hemolytic streptococci			—	—	≥24
Aziocillin	AZ-75	75 μg			
P. aeruginosa			≤17	—	≥18
Bacitracin	B-10	10 units	≤8	9–12	≥13
Carbenicillin	CB-100	100 μg			
Enterobacteriaceae and					
Acinetobacter			≤19	20–22	≥22
P. aeruginosa			≤13	14–16	≥17
Cefaclor	CEC-30	30 μg			
Enterobactiaceae and					
staphylococci			≤14	15–17	≥18
Haemophilus spp.			≤16	17–19	≥20
Cefazolin	CZ-30	30 μg			
Enterobacteriaceae and					
staphylococci			≤14	15–17	≥18
Cephalothin	CF-30	30 μg			
Enterobacteriaceae, and					
staphylococci			≤14	15–17	≥18
Chloramphenicol	C-30	30 μg			
Enterobacteriaceae,					
P. aeruginosa, Acinetobactor,					
staphylococci, enterococci,					
and *V. cholerae*			≤12	13–17	≥18
Haemophilus spp.			≤25	26–28	≥29
S. pneumoniae			≤20	—	≥21
Streptococci			≤17	18–20	≥21

(continued)

Table 36.1 Zones of Inhibition in the Kirby-Bauer Method of Antimicrobic Sensitivity Testing (Cont.)

ANTIBIOTIC	CODE	POTENCY	Zone of Inhibition (mm)		
			RESISTANT	INTERMEDIATE	SENSITIVE
Ciprofloxacin *Enterobacteriaceae, P. aeruginosa, Acinetobacter,* staphylococci and enterococci *Haemophilus* spp. *N. gonorrhoeae*	CIP-5	5 μg	≤15 — ≤27	16–20 — 28–40	≥21 ≥21 ≥41
Clarithromycin *Staphylococcus* spp. *Haemophilus* spp. *S. pneumoniae* and other streptococci	CLR-15	15 μg	≤13 ≤10 ≤16	14–17 11–12 17–20	≥18 ≥13 ≥21
Clindamycin *Staphylococcus* spp. *S. pneumoniae* and other streptococci	CC-2	2 μg	≤14 ≤15	15–20 16–18	≥21 ≥19
Doxycycline *Enterobacteriaceae, P. aeruginosa, Acinetobactor,* staphylococci and enterococci	D-30	30 μg	≤12	13–15	≥13
Erythromycin *Staphylococcus* spp. and enterococci *S. pneumoniae* and other streptococci	E-15	15 μg	≤13 ≤15	14–22 16–20	≥23 ≥21
Gentamicin *Enterobacteriaceae P. aeruginosa, Acinetobacter,* and staphylococci	GM-120	120 μg	≤12	13–14	≥15
Imipenem *Enterobacteriaceae, P. aeruginosa, Acinetobacter,* and staphylococci *Haemophilus* spp.	IPM-10	10 μg	≤13 —	14–15 —	≥16 ≥16
Kanamycin *Enterobacteriaceae* and staphylococci	K-30	30 μm	≤13	13–17	≥18
Lomefloxacin *Enterobacteriaceae, P. aeruginosa, Acinetobacter,* and staphylococci *Haemophilus* spp. *N. gonorrhoeae*	LOM-10	10 μg	≤18 — ≤26	19–21 — 27–37	≥22 ≥22 ≥38
Loracarbef *Enterobacteriaceae* and staphylococci *Haemophilus* spp.	LOR-30	30 μg	≤14 ≤15	15–17 16–18	≥18 ≥19
Meziocillin *Enterobacteriaceae* and *Acinetobacter* *P. aeruginosa*	MZ-75	75 μg	≤17 ≤15	18–20 —	≥21 ≥16

Table 36.1 Zones of Inhibition in the Kirby-Bauer Method of Antimicrobic Sensitivity Testing (Cont.)

ANTIBIOTIC	CODE	POTENCY	Zone of Inhibition (mm)		
			RESISTANT	INTERMEDIATE	SENSITIVE
Minocycline *Enterobacteriaceae,* *P. aeruginosa, Acinetobacter,* staphylococci, and enterococci	MI-30	30 µg	≤14	15–18	≥19
Moxalactam *Enterobacteriaceae,* *P. aeruginosa, Acinetobacter,* and staphylococci	MOX-30	30 µg	≤14	15–22	≥23
Nafcillin *Staphylococcus aureus*	NF-1	1 µg	≤10	11–12	≥13
Nalidixic Acid *Enterobacteriaceae*	NA-30	30 µg	≤13	14–18	≥19
Neomycin	N-30	30 µg	≤12	13–16	≥17
Netilmicin *Enterobacteriaceae,* *P. aeruginosa, Acinetobacter* and staphylococci	NET-30	30 µg	≤12	13–14	≥15
Norfloxacin *Enterobacteriaceae,* *P. aeruginosa, Acinetobacter,* staphylococci and entercocci	NCR-10	10 µg	≤12	13–16	≥17
Novobiocin	NB-30	30 µg	≤17	18–21	≥22
Oxacillin *Staphylococcus aureus* staphylococcus (coagulase negative)	OX-1	1 µg	≤10 ≤17	11–12 —	≥13 ≥18
Penicillin *Staphylococcus* spp. *Enterococcus* spp. *L. monocytogenes* *N. gonorrhoeae* β-hemolytic streptococci	P-10	10 units	≤28 ≤14 ≤19 ≤26 —	— — 20–27 27–46 —	≥29 ≥15 ≥28 ≥47 ≥24
Piperacillin *Enterobacteriaceae,* and *Acinetobacter* *P. aeruginosa*	PIP-100	100 µg	≤17 ≤17	18–20 —	≥21 ≥18
Polymyxin B	PB-300	300 U	≤8	9–11	≥12
Rifampin *Staphylococcus* spp. *Enterococcus* spp. and *Haemophilis* spp. *S. pneumoniae*	RA-5	5 µg	≤16 ≤16	17–19 17–18	≥20 ≥19
Spectinomycin *N. gonorrhoeae*	SPT-100	100 µg	≤14	15–17	≥18
Streptomycin *Enterobacteriaceae*	S-300	300 µg	≤11	12–14	≥15

(continued)

Table 36.1 Zones of Inhibition in the Kirby-Bauer Method of Antimicrobic Sensitivity Testing (Cont.)

ANTIBIOTIC	CODE	POTENCY	Zone of Inhibition (mm)		
			RESISTANT	INTERMEDIATE	SENSITIVE
Sulfisoxazole *Enterobacteriaceae,* *P. aeruginosa, Acinetobacter,* *V. cholerae,* and staphylococci	G-25	25 μg	≤12	13–16	≥17
Tetracycline *Enterobacteriaceae,* *P. aeruginosa, Acinetobacter,* *V. cholerae,* staphylococci and enterococci *Haemophilus* spp. *N. gonorrhoeae* *S. pneumoniae* and other streptococci	Te-30	30 μm	≤14 ≤25 ≤30 ≤18	15–18 26–28 31–37 19–22	≥19 ≥29 ≥38 ≥23
Tobramycin *Enterobacteriaceae,* *P. aeruginosa, Acinetobacter,* and staphylococci	NN-10	10 μg	≤12	13–14	≥15
Trimethoprim *Enterobacteriaceae* and staphylococci	TMP-5	5 μg	≤10	11–15	≥16
Vancomycin *Staphylococcus* spp. *Enterococcus* spp. *S. pneumoniae* and other streptococci	Va-30	30 μg	— ≤14 —	— 15–16 —	≥15 ≥17 ≥17

Courtesy and © Becton-Dickinson and Company

a sterile forceps or inoculating loop to secure it to medium.

6. Invert and incubate the plate for 16 to 18 hours at 37° C.

Second Period

(Interpretation)

After incubation, measure the zone diameters with a metric ruler to the nearest whole millimeter. The zone of complete inhibition is determined without magnification. Ignore faint growth or tiny colonies that can be detected by very close scrutiny. Large colonies growing within the clear zone might represent resistant variants or a mixed inoculum and may require reiden-

tification and retesting in clinical situations. Ignore the "swarming" characteristics of *Proteus,* measuring only to the margin of heavy growth.

Record the zone measurements on the table of Laboratory Report 36 and on the chart on the demonstration table, which has been provided by the instructor.

Use table 36.1 for identifying the various disks.

To determine which antibiotics your organism is sensitive to (S), or resistant to (R), or intermediate (I), consult the above table. It is important to note that the significance of a zone of inhibition varies with the type of organism. If you cannot find your antibiotic on the chart, consult a chart that is supplied on the demonstration table or bulletin board. Table 36.1 is incomplete.

Laboratory Report

Student: _____

Date: _____ Section: _____

36 Antimicrobic Sensitivity Testing: The Kirby-Bauer Method

A. Results

1. List the antimicrobics that were used for each organism. Consult table 36.1 to identify the various disks. After measuring and recording the zone diameters, consult table 36.1 for interpretation. Record the degrees of sensitivity (R, I, or S) in the sensitivity column. Exchange data with other class members to complete the entire chart.

	ANTIMICROBIC	ZONE DIA.	RATING (R, I, S)	ANTIMICROBIC	ZONE DIA.	RATING (R, I, S)
S. aureus						
P. aeruginosa						
Proteus vulgaris						
E. coli						

2. Which antimicrobics would be suitable for the control of the following organisms?

    S. aureus: _____

    E. coli: _____

    P. vulgaris: _____

    P. aeruginosa: _____

B. Short Answer Questions

1. Differentiate between the following and provide one example of each:

    a. antibiotics and antimicrobial drugs.

    _____

    b. broad- and narrow-spectrum antibiotics.

    _____

2. What factors influence the size of the zone of inhibition for an antibiotic?

    _____

    _____

3. Two antibiotics are tested for their efficacy against a single bacterial species. If antibiotic A and antibiotic B produce zones of inhibition with the same diameter, how can the bacterium be considered resistant to antibiotic A, but sensitive to antibiotic B?

    _____

    _____

4. Why are certain gram-negative bacteria more resistant than gram-positive bacteria to antibiotics that attack cytoplasmic targets?

    _____

    _____

5. Why are gram-positive bacteria typically more resistant than gram-negative bacteria to antibiotics, that disrupt plasma membranes, such as polymyxin B?

    _____

    _____

6. If a bacterial isolate shows intermediate to moderate resistance to an antibiotic, how might this antibiotic still be successfully used in the treatment of this microbe?

    _____

    _____

Evaluation of Antiseptics:
The Filter Paper Disk Method

37

Everyday, we use a number of different chemical agents to control or kill microorganisms. Hospitals disinfect contaminated areas with strong oxidizing agents such as sodium hypochlorite, which is identical to common household bleach. Alcohol is applied by a medical professional to the skin before giving an injection, and betadine—an organic form of iodine—is swabbed onto the skin before surgery to ensure that skin bacteria such as *Staphylococcus aureus* will not cause postsurgical infections. Cities and municipalities add chlorine to water supplies to prevent the spread of potential pathogens such as *Salmonella typhi* in the drinking water. Chemical agents are added to food to retard spoilage by microbes, to increase the shelf-life of food products in the supermarket, and to prevent the growth of pathogenic bacteria such as *Clostridium botulinum,* which causes a deadly form of food poisoning.

Chemical agents that are used to control microorganisms can be defined as antiseptics or disinfectants. Antiseptics are substances such as alcohol or betadine that inhibit microbial growth or kill microorganisms and are gentle enough to be applied to living tissue. However, these compounds do not destroy endospores. They are used for hand washing, treating surface wounds on the skin, and preparing the skin for invasive procedures such as incisions and inoculations prevent infections by opportunistic skin bacteria. Disinfectants are chemical agents that are applied to inanimate objects such as floors, walls and tabletops to kill microorganisms. They are usually more harsh than antiseptics and are therefore damaging to living tissue. Some disinfectants are classified as **steriliants** or **sporocides,** which means that they destroy all microbial life, including endospores. An example is ethylene oxide, a gas that is used to sterilize heat-sensitive objects such as plastic petri plates, plastic pipettes, plastic syringes, and some surgical instruments that cannot be exposed to high temperatures. Agents that reduce microbial numbers to a safe level but do not completely eliminate all microbes are defined as **sanitizers.** These agents are used in the food industry to treat cooking equipment such as dishes and utensils. If a particular agent only inhibits the growth of bacterial cells but does not kill them, we say that the agent is **bacteriostatic.** Were the agent to be removed, growth of the

bacterial cells would resume. In contrast, agents that kill bacterial cells are termed **bacteriocidal.** Agents can also be specific for certain groups of microorganisms; thus, there are bactericides, fungicides, and viricides, and their use would depend on which group needed to be controlled. Some antiseptics are classified as drugs and are therefore regulated by the Food and Drug Administration.

It is difficult to compare the relative effectiveness of disinfectants and antiseptics because they can have vastly different modes of action and chemical properties such as solubility, which can influence the effectiveness of the agent under test conditions. However, by modifying the method used in the Kirby-Bauer technique for evaluating antibiotics, we can use this procedure to determine if an antiseptic or disinfectant inhibits the growth of a test organism and compare the relative efficacy of one agent against another. In this procedure, a plate of suitable medium is streaked with the test organism. Filter paper disks, the same size as antibiotic disks, are impregnated with the agent by dipping the disk in a germicide or disinfectant and placing the disk on an inoculated plate. The plate is incubated for 48 hours. The agent will diffuse from the disk into the agar, forming a concentration gradient. If the substance is inhibitory, a clear zone of inhibition will surround the disk where no growth has occurred. The size of the zone can be used to quantitatively compare one agent's effectiveness against other chemical substances.

In this exercise, you will measure the relative effectiveness of various disinfectants and antiseptics against *Staphylococcus aureus,* a gram-positive bacterium, and *Pseudomonas aeruginosa,* a gram-negative bacterium. Suggested chemical agents to be tested are: bleach (1/10 dilution), 5% phenol, 5% aqueous iodine, betadine, commercial mouth washes, amphyl, and 5% formaldehyde.

First Period

(Disk Application)

Materials

per student:
- 1 nutrient agar pour and 1 petri plate
- broth culture of *S. aureus* and *P. aeruginosa on* demonstration table

Table 37.1 Student Assignments

Chemical Agent		Student Number	
		S. aureus	*P. aeruginosa*
5% Phenol		1, 15, 29	2, 16, 30
5% Formaldehyde		3, 17, 31	4, 18, 32
5% Iodine		5, 19, 33	6, 20, 34
1:10 Bleach		7, 21	8, 22
Mouthwash		9, 23	10, 24
Betadine		11, 25	12, 26
Amphyl		13, 27	14, 28

- petri dish containing sterile disks of filter paper (1/2″ dia)
- forceps and Bunsen burner
- chemical agents in small beakers (5% phenol, 5% formaldehyde, 5% aqueous iodine)

1. Consult table 37.1 to determine your assignment.
2. Liquefy a nutrient agar pour in a water bath and cool to 50° C.
3. Label the bottom of a petri plate with the names of the organism and chemical agent.
4. Inoculate the agar pour with a loopful of *S. aureus* and aseptically pour it into a sterile petri plate, allowing the medium to solidify. (Alternatively a plate of solidified medium can be uniformly streaked with the test organism.)
5. Sterilize the forceps by dipping it in alcohol and flaming (*careful: flames and alcohol*). Using the flamed forceps, pick up a sterile filter paper disk and dip it halfway into a disinfectant. Place the disk on the inoculated plate and gently press the disk onto the agar surface. Place no more than four disks on a 150-mm plate.
6. Repeat the same procedure for *Pseudomonas aeruginosa.*
7. Incubate the plates for 48 hours at 37° C.

Second Period

(Evaluation)

1. Measure the zone of inhibition from the edge of the disk to the edge of the growth for the test organism (see illustration 5, figure 37.1).
2. Evaluate the antiseptics and disinfectants based on the zones of inhibition for each agent. Complete the table in Laboratory Report 37.

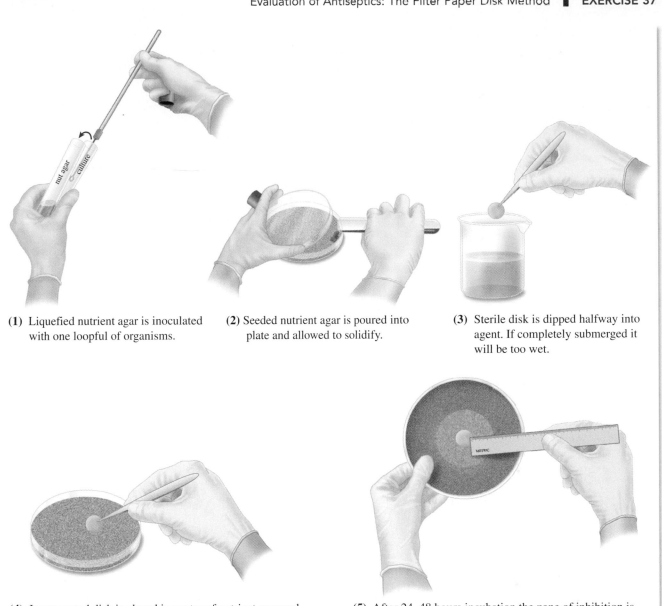

(1) Liquefied nutrient agar is inoculated with one loopful of organisms.

(2) Seeded nutrient agar is poured into plate and allowed to solidify.

(3) Sterile disk is dipped halfway into agent. If completely submerged it will be too wet.

(4) Impregnated disk is placed in center of nutrient agar and pressed down lightly to secure it.

(5) After 24–48 hours incubation the zone of inhibition is measured on bottom of plate. Note that measurement is between disk edge and growth.

Figure 37.1 Filter paper disk method of evaluating an antiseptic.

Laboratory Report

Student: _____

Date: _____ Section: _____

37 Evaluation of Antiseptics: The Filter Paper Disk Method

A. Results

1. With a millimeter scale, measure the zones of inhibition between the edge of the filter paper disk and the organisms. Record this information. Exchange your plates with other students' plates to complete the measurements for all chemical agents.

DISINFECTANT/ANTISEPTIC	MILLIMETERS OF INHIBITION	
	Staphylococcus aureus	*Pseudomonas aeruginosa*
5% phenol		
5% formaldehyde		
5% iodine		
1:10 bleach		
mouthwash		
betadine		
amphyl		

2. Which chemical was the most effective for inhibiting growth of *S. aureus?* Of *P. aeruginosa?*

   _____

   _____

3. Which chemical was the least effective for inhibiting growth of *S. aureus?* Of *P. aeruginosa?*

   _____

   _____

4. What do your results indicate about the relative chemical resistances of these two species?

   _____

   _____

B. Short Answer Questions

1. Differentiate between antiseptic and disinfectant. Include examples of each in your answer. Indicate whether any chemicals can be used as both.

   _____

   _____

2. What factors influence the size of the zone of inhibition produced by a chemical?

_____

_____

3. How might the physical differences between gram-positive and gram-negative bacteria contribute to differences in chemical resistances?

_____

_____

Effectiveness of Hand Scrubbing

The importance of hand washing in preventing the spread of disease is accredited to the observations of Ignaz Semmelweis at the Lying-In Hospital in Vienna in 1846 and 1847. Semmelweis was the head of obstetrics and he noted that the number of cases of childbirth fever (puerperal sepsis) was primarily the result of the lack of sanitary practices. He observed that medical students and physicians would go directly from dissection and autopsy rooms to a patient's bedside and assist in deliveries without washing their hands. In these wards, the death rate from childbirth fever was very high, approaching 20% in some cases. In contrast, some women were attended to by midwives and nurses who were not allowed in autopsy rooms and who were more sanitary in handling patients. When women were assisted in deliveries by midwives and nurses, the death rate was considerably less. As a result of his observations, Semmelweis instituted a policy whereby physicians and medical students had to disinfect their hands with a solution of chloride of lime (bleach) prior to examining obstetric patients or assisting in deliveries. His efforts resulted in a significant decrease of puerperal sepsis, down to about 1% for women treated by physicians and medical students. But this success was short lived, as complaints by doctors and medical students to the director of the hospital forced Semmelweis to abandon this practice, and death rates for childbirth fever once again increased. Despondent, Semmelweis resigned his post and returned to his native Hungary.

Today, it is routine practice to wash one's hands prior to the examination of a patient and to do a complete surgical scrub prior to surgery. Scrubbing is the most important way to prevent infections in hospitals and physician's offices, and is an effective way to remove some opportunistic pathogens that occur on the skin, such as *Staphylococcus aureus.* The failure of medical professionals to wash their hands before examining patients has resulted in a serious increase in infections in patients, especially **nosocomial,** or hospital-acquired infections. Hand washing is also important in day care centers and for food preparers who work in private and public kitchens. There individuals can transmit enteric pathogens such as bacteria and viruses to susceptible individuals.

The human skin is inhabited by a diverse group of microorganisms. They protect us from invasion by pathogens and hence contribute to our overall health. The normal flora that dwells on the skin can be placed into three main groups:

Diphtheroids: these are gram-positive bacteria that are similar to *Corynebacterium diphtheriae* owing to their variable morphology. However, they are nonpathogenic. An example is *Propionibacterium acnes,* an anaerobic diphtheroid that lives in hair follicles and breaks down sebum, the oily secretion of the follicle that prevents the skin from drying out.

Staphylococci: a second major group of bacteria on the skin is the staphylococci. An example is *Staphylococcus epidermidis,* a nonpathogenic, coagulase-negative staphylococcus. These organisms inhibit pathogens from establishing a presence on the skin because they effectively compete for nutrients on the skin and produce inhibitory substances. Some individuals carry *Staphylococcus aureus* in the nose, on various parts of the skin, and on the hands. This represents a means for how this opportunistic pathogen can be transmitted to susceptible individuals by medical personnel working with patients and by food handlers.

Yeasts and **Fungi:** various yeast and fungi can be found on the skin. Some are normal inhabitants of the skin that degrade lipid secretions from the secretory glands. The spores of transient, saprophytic fungi from the environment can be deposited on human skin and grow out as fungal colonies when the skin is sampled for microorganisms. Normally, the fungi and yeast found on the skin are nonpathogenic. However, some fungi and yeasts can cause opportunistic infections, especially in immunocompromised patients who are taking chemotherapy or immunosuppressive drugs. Dermatophytes are fungi that infect the hair, skin, and nails, causing athlete's foot and related infections in humans.

In addition to the normal flora, there are transient bacteria that can occur on the skin. Organisms such as endospore formers may be cultured by swabbing the skin because their endospores are present. However, they are not part of the persistent population that inhabits the skin. Other bacteria may be present because the skin has become temporarily contaminated with them. Most are easily removed by washing because they are contaminants on the skin and antiseptic soaps are effective in killing them. The normal flora is much more difficult to remove by washing because these organisms reside in hair follicles and are entrenched in the skin, making them very difficult to remove or kill.

In this exercise, the class will evaluate the effectiveness of length of time in removal of organisms from the hands using a surgical scrub technique. One member of the class will be selected to perform the scrub. Another student will assist by supplying the soap, brushes, and basins, as needed. During the scrub, at 2-minute intervals, the hands will be scrubbed into a basin of sterile water. Bacterial counts will be made of these basins to determine the effectiveness of the previous 2-minute scrub in reducing the bacterial flora of the hands. Members of the class not involved in the scrub procedure will make the inoculations from the basins for the plate counts.

Scrub Procedure

The two members of the class who are chosen to perform the surgical scrub will set up their materials near a sink for convenience. As one student performs the scrub, the other will assist in reading the instructions and providing materials as needed. The basic steps, which are illustrated in figure 38.1, are also described in detail below. Before beginning the scrub, both students should read all the steps carefully.

Materials

- 5 sterile surgical scrub brushes, individually wrapped
- 5 basins (or 2000 ml beakers), containing 1000 ml each of sterile water. These basins should be covered to prevent contamination.
- 1 dispenser of green soap
- 1 tube of hand lotion

Step 1 To get some idea of the number of transient organisms on the hands, the scrubber will scrub all surfaces of each hand with a sterile surgical scrub brush for 30 seconds in basin A. No green soap will be used for this step. The successful performance of this step will depend on:

- spending the same amount of time on each hand (30 seconds),
- maintaining the same amount of activity on each hand, and
- scrubbing under the fingernails, as well as working over their surfaces.

After completion of this 60-second scrub, notify Group A that their basin is ready for the inoculations.

Step 2 Using the *same* brush as above, begin scrubbing with green soap for 2 minutes, using cool tap water to moisten and rinse the hands. One minute is devoted to each hand.

The assistant will make one application of green soap to each hand as it is being scrubbed.

Rinse both hands for 5 seconds under tap water at the completion of the scrub.

Discard the brush.

Note: This same procedure will be followed exactly in steps 4, 6, and 8 of figure 38.1.

Step 3 With a *fresh* sterile brush, scrub the hands in basin B in a manner that is identical to step 1. Don't use soap. Notify Group B when this basin is ready.

Note: Exactly the same procedure is used in steps 5, 7, and 9 of figure 38.1, using Basins C, D, and E.

Remember: It is important to use a fresh sterile brush for the preparation of each of these basins.

After Scrubbing After all scrubbing has been completed, the scrubber should dry his or her hands and apply hand lotion.

Making the Pour Plates

While the scrub is being performed, the rest of the class will be divided into five Groups (A, B, C, D, and E) by the instructor. Each group will make six plate inoculations from one of the five basins (A, B, C, D, or E). It is the function of these groups to determine the bacterial count per milliliter in each basin. In this way, we hope to determine, in a relative way, the effectiveness of scrubbing in bringing down the total bacterial count of the skin.

Materials

- 30 veal infusion agar pours—6 per group
- 1 ml pipettes
- 30 sterile petri plates—6 per group
- 70% alcohol
- L-shaped glass stirring rod (optional)

(1) Sixty-second hand scrub into Basin A. No soap.

(2) Two-minute soap scrub with running water.

(3) Sixty-second hand scrub into Basin B. No soap.

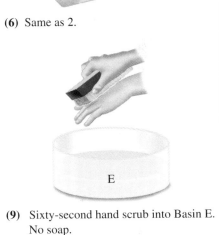

(4) Same as 2.

(5) Sixty-second hand scrub into Basin C. No soap.

(6) Same as 2.

(7) Sixty-second hand scrub into Basin D. No soap.

(8) Same as 2.

(9) Sixty-second hand scrub into Basin E. No soap.

Figure 38.1 Hand scrubbing routine.

1. Liquefy six pours of veal infusion agar and cool to 50° C. While the medium is being liquefied, label two sets of plates each: 0.1 ml, 0.2 ml, and 0.4 ml. Also, indicate your group designation on the plate.

2. As soon as the scrubber has prepared your basin, take it to your table and make your inoculations as follows:

   a. Stir the water in the basin with a pipette or an L-shaped stirring rod for 15 seconds. If the stirring rod is used (figure 38.2), sterilize it before using by immersing it in 70% alcohol and flaming. *For consistency of results all groups should use the same method of stirring.*

   b. Deliver the proper amounts of water from the basin to the six petri plates with a sterile serological pipette. Refer to figure 38.3. If a pipette was used for stirring, it may be used for the deliveries.

Figure 38.2 An alternative method of stirring utilizes an L-shaped glass stirring rod.

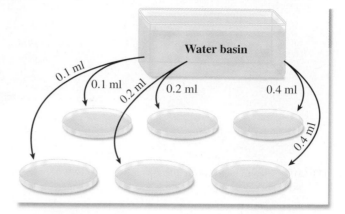

Figure 38.3 Scrub water for count is distributed to six petri plates in amounts as shown.

c. Pour a tube of veal infusion agar, cooled to 50°C, into each plate, rotate to get good distribution of organisms, and allow to cool.

d. Incubate the plates at 37° C for 24 hours.

3. After the plates have been incubated, select the pair that has the best colony distribution with no fewer than 30 or more than 300 colonies. Count the colonies on the two plates and record your counts on the chart on the chalkboard.

4. After all data are on the chalkboard, complete the table and graph on Laboratory Report 38.

Laboratory Report

Student: _____

Date: _____ Section: _____

38 Effectiveness of Hand Scrubbing

A. Results

1. The instructor will draw a table on the chalkboard similar to the one below. Examine the six plates that your group inoculated from the basin of water. Select the two plates of a specific dilution that have approximately 30 to 300 colonies and count all of the colonies of each plate with a Quebec colony counter. Record the counts for each plate and their averages on the chalkboard. Once all the groups have recorded their counts, record the dilution factors for each group in the proper column. To calculate the organisms per milliliter, multiply the average count by the dilution factor.

| GROUP | 0.1 ml COUNT | | 0.2 ml COUNT | | 0.4 ml COUNT | | DILUTION FACTOR* | ORGANISMS PER MILLILITER |
	Per Plate	Average	Per Plate	Average	Per Plate	Average		
A								
B								
C								
D								
E								

*Dilution factors: 0.1 ml = 10; 0.2 ml = 5; 0.4 ml = 2.5

2. Graph

   After you have completed this tabulation, plot the number of organisms per milliliter that was present in each basin.

MINUTES:	1	3	6	9	12
BASIN:	A	B	C	D	E

3. What conclusions can be derived from this exercise?

   _____

   _____

   _____

4. What might be an explanation of a higher count in basin D than in B, ruling out contamination or faulty techniques?

   _____

   _____

B. Short Answer Questions

1. Name the three types of microbes most commonly associated with skin.

   _____

   _____

2. How are normal skin flora beneficial to the host?

   _____

   _____

3. What important opportunistic pathogen is associated with skin?

    _____

    _____

4. Differentiate between normal flora and transient bacteria found on skin. Which type is more difficult to remove? Explain.

    _____

    _____

5. Why is it so important that surgeons scrub their hands prior to surgery even though they wear rubber gloves?

    _____

    _____

Identification of Unknown Bacteria

One of the most interesting experiences in introductory microbiology is to attempt to identify an unknown microorganism that has been assigned to you as a laboratory problem. The next six exercises pertain to this phase of microbiological work. You will be given one or more cultures of bacteria to identify. The only information that might be given to you about your unknowns will pertain to their sources and habitats. All the information needed for identification will have to be acquired by you through independent study.

Although you will be engrossed in trying to identify an unknown organism, there is a more fundamental underlying objective of this series of exercises that goes far beyond simply identifying an unknown. That objective is to gain an understanding of the cultural and physiological characteristics of bacteria. Physiological characteristics will be determined with a series of biochemical tests that you will perform on the organisms. Although correctly identifying the unknowns that are given to you is very important, it is just as important that you thoroughly understand the chemistry of the tests that you perform on the organisms.

The first step in the identification procedure is to accumulate information that pertains to the organisms' morphological, cultural, and physiological (biochemical) characteristics. This involves making different kinds of slides for cellular studies and the inoculation of various types of media to note the growth characteristics and types of enzymes produced. As this information is accumulated, it is recorded in an orderly manner on descriptive charts, which are located in the back of the manual.

After sufficient information has been recorded, the next step is to consult a taxonomic key, which enables one to identify the organism. For this final step, *Bergey's Manual of Systematic Bacteriology* will be used. Copies of volumes 1 and 2 of this book will be available in the laboratory, library, or both.

Success in this endeavor will require meticulous techniques, intelligent interpretation, and careful record keeping. Your mastery of aseptic methods in the handling of cultures and the performance of inoculations will show up clearly in your results. Contamination of your cultures with unwanted organisms will yield false results, making identification hazardous speculation. If you have reason to doubt the validity of the results of a specific test, repeat it; *don't rely on chance!* As soon as you have made an observation or completed a test, record the information on the descriptive chart. Do not trust your memory—record data immediately.

Morphological Study
of Unknown Bacterium

The first step in the study of your unknown bacterium is to set up stock cultures that will be used in the subsequent exercises. Refer to Exercise 20 for the importance of establishing both a reserve stock culture and a working stock culture. Your reserve stock culture will not be used for making slides or inoculating tests. It will be stored in the refrigerator in case your working stock becomes contaminated and you need to make a fresh working stock. The working stock will be used to inoculate the various tests that you will perform to identify your unknown bacterium. It is crucial that you practice good aseptic technique when inoculating from your working stock in order to avoid contaminating the culture. If it becomes contaminated or loses viability, you can prepare a fresh culture from the reserve stock culture that you have maintained in the refrigerator.

Identifying your unknown will be a kind of "microbiological adventure" that will test the skills and knowledge that you have acquired thus far. You will gather a great deal of information regarding your unknown by performing staining reactions and numerous metabolic tests. The Gram stain will play a very critical role in the process because it will eliminate thousands of possible organisms. The results of these tests will be compared to flow charts provided in this manual and to information in *Bergey's Manual*. From your "detective" work, you will be able to ascertain the identity of the unknown that you were given. To set up the stock cultures, proceed in the following way (see figure 39.1).

Stock Cultures

You will receive a broth culture or an agar slant of your unknown bacterium. From this culture you will prepare your working stock and your reserve stock cultures. From the working stock, you will be able to determine such things as cell morphology, the Gram reaction of the unknown, and, in some cases, whether the culture forms any pigment. You can also determine other morphological characteristics such as the presence of a glycocalyx, endospores, or cytoplasmic granules.

Materials

First Period

- nutrient agar or tryptone agar slants

1. Label the agar slants with the code number of the unknown, your name, lab section, and date.
2. Inoculate both slants with your unknown organism. Begin your streak at the bottom of the slant and move the inoculating loop toward the top of the slant in a straight motion. Remember to practice good aseptic technique.
3. Place the respective tubes in the appropriate baskets labeled with the two incubation temperatures, 20° C and 37° C (figure 39.1). Incubate the slants for 18–24 hours.

Second Period

1. Examine the slants. Look for growth. Some organisms produce sparse growth and you must examine the cultures closely to determine if growth is present. Is either culture producing a pigment and, if so, is the pigment associated with the cells or has it diffused into the agar? Remember, however, that pigment production could require longer incubation times.
2. Determine which incubation temperature produced the best growth. If no growth occurred on either slant, your original culture could be nonviable or more time is needed for growth of the culture to occur. A third possibility is that neither temperature supported growth. Think through the possibilities and decide what course of action you need to take.
3. If growth occurred on the slant, pick the tube with the best growth and designate it as your **reserve stock culture.** Store the reserve stock in the refrigerator. Cultures stored in this manner are viable for several weeks. Do not use the reserve stock to make inoculations of the various media you will employ or to make stains. **Do not store the culture in your desk.**

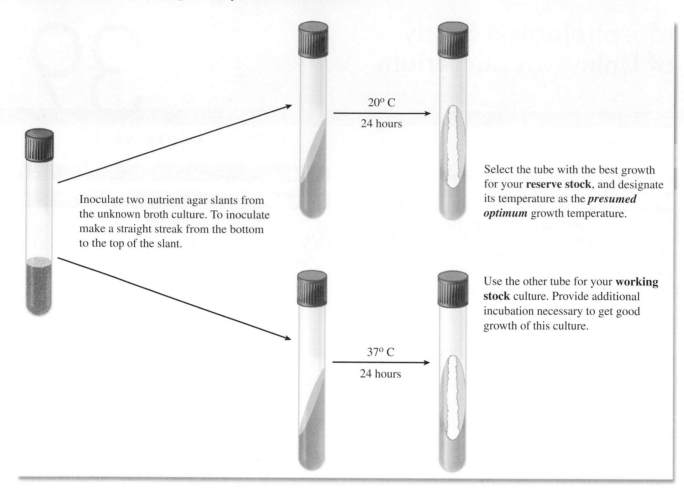

Figure 39.1 Stock culture procedure.

Inoculate two nutrient agar slants from the unknown broth culture. To inoculate make a straight streak from the bottom to the top of the slant.

20° C
24 hours

Select the tube with the best growth for your **reserve stock**, and designate its temperature as the *presumed optimum* growth temperature.

Use the other tube for your **working stock** culture. Provide additional incubation necessary to get good growth of this culture.

37° C
24 hours

4. Designate the second culture as your **working stock culture.** This culture will be the source of the inoculum for the various tests and stains that you will perform (figure 39.2). If the culture is 18–24 hours old, it can be used to perform the Gram stain. If not, you will have to prepare a fresh slant from the working stock to do the Gram stain.

As soon as morphological information is acquired, be sure to record your observations on the descriptive chart on page 263. Proceed as follows:

Materials

- Gram-staining kit
- spore-staining kit
- acid-fast staining kit
- Loeffler's methylene blue stain
- nigrosin or india ink
- tubes of nutrient broth and nutrient agar
- gummed labels for test tubes

New Inoculations

For all of these staining techniques, you will need 24–48 hour cultures of your unknown. If your working stock slant is a fresh culture, use it. If you don't have a fresh broth culture of your unknown, inoculate a tube of nutrient broth and incubate it at its estimated optimum temperature for 24 hours.

Gram's Stain

Once you have a good Gram stain of your organism, you can determine several characteristics of your unknown. First, you should be able to determine the morphology of your organism: rod, coccus, spiral, etc. Furthermore, if you have prepared a thin smear, you can ascertain something about the arrangement of the cells. Note whether the cells occur singly, in pairs, in masses, or in chains. For example, the streptococci occur in chains whereas the staphylococci occur in masses much like a bunch of grapes. The Gram stain can also be used to determine the size of your organism. Refer to Exercise 5.

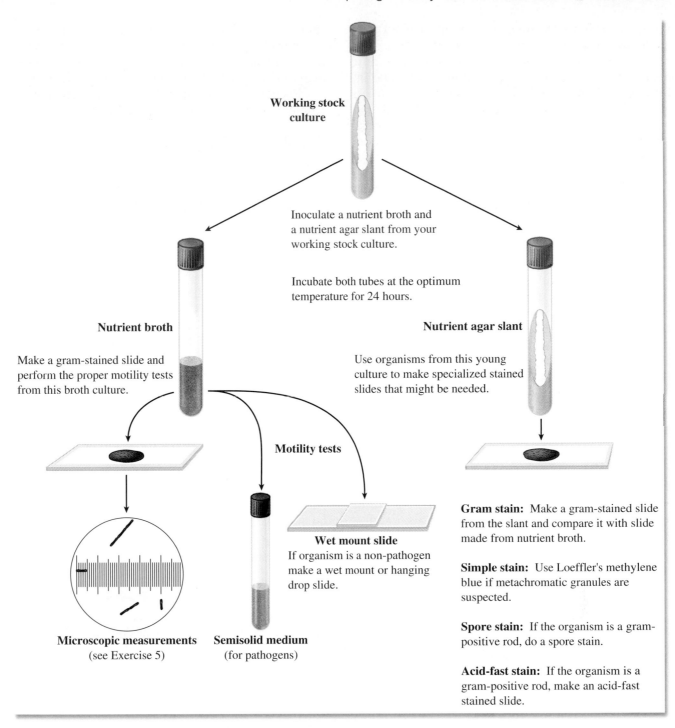

Figure 39.2 Procedure for morphological study.

Note: The results from the Gram stain can be verified by performing the following test. Place one drop of 3% KOH on a microscope slide and transfer a loopful of your unknown cells to the KOH solution. While observing the slide edge-on at eye level, mix the cells and KOH solution and slowly raise the loop from the cells. Gram-negative cells will lyse in the KOH solution, releasing their DNA and causing the liquid to become very viscous. Often, "strings" of DNA can be seen adhering to the loop as it is raised from the slide. Gram-positive cells do not undergo lysis in the KOH solution, and hence an increase in viscosity or DNA strings will not be seen in these cells.

Keep in mind that short rods with round ends (coccobacilli) look like cocci. If you have what seems to be a coccobacillus, examine many cells before you make a final decision. Also, keep in mind that *while rod-shaped organisms frequently appear as cocci under certain growth conditions, cocci rarely appear as rods.* (*Streptococcus mutans* is unique in forming rods under certain conditions.) Thus, it is generally safe to assume that if you have a slide on which you see both coccus-like cells and short rods, the organism is probably rod-shaped. This assumption is valid, however, only if you are not working with a contaminated culture!

Record the shape of the organism and its reaction to the stain on the descriptive chart on page 263.

Motility

For nonpathogens, the wet mount or hanging drop prepared from a broth culture is the preferred way to determine motility. Refer to Exercise 18. For pathogens, SIM medium can be used to ascertain motility (Exercise 18). Inoculate the culture by stabbing the "deep" with an inoculating needle. If growth occurs only along the line of inoculation, the organism is non-motile. In contrast, turbidity throughout the tube would indicate that your organism is motile. In general, cocci are non-motile whereas rods can be either motile or non-motile.

Endospores

If your unknown is a gram-positive rod, it may be an endospore former. Endospores, however, do not usually occur in cocci or in gram-negative rods. Examination of your gram-stained slide made from the agar slant should provide a clue since endospores show up as transparent oval structures in gram-stained preparations. Endospores can also be seen on unstained organisms if studied with phase-contrast optics.

If there seems to be evidence that the organism is a spore-former, make a slide using one of the spore-staining techniques you used in Exercise 16. *Since some spore-formers require at least a week's time of incubation before forming spores, it is prudent to double-check for spores in older cultures.*

Record on the descriptive chart whether the spore is terminal, subterminal, or in the middle of the rod.

Acid-Fast Staining

The mycobacteria and some species of *Nocardia* are acid-fast. For these bacteria, the presence of acid-fastness can interfere with the Gram stain, causing these bacteria to stain gram-negative. Performing the acid-fast stain will sort out part of this problem. Do not depend solely on the Gram stain as the results can be misleading, especially for the acid-fast bacteria.

If your unknown is a gram-positive, non-spore-forming rod, it could be an acid-fast bacterium. Acid-fastness can vary with culture age, but most cultures display this property after 2 days of incubation. For best results, do not do this stain on old cultures. Refer to Exercise 17 for the staining procedure.

Other Structures

If the cytoplasm in the gram-stained cells appears uneven, you may want to do a simple stain with Loeffler's methylene blue (Exercise 12) to determine the presence of metachromatic granules (volutin), which are storage granules of polyphosphate.

Although a capsule stain (Exercise 14) may be performed at this time, it might be better to wait until a later date when you have the organism growing on blood agar. Capsules usually are more apparent when the organisms are grown on this medium.

Laboratory Report

There is no Laboratory Report to fill out for this exercise. All information is recorded on the descriptive chart.

Descriptive Chart

STUDENT: _____

LAB SECTION: _____

Habitat: _____ Culture No.: _____

Source: _____

Organism: _____

MORPHOLOGICAL CHARACTERISTICS

Cell shape:

Arrangement:

Size:

Spores:

Gram's Stain:

Motility:

Capsules:

Special Stains:

CULTURAL CHARACTERISTICS

Colonies:

Nutrient Agar:

Blood Agar:

Agar Slant:

Nutrient Broth:

Gelatin Stab:

Oxygen Requirements:

Optimum Temp.:

PHYSIOLOGICAL CHARACTERISTICS

	TESTS	RESULTS
Fermentation	Glucose	
	Lactose	
	Sucrose	
	Mannitol	
Hydrolysis	Gelatin Liquefaction	
	Starch	
	Casein	
	Fat	
IMViC	Indole	
	Methyl Red	
	V–P (acetylmethylcarbinol)	
	Citrate Utilization	
	Nitrate Reduction	
	H_2S Production	
	Urease	
	Catalase	
	Oxidase	
	DNase	
	Phenylalanase	

	REACTION	TIME
Litmus Milk	Acid	_____
	Alkaline	_____
	Coagulation	_____
	Reduction	_____
	Peptonization	_____
	No Change	_____

Cultural Characteristics

The cultural characteristics of an organism pertain to its macroscopic appearance on different kinds of media. In *Bergey's Manual,* you will find descriptive terms used by bacteriologists for recording cultural characteristics. For the general description of colonies, nutrient agar or any complex, rich medium is useful for this purpose. The nature of the growth in a nutrient broth can vary, and this too can be a source of certain information about an organism. Thioglycollate medium (Exercise 21) can be used to determine the oxygen requirements of an organism: Do the cells grow in a tube of this medium? Some media, such as blood agar are "differential," demonstrating the hemolytic capability of an organism. In the following exercise, you will inoculate your unknown into different media to determine its cultural characteristics in the various media.

First Period

(Inoculations)

During this period, one nutrient agar plate, one nutrient gelatin deep, two nutrient broths, and one tube of fluid thioglycollate medium will be inoculated. Inoculations will be made with the original broth culture of your unknown. The reason for inoculating two tubes of nutrient broth here is to recheck the optimum growth temperature of your unknown. In Exercise 39, you incubated your nutrient agar slants at 20° C and 37° C. It may well be that the optimum growth temperature is closer to 30° C. It is to check out this intermediate temperature that an extra nutrient broth is being inoculated. Proceed as follows:

Materials

for each unknown:
- 1 nutrient agar pour
- 1 nutrient gelatin deep
- 2 nutrient broths
- 1 fluid thioglycollate medium (FTM)
- 1 petri plate

1. Pour a petri plate of nutrient agar for each unknown and streak it with a method that will give good isolation of colonies. Use the original broth culture for streaking.
2. Inoculate the tubes of nutrient broth with a loop.
3. Make a stab inoculation into the gelatin deep by stabbing the inoculating needle (straight wire) directly down into the medium to the bottom of the tube and pulling it straight out. The medium must not be disturbed laterally.
4. Inoculate the tube of FTM with a loopful of your unknown. Mix the organisms throughout the tube by rolling the tube between your palms.
5. Place all tubes except one nutrient broth into a basket and incubate for 24 hours at the temperature that seemed optimal in Exercise 39. Incubate the remaining tube of nutrient broth separately at 30° C. Incubate the agar plate, inverted, at the presumed best temperature.

Second Period

(Evaluation)

After the cultures have been properly incubated, *carry them to your desk in a careful manner* to avoid disturbing the growth pattern in the nutrient broths and FTM. Before studying any of the tubes or plates, place the tube of nutrient gelatin in an ice water bath. It will be studied later. Proceed as follows to study each type of medium and record the proper descriptive terminology on the descriptive chart on page 263.

Materials

- reserve stock agar slant of unknown
- spectrophotometer and cuvettes
- hand lens
- ice water bath near sink

Nutrient Agar Slant (Reserve Stock)

Examine your reserve stock agar slant of your unknown that has been stored in the refrigerator since the last laboratory period. Evaluate it in terms of the following criteria:

Amount of Growth The abundance of growth may be described as *none, slight, moderate,* and *abundant.*

Color Pigments can be associated with a colony, for example, prodigiosin, the red pigment made by *Serratia marcescens* when grown at 27° C. However, pigments can be produced by an organism that diffuse

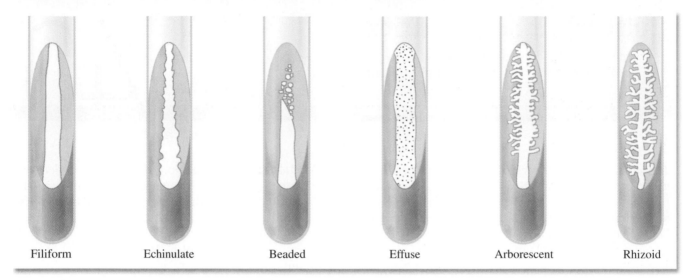

| Filiform | Echinulate | Beaded | Effuse | Arborescent | Rhizoid |

Figure 40.1 Types of bacterial growth on nutrient agar slants.

into the medium, causing the medium to be colored, such as the case for the green fluorescent pigment produced by *Pseudomonas fluorescens*. To check for diffusable pigments, hold your plate up to the light and observe the color of the medium in the plate. Most bacteria, however, do not produce pigments, and their colonies are white or buff colored.

Opacity Organisms that grow prolifically on the surface of a medium will appear more opaque than those that exhibit a small amount of growth. Degrees of opacity may be expressed in terms of *opaque, transparent,* and *translucent* (partially transparent).

Form The gross appearance of different types of growth are illustrated in figure 40.1. The following descriptions of each type will help in differentiation:

> *Filiform:* characterized by uniform growth along the line of inoculation
> *Echinulate:* margins of growth exhibit toothed appearance
> *Beaded:* separate or semiconfluent colonies along the line of inoculation
> *Effuse:* growth is thin, veil-like, unusually spreading
> *Arborescent:* branched, treelike growth
> *Rhizoid:* rootlike appearance

Nutrient Broth

The nature of growth on the surface, subsurface, and bottom of the tube is significant in nutrient broth cultures. Describe your cultures as thoroughly as possible on the descriptive chart with respect to these characteristics:

Surface Figure 40.2 illustrates different types of surface growth. A *pellicle* type of surface differs from the *membranous* type in that the latter is much thinner. A *flocculent* surface is made up of floating adherent masses of bacteria.

Subsurface Below the surface, the broth may be described as *turbid* if it is cloudy, *granular* if specific small particles can be seen, *flocculent* if small masses are floating around, and *flaky* if large particles are in suspension.

Sediment The amount of sediment in the bottom of the tube may vary from none to a great deal. To describe the type of sediment, agitate the tube, putting the material in suspension. The type of sediment can be described as *granular, flocculent, flaky,* and *viscid.* Test for viscosity by probing the bottom of the tube with a sterile inoculating loop.

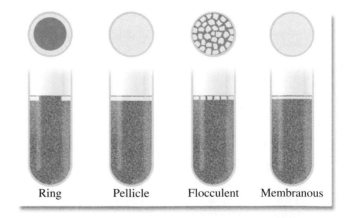

| Ring | Pellicle | Flocculent | Membranous |

Figure 40.2 Types of surface growth in nutrient broth.

Amount of Growth   To determine the amount of growth, it is necessary to shake the tube to disperse the organisms. Terms such as *slight* (scanty), *moderate*, and *abundant* adequately describe the amount.

Temperature Requirements   To determine which temperature produces better growth, transfer the contents of the nutrient broth tubes to separate cuvettes and measure the optical density (absorbance) with a spectrophotometer. Because the cultures may be too turbid to measure, you may have to dilute the cultures with water before taking the readings. Record in the descriptive chart which temperature produces better growth for your organism. This temperature will be closer to the one needed for optimum growth of your organism.

Fluid Thioglycollate Medium

The growth pattern of your bacterium in fluid thioglycollate medium will give some indication of the oxygen requirement of your organism. Examine your FTM tube and compare the growth pattern of your organism with that of figure 21.5 on page 148. More than likely, your bacterium will be either aerobic, microaerophilic, or a facultative anaerobe. Strict anaerobes such as *Clostridium* require special culture conditions for growth.

Gelatin Stab

Some bacteria produce **proteases,** enzymes that degrade proteins. Determine if your unknown produces proteases by examining the nutrient gelatin tube that you inoculated with your unknown. After incubation, place the culture in an ice bath and allow it to stand for several minutes. Remove the tube and tilt it several times from side to side to ascertain if liquefaction has occurred. Any degraded gelatin will remain liquid after being placed in the ice bath. If liquefaction has not occurred, the contents of the tube will be a solid. Also be sure to note if your organism can grow in gelatin since some bacteria are unable to do so. Check the configuration with figure 40.3 to see if any of the illustrations match your tube. A description of each type follows:

> *Crateriform:* saucer-shaped liquefaction
> *Napiform:* turnip-like
> *Infundibular:* funnel-like or inverted cone
> *Saccate:* elongate sac, tubular, cylindrical
> *Stratiform:* liquefied to the walls of the tube in the upper region

Note: The configuration of liquefaction is not as significant as the mere fact that liquefaction takes place. If your organism liquefies gelatin, but you are unable to

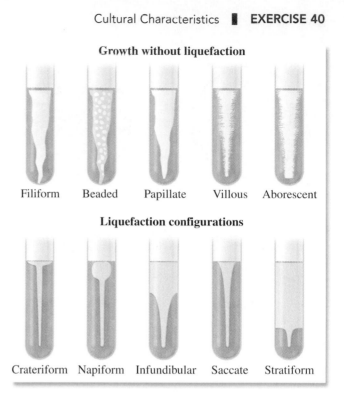

Growth without liquefaction

Filiform   Beaded   Papillate   Villous   Aborescent

Liquefaction configurations

Crateriform   Napiform   Infundibular   Saccate   Stratiform

Figure 40.3 Growth in gelatin stabs.

determine the exact configuration, don't worry about it. However, be sure to record on the descriptive chart the *presence* or *absence* of protease production.

Another important point: Some organisms produce protease at a very slow rate. Tubes that are negative should be incubated for another 4 or 5 days to see if protease is produced slowly.

Type of Growth   (No Liquefaction) If no liquefaction has occurred, check the tube to see if the organism grows in nutrient gelatin (some do, some don't). If growth has occurred, compare the growth with the top of the illustration in figure 40.3. It should be pointed out, however, that, from an identification standpoint, the nature of growth in gelatin is not very important.

Nutrient Agar Plate

Colonies grown on plates of nutrient agar should be studied with respect to size, color, opacity, form, elevation, and margin. With a dissecting microscope or hand lens study individual colonies carefully. Refer to figure 40.4 for descriptive terminology. Record your observations on the descriptive chart.

▐ **Laboratory Report**

There is no Laboratory Report for this exercise. Record all information on the descriptive chart on page 263.

CONFIGURATIONS

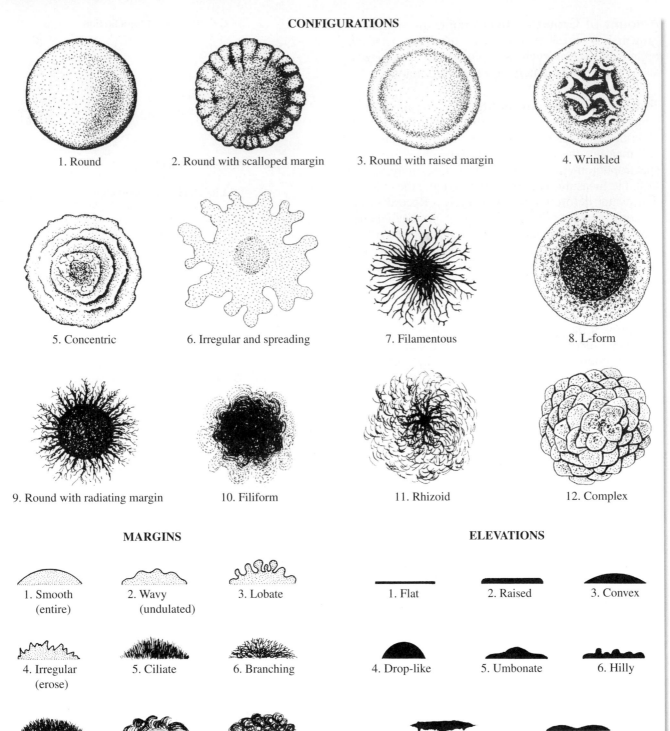

1. Round
2. Round with scalloped margin
3. Round with raised margin
4. Wrinkled

5. Concentric
6. Irregular and spreading
7. Filamentous
8. L-form

9. Round with radiating margin
10. Filiform
11. Rhizoid
12. Complex

MARGINS

1. Smooth (entire)
2. Wavy (undulated)
3. Lobate
4. Irregular (erose)
5. Ciliate
6. Branching
7. Wooly
8. Thread-like
9. "Hair-Lock"-like

ELEVATIONS

1. Flat
2. Raised
3. Convex
4. Drop-like
5. Umbonate
6. Hilly
7. Ingrowing into medium
8. Crateriform

Figure 40.4 Colony characteristics.

Physiological Characteristics:
Oxidation and Fermentation Tests

The sum total of the chemical reactions that occur in a cell are referred to as metabolism, and the individual chemical reactions that make up the metabolic pathways in a cell are catalyzed by protein molecules called **enzymes.** Most enzymes function inside the cell where metabolic pathways carry out the breakdown **(catabolism)** of food materials and the biosynthesis of cell constituents **(anabolism).** Because bacteria cannot carry out phagocytosis owing to their rigid cell walls, they excrete **exoenzymes** that function outside the cell to degrade large macromolecules. For example, exoenzymes break down proteins and polysaccharides into amino acids and monosaccharides, respectively, which are then transported into the cell for metabolic needs. Protease, DNase, and amylase are examples of exoenzymes (figure 41.1).

Some enzymes are assisted in catalytic reactions by **coenzymes.** The latter transfer small molecules from one molecule to another. For example, the coenzymes NAD^+ and FAD transfer protons and coenzyme A transfers acetate groups. Most coenzymes are derivatives of vitamins. As examples, NAD^+ is synthesized from niacin, and FAD comes from folic acid. Coenzymes are only required by a cell in catalytic amounts, however, and when an enzymatic reaction catalyzes an oxidation step that converts NAD^+ to NADH, the coenzyme must be converted back into its oxidized form if the metabolic pathway is to continue

to function. Many of the reactions that define respiration and fermentation are concerned with regenerating coenzymes such as NAD^+ and FAD.

The primary goal of catabolism is the production of energy, which is needed for biosynthesis and growth. Bacteria can obtain their energy needs by two different metabolic means, respiration or fermentation. In respiration, organic molecules are completely degraded to carbon dioxide and water. ATP is generated by the energy created from a proton gradient that is established across the cell membrane when protons are transported from the cytoplasm to the outside of the cell. The shuttling of electrons down an electron transport chain involving cytochromes facilitates the movement of the protons to the outside of the cell. This process is called **oxidative phosphorylation** and, in the process, reduced coenzyme NADH generated in metabolic reactions is converted back to NAD^+ because oxygen acts as the terminal electron acceptor and is converted to water. In contrast, fermentation is the partial breakdown of organic molecules to alcohols, aldehydes, acids, and gases such as carbon dioxide and hydrogen. In this process, organic molecules in metabolic pathways serve as terminal electron acceptors and become the end products in a fermentation pathway. Reactions that carry out oxidation steps and utilize NAD^+ in metabolic pathways are coupled to reactions that use NADH to reduce the

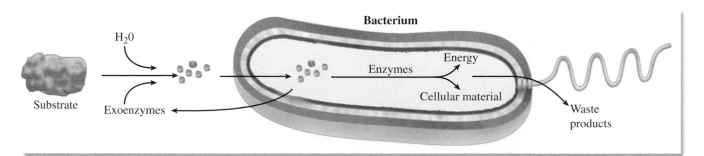

Figure 41.1 Note that the hydrolytic exoenzymes split larger molecules into smaller ones, utilizing water in the process. The smaller molecules are then assimilated by the cell to be acted upon by endoenzymes to produce energy and cellular material.

metabolic intermediates. An example is the oxidation of glyceraldehyde-3-phosphate in glycolysis being coupled with the formation of lactate from pyruvate when *Streptococcus lactis* ferments glucose.

In fermentation, ATP is synthesized by **substrate level phosphorylation** in which metabolic intermediates in pathways directly transfer high-energy phosphates to ADP to synthesize ATP. In the glycolytic fermentation of glucose, ATP is formed when phosphoenol pyruvate transfers a high-energy phosphate to ADP and pyruvate is formed. In general, fermentation is much less efficient in producing energy relative to respiration because the use of metabolic intermediates as electron acceptors leaves most of the available energy in molecules that form the end products. Some bacteria are capable of growing both by respiration and fermentation. *Escherichia coli* is a facultative aerobe that will grow by respiratory means if oxygen is present but will switch metabolic gears in anaerobic conditions and grow by fermentation.

Sugars, particularly glucose, are compounds most widely used by fermenting organisms. However, other compounds such as organic acids, amino acids, and fats are fermented by bacteria. Butter becomes rancid because bacteria ferment butter fat producing volatile and odoriferous organic acids. The end products of a particular fermentation are like a "fingerprint" for an organism and can be used in its identification. For example, *Escherichia coli* can be differentiated from *Enterobacter aerogenes* because the primary fermentation end products for *E. coli* are mixed organic acids, whereas *E. aerogenes* produces acetylmethylcarbinol, a neutral end product.

Tests To Be Performed

Two different kinds of tests will be performed in this exercise: (1) **fermentation tests** to determine if your unknown is capable of carrying out various fermentation reactions and (2) **oxidative tests** to determine if your unknown carries out respiratory metabolism. One test, the O/F glucose test is designed to differentiate between these two modes of metabolism and ascertain if the organism is oxidative, fermentative, or capable of both kinds of metabolism. The fermentation tests to be done are the O/F glucose, specific sugar fermentations, mixed-acid fermentation (methyl red [MR] test), butanediol fermentation (Voges-Proskauer [VP] test), and citrate test (figure 41.2). The oxidative tests to be performed are: the oxidase, catalase, and nitrate tests (figure 41.3). If the O/F glucose test determines

that your organism is oxidative and not capable of fermenting sugars, then your bacterium cannot be identified by fermentation tests, and you will have to rely on other tests to identify your unknown.

The performance of these tests on your unknown may involve a considerable number of inoculations because a set of positive test controls are needed to which you will compare your unknown bacterium (figure 41.4). Although photographs of the various tests are provided this in manual, seeing the actual test results will be much more meaningful. Also keep in mind that some bacteria may not give the same exact results as listed in *Bergey's Manual* as an isolate can often differ from its description in the manual.

As you perform the various tests, attempt to determine which tests may define a specific group of organisms. Some tests may be specific in an identification of an unknown while others may not be specific and therefore not useful in determining the identify of your unknown. Keep in mind that *it is not routine practice to perform all the tests in identifying an unknown.* Although your goal is to identify your unknown, it is also an important for you to learn how to perform the various tests and how to interpret them. The use of an unknown bacterium to identify simply makes it more of a challenge. In actual practice in hospitals and clinical laboratories, biochemical tests are used very selectively. The "shotgun" method of using all the tests is to be avoided because it is wasteful and can lead to confusing results.

Fermentation Tests

First Period

Inoculation should be set up for positive test controls and for your unknown. The media for each set of inoculations are listed separately under each heading.

Unknown Inoculations

The first biochemical test in determining the identity of your organism will be to ascertain whether your organism is oxidative or fermentative. For this, you will inoculate O/F glucose tubes with your unknown, a fermentative and oxidative organism, *Escherichia coli,* and an oxidative organism, *Pseudomonas aeruginosa.*

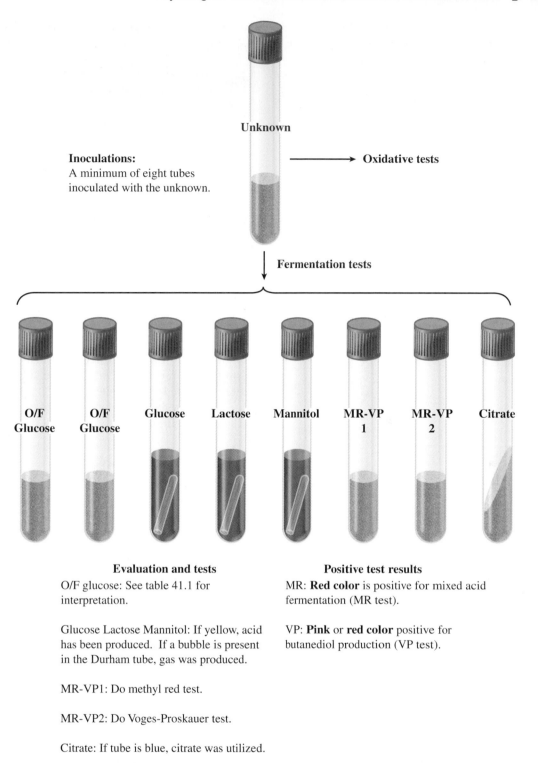

Unknown

Inoculations:
A minimum of eight tubes
inoculated with the unknown.

→ **Oxidative tests**

Fermentation tests

| O/F Glucose | O/F Glucose | Glucose | Lactose | Mannitol | MR-VP 1 | MR-VP 2 | Citrate |

Evaluation and tests

O/F glucose: See table 41.1 for interpretation.

Glucose Lactose Mannitol: If yellow, acid has been produced. If a bubble is present in the Durham tube, gas was produced.

MR-VP1: Do methyl red test.

MR-VP2: Do Voges-Proskauer test.

Citrate: If tube is blue, citrate was utilized.

Positive test results

MR: **Red color** is positive for mixed acid fermentation (MR test).

VP: **Pink** or **red color** positive for butanediol production (VP test).

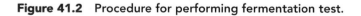

Figure 41.2 Procedure for performing fermentation test.

Oxidative Test

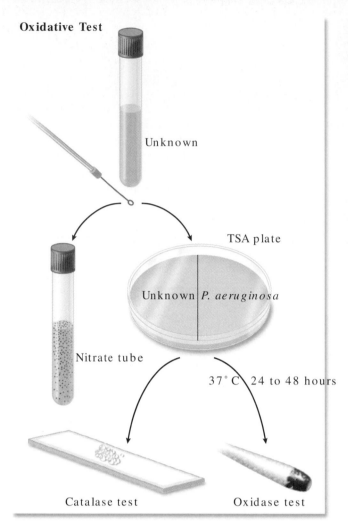

Figure 41.3   Procedure for performing oxidative tests.

Labels in figure: Unknown, TSA plate, Unknown | *P. aeruginosa*, Nitrate tube, 37° C   24 to 48 hours, Catalase test, Oxidase test

Materials

- 6 O/F glucose tubes
- sterile mineral oil

1. Each unknown organism and each test organism will be inoculated into two tubes of O/F glucose by stabbing with an inoculating needle.
2. To one of the tubes for each organism, aseptically deliver about 1 ml of sterile mineral oil after you have inoculated the tube. The mineral oil will establish anaerobic conditions in the tube. The tube without the mineral oil will be aerobic; therefore, be sure to loosen the cap about a quarter of a turn to allow access to the air.
3. Incubate the tubes at 37° C for 24 hours.
4. Record the results and compare them to the data in table 41.1 and the results in figure 41.5. **Note:** If your tubes do not show any color change from the uninoculated control at 24 hours, incubate them for an additional 48 hours and read them again.

Table 41.1   Interpretation of the O/F Glucose Test

Result		Interpretation
ANAEROBIC	AEROBIC	
Yellow	Yellow	Oxidative and fermentative metabolism
Green	Yellow	Oxidative metabolism
Green	Green	Sugar not metabolized (nonsaccharolytic)

Specific Fermentation Reactions

If your organism was found to be fermentative in its metabolism, it will be important to determine which specific sugars are fermented or which fermentation pathways are used for growth. Testing for the fermentation of specific sugars or the end products of fermentation pathways are important phenotypic characteristics used to identify bacteria in *Bergey's Manual*. The following fermentation tests will be studied in this exercise: (1) Durham tube sugar fermentations, (2) mixed-acid fermentation (methyl red test), (3) 2,3-butanediol fermentation (Voges-Proskauer test), and (4) citrate test.

Unknown Inoculations

First Period   Figure 41.2 illustrates the procedure for inoculating the fermentation tests with your unknown. Your instructor may suggest other carbohydrates to be tested, and therefore blanks have been provided in the materials section for this purpose. *Different colored tube caps may be used to distinguish the different carbohydrates and so be sure to record the cap color with the sugars given below.*

Materials: (*for each unknown*)

- carbohydrate broths with Durham tubes and phenol red indicator with the following sugars:
  1 glucose tube
  1 lactose tube
  1 mannitol tube
- 2 MR-VP broth tubes
- 1 Simmon's citrate tube
- 1 trypticase soy agar (TSA) plate

1. Label each tube with the number of your unknown and an identifying letter as shown in figure 41.2.
2. Label one-half of a TSA plate with your unknown and the other half with *Pseudomonas aeruginosa*. This plate will be used in the next section for oxidative tests to determine if your organism produces the respiratory enzyme, cytochrome oxidase (figure 41.3).

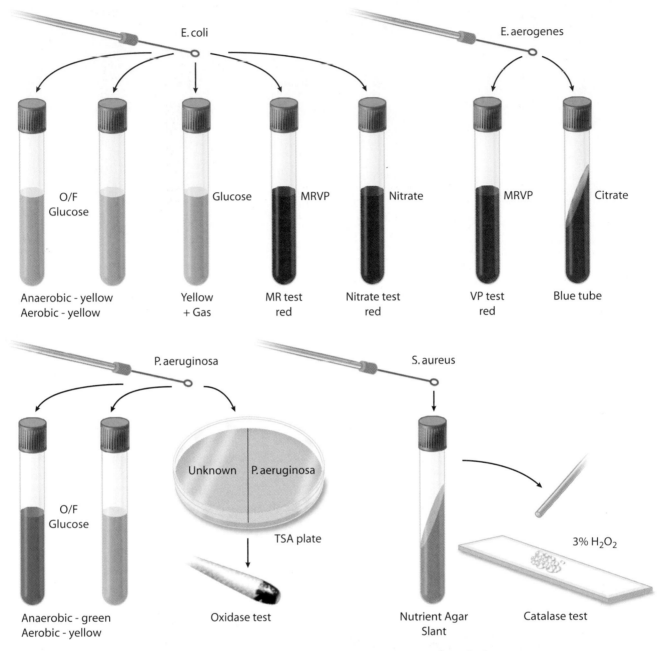

Figure 41.4 **Procedure for doing positive test controls for fermentation and oxidative tests**

3. Inoculate the Durham sugar tubes and the MR-VP broth tubes with your unknown.
4. Using an inoculating needle, first inoculate the Simmon's citrate slant by streaking the slant, and then stab the center of the slant about three-quarters of the way down into the butt of the tube.
5. Incubate the carbohydrate tubes, the Simmon's citrate tube, and the TSA plate for 24 hours.
6. Incubate the MR-VP broth tubes for 3 to 5 days.

Test Control Inoculations

Figure 41.4 above illustrates the procedure for inoculating the test control tubes.

Materials

- 4 O/F glucose deeps
- 1 glucose broth with Durham tube and phenol red indicator

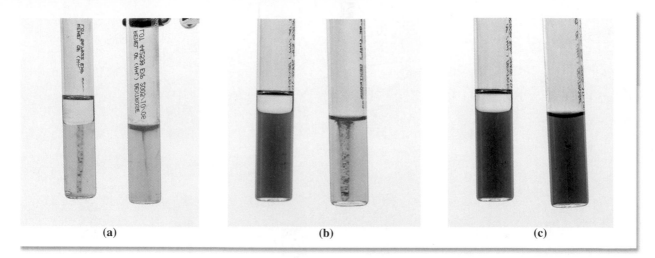

Figure 41.5 O/F glucose test. (a) Fermentative and oxidative; (b) oxidative; (c) glucose not metabolized or inert. © The McGraw-Hill Companies/Auburn University Photographic Services

- 2 MR-VP broth tubes
- 1 Simmon's citrate tube

1. Label each tube with the organism inoculated:

   O/F glucose deeps
   glucose broth
   MR-VP broth
   MR-VP broth
   Simmon's citrate

2. Inoculate each of the test tubes with the appropriate organism as listed.
3. Incubate the glucose tube and the Simmon's citrate tube at 37° C for 24 hours.
4. Incubate the MR-VP broth tubes at 37° C for 3 to 5 days.

Second Period

(Test Evaluations)

After 24 to 49 hours incubation, arrange all your tubes with the unknown tubes in one row and the test controls in another. As you interpret the results, record the information in the descriptive chart on page 257. Do not trust your memory. Any result that is not properly recorded will have to be repeated.

Carbohydrates in Durham Tubes

If an organism ferments a sugar, acid is usually produced, and gas may also be an end product of the fermentation. The presence of acid is indicated by a color change in the pH indicator, phenol red, from red at alkaline pH values to yellow at acidic pH values. The production of gas such as hydrogen and carbon dioxide is revealed by the displacement of medium from the Durham tube (figure 41.6).

Note: Positive gas production should only be recorded when at least 10% of the medium has been displaced from the Durham tube.

Each sugar broth is supplemented with a specific carbohydrate at a concentration of 0.5% as well as beef extract or peptone to satisfy the nitrogen

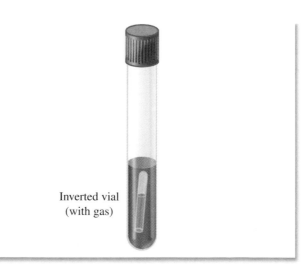

Inverted vial
(with gas)

Figure 41.6 Durham fermentation tube.

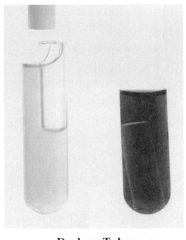

Durham Tubes

Tube on left is positive;
tube on right is negative.

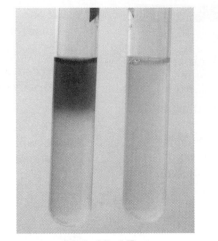

Methyl Red Test

Tube on left is positive *(E. coli)*;
tube on right is negative.

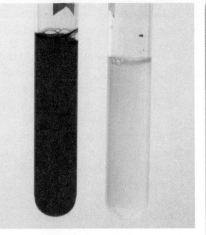

Voges-Proskauer Test

Tube on left is positive *(E. aerogenes)*;
tube on right is negative.

Figure 41.7 Durham tubes, mixed-acid, and butanediol fermentation tests.
© The McGraw-Hill Companies/Auburn University Photographic Service

requirements of most bacteria. It is reasonable to assume that your unknown may ferment other sugars, but glucose, lactose, and mannitol are reasonable choices to start with as they are important in differentiating some of the medically important bacteria. Your instructor may suggest additional carbohydrates to be tested now or later that will assist in the identification of your organism.

Interpretation of the Results

Examine the glucose tube inoculated with *E. coli.* Note that the phenol red has turned from red to yellow, indicating the presence of acids from the fermentation of the glucose. Also note if medium has been displaced from the Durham tube. If at least 10% of the liquid has been displaced, it means that gas has been formed from the fermentation of the sugar. Figure 41.7 illustrates the difference between a positive and a negative test for both acid and gas production.

Now examine the test tubes with the test sugars, glucose, lactose, and mannitol that you inoculated with your unknown organism. Record the results for acid and gas production, comparing them to the positive control tubes. If there was no color change, record "none" in the descriptive chart. No color change is usually consistent with an oxidative organism. Keep in mind that a negative test result for your unknown is just as important as a positive result.

Mixed-Acid Fermentation

(Methyl Red Test)

An important test in differentiating some of the gram-negative intestinal bacteria is that of mixed-acid fermentation. Genera such as *Escherichia, Salmonella, Proteus,* and *Aeromonas* ferment glucose to produce a number of organic acids such as lactic, acetic, succinic, and formic acids. In addition CO_2, H_2, and ethanol are also produced in this fermentation. The amount of acid produced is sufficient to lower the pH of the MR-VP broth to 5.0 or less.

To test for the presence of these acids, the pH indicator, methyl red, is added to the medium, which turns red if acid is present. A positive methyl red test indicates that the organism has carried out *mixed-acid fermentation.* The bacteria that are mixed-acid fermenters also generally produce gas because they elaborate the enzyme **formic hydrogenlyase,** which splits formic acid to produce CO_2 and H_2.

Medium MR-VP medium is a glucose broth that is buffered with peptone and dipotassium phosphate.

Test Procedure Perform the methyl red test first on the control, *E. coli* and then on your unknown.

Materials

• dropping bottle of methy red indicator

1. Add 3–4 drops of methyl red indicator to one of the MR-VP control tubes inoculated with *E. coli*. The tube should become red immediately. A positive tube is shown in the middle illustration of figure 41.7.

2. Repeat this procedure for one of the MR-VP broth tubes inoculated with your unknown organism. If your tube does not become red but remains unchanged (figure 41.7), your unknown is methyl red negative.

3. Record your results in the descriptive chart on page 263.

2,3-Butanediol Fermentation

(Voges-Proskauer Test)

Some of the gram-negative intestinal bacteria do not carry out mixed-acid fermentation, but rather they ferment glucose to produce limited amounts of some organic acids and primarily a more neutral end product, 2,3-butanediol. All species of *Enterobacter* and *Serratia* as well as some species of *Erwinia* and *Aeromonas* carry out the butanediol fermentation. There are also some species of *Bacillus* that produce butanediol when grown on glucose. If an organism produces butanediol and is positive for the Voges-Proskauer (VP) test, it is usually negative for the methyl red test. The methyl red test and the Voges-Proskauer test are important tests for differentiating the gram-negative bacteria.

The neutral end product, 2,3-butanediol, is not detected directly but must be converted to acetoin by oxidation of the 2,3-butanediol. The acetoin reacts with Barritt's reagent, which consists of α naphthol and KOH. The reagent is added to a **3 to 5 day old culture** grown in MR-VP medium and vigorously shaken to oxidize the 2,3-butanediol to acetoin. The tube is allowed to stand at room temperature for 30 minutes, during which time the tube will turn pink to red if acetoin is present (figure 41.7, right illustration).

Test Procedure Perform the VP test on the control MR-VP broth tube inoculated with *Enterobacter aerogenes* and on the second MR-VP broth tube inoculated with your unknown organism. Follow this procedure:

Materials

• Barritt's reagent
• 2 pipettes (1 ml size)
• 2 empty test tubes

1. Lable one empty test tube for your unknown and the other for *E. aerogenes* (positive control).

2. Pipette 1 ml of culture from your unknown to its respective tube and 1 ml of *E. aerogenes* to its respective tube. Use separate pipettes for each transfer.

3. Add 18 drops (about 0.5 ml) of Barritt's reagent A (alpha-naphthol) to each of the tubes containing 1 ml of culture.

4. Add 18 drops (0.5 ml) of Barritt's reagent B (KOH) to each of the test tubes.

5. Cap or cover the mouth of each test tube and shake the tubes vigorously. Allow the tubes to stand for 30 minutes. In this time, the tube with *E. aerogenes* should turn pink to red. compare this to your unknown. *Vigorous shaking is necessary to oxidize the 2,3-butanediol to acetoin, which reacts with Barritt's reagents to give the red color.*

The left-hand tube in the right-hand illustration of figure 41.7 shows a positive VP result, which is pink to red.

6. Record your results on the descriptive chart on page 263.

Citrate Test

Some bacteria are capable of using citrate as a sole carbon source. Normally citrate is oxidatively metabolized by the Kreb's cycle. However, some bacteria such as *Enterobacter aerogenes* and *Salmonella thyphimurium* can cleave citrate to produce oxalo-acetate and pyruvate. These intermediates are then fermented to produce several end products such as formate, acetate, lactate, acetoin, and CO_2. The medium also contains ammonium salts that serve as a sole nitrogen source for growth. Organisms degrading citrate must also use the ammonium salts, and in the process, they produce ammonia that causes the medium to become alkaline. Under alkaline conditions, the pH indicator in the medium turns from dark green to a deep Prussian blue, indicating the utilization of citrate.

Materials

• Simmon's citrate tubes

1. Label one tube of Simmon's citrate with *Enterobacter aerogenes* and another tube with your unknown number. Using an inoculating needle, first streak the surface of the slant and then stab the needle into the middle of the slant.

2. Incubate the tubes with *E. aerogenes* and your unknown at 37° C for 24 to 48 hours.

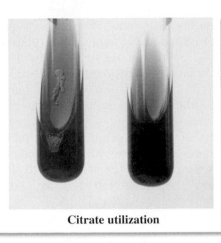

Citrate utilization

Figure 41.8 Left-hand tube exhibits citrate utilization (Prussian blue color). Right-hand tube is uninoculated or negative (green).
© The McGraw-Hill Companies/Auburn University Photographic Services

3. Examine the slants. The slant inoculated with *E. aerogenes* will be a deep Prussian blue because the organism has utilized citrate. Compare this tube to your unknown. If the tube for your unknown has remained green, citrate was not utilized (figure 41.8).
4. Record your results in the descriptive chart.

Oxidative Tests (Refer to Figure 41.3)

Oxidase Test

The oxidase test assays for the presence of cytochrome oxidase, an enzyme in the electron transport chain. This enzyme catalyzes the transfer of electrons from reduced cytochrome *c* to molecular oxygen, producing oxidized cytochrome *c* and water. Cytochrome oxidase occurs in bacteria that carry out respiration where oxygen is the terminal electron acceptor; hence, the test differentiates between those bacteria that have cytochrome oxidase and use oxygen as a terminal electron acceptor from those that can use oxygen as a terminal electron acceptor but have other types of terminal oxidases. The enzyme is detected by the use of an artificial electron acceptor, N,N,N′,N′-tetramethyl-*p*-phenylenediamine, which changes from yellow to purple when electrons are transferred from reduced cytochrome *c* to the artificial acceptor.

The oxidase test will differentiate most species of oxidase-positive *Pseudomonas* from the Enterobacteriaceae, which are oxidase negative. The artificial acceptor is somewhat unstable and can oxidize if left exposed to air for prolonged periods of time. In this exercise, you will use a commercially prepared reagent stored in glass ampules that are broken just prior to use.

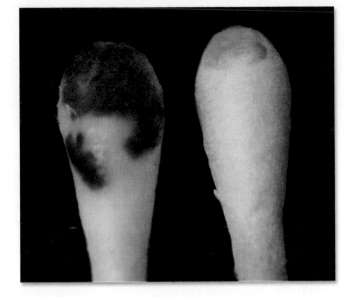

Figure 41.9 Left-hand swab shows a purple reaction due to oxidase production. Right-hand swab shows a culture that is oxidase negative.
© The McGraw-Hill Companies/Auburn University Photographic Service

Materials

- TSA plate streaked with your unknown on one-half and *Pseudomonas aeruginosa* streaked on the other half (figure 41.3).
- ampule of 1% oxidase reagent, N,N,N′,N′-tetramethyl-*p*-phenylenediamine dihydrochloride (Difco)
- Sterile swabs
- Whatman no. 2 filter paper

1. Grasp an ampule of oxidase reagent between your thumb and forefinger. Hold the ampule so that it is pointed away from you and squeeze the ampule until the glass breaks. Tap the ampule gently on the tabletop several times.
2. Touch a sterile swab to the growth of *Pseudomonas aeruginosa* on the TSA plate. Deliver several drops of oxidase reagent to the cells on the swab. (**Note:** You do not have to remove the cap of the oxidase reagent as it has a small hole for delivery of the reagent.)

    Alternatively: Transfer growth from the TSA plate to a piece of filter paper and add several drops of reagent to the cells on the paper.
3. A positive culture will cause the reagent to turn from yellow to purple in 10 to 30 seconds. A change after 30 seconds is considered a negative reaction (figure 41.9).
4. Repeat the test procedure for your unknown organism and record the results in the descriptive chart.

Catalase Test

When aerobic bacteria grow by respiration, they use oxygen as a terminal electron acceptor, converting it to water. However, they also produce hydrogen peroxide as a by-product of this reaction. Hydrogen peroxide is a highly reactive oxidizing agent that can damage enzymes, nucleic acids, and other essential molecules in the bacterial cell. To avoid this damage, aerobes produce the enzyme **catalase,** which degrades hydrogen peroxide into harmless oxygen and water.

$$2H_2O_2 \xrightarrow{\text{catalase}} 2H_2O + O_2$$

Strict anaerobes and aerotolerant bacteria such as *Streptococcus* lack this enzyme, and hence they are unable to deal with the hydrogen peroxide produced in aerobic environments. The presence of catalase is one way to differentiate these bacteria from aerobes or facultative aerobes, both of which produce catalase. For example, catalase production can be used to differentiate aerobic staphylococci from streptococci and enterococci, which lack this enzyme.

Test Procedure   To determine if catalase is produced, a small amount of growth is transferred from a plate or slant, using a wooden stick, to a clean microscope slide. A couple of drops of 3% hydrogen peroxide are added to the cells on the slide. If catalase is produced, there will be vigorous bubbling due to the breakdown of hydrogen peroxide and the production of oxygen gas.

   Note: Do not use a wire loop to transfer and mix the cells as iron can cause the hydrogen peroxide to break down, releasing oxygen. Also, do not perform the catalase test on cells growing on blood agar since blood contains catalase.

Materials

- 3% hydrogen peroxide
- nutrient agar slant tube with *Staphylococcus aureus* and your unknown on the TSA plate

1. Using the end of a wooden swab, transfer some cells from the *S. aureus* culture to the surface of a clean microscope slide.
2. Add 2 to 3 drops of 3% hydrogen peroxide to the cells, mix with the wooden stick, and observe for vigorous bubbling (figure 41.4).
3. Repeat the same procedure for your test organism and record your results in the descriptive chart.

Nitrate Reduction

Some facultative anaerobes can use nitrate as a terminal electron acceptor in a type of anaerobic respiration called **nitrate respiration.** Bacteria such

as *Paracoccus* and some *Pseudomonas* and *Bacillus* reduce nitrate to a gaseous end products such as N_2O or N_2. Other bacteria such as *Escherichia coli* partially reduce nitrate to nitrite. Several enzymes are involved in the reduction of nitrate, one of which is nitrate reductase, which catalyzes the transfer of electrons from cytochrome *b* to nitrate, reducing it to nitrite. The enzymes involved in nitrate reduction are inducible and are only produced if nitrate is present and anaerobic conditions exist for growth. The chemical reaction for the reduction of nitrate to nitrite is as follows:

$$NO_3^- + 2e^- + 2H^+ \xrightarrow{\text{nitrate reductase}} NO_2^- + H_2O$$

Test Procedure   The ability of bacteria to reduce nitrate can be determined by assaying for the end products of nitrate reduction: gas or nitrite. Cultures are grown in beef extract medium containing potassium nitrate. Gases produced from nitrate reduction are captured in Durham tubes placed in the nitrate medium. Partial reduction of nitrate to nitrite is assayed for by adding sulfanilic acid (reagent A) followed by dimethyl-alpha-naphthylamine (reagent B). If nitrite is produced by reduction, it will form a chemical complex with the sulfanilic acid and the dimethyl-alpha-naphthylamine to give a dark red color (figure 41.10). A negative test could mean that nitrate was not reduced or that a some other reduced form nitrogen was produced, such as ammonia or hydroxlyamine. As a check, zinc powder is added to the test medium. Zinc metal will chemically reduce nitrate to nitrite, causing the medium to turn dark red as result of the formation of the chemical complex. If a nongaseous product such as ammonia was produced, no color will develop after the addition of the zinc metal.

Materials

- nitrate broth cultures with Durham tubes of the unknown organism and the test control *E. coli*
- nitrate test reagents: reagent A—sulfanilic acid; reagent B—dimethyl-alpha-naphthylamine
- zinc powder

1. Examine the nitrate broth of your unknown. If gas has been displaced in the Durham tube, it means that your organism has reduced nitrate to a gaseous end product, such as nitrogen gas. If no gas is present, reduction may have resulted in the formation of nitrite or the formation of a nongaseous end product.

Figure 41.10 Nitrate test: Left-hand tube shows red color due to nitrate reduction. Middle tube shows reduction of nitrate to nitrogen gas that is trapped in the Durham tube. Right tube is an uninoculated control.

© The McGraw-Hill Companies/Auburn University Photographic Service

1. To test for the presence of nitrite, first assay the test control *E. coli* culture by adding 2 to 3 drops of reagent A and 2 to 3 drops of reagent B to the nitrate broth culture of the organism. A deep red color will develop immediately (figure 41.10).

> **CAUTION:** Avoid skin contact with solution B. Dimethyl-alpha-naphthylamine is carcinogenic.

2. Repeat this same test procedure for your unknown bacterium. If a red color fails to appear, your organism did not reduce nitrate or it may have produced a nongaseous end product of nitrate reduction.

   Zinc Test: To the negative culture, add a pinch of zinc powder and shake the tube vigorously. If a red color develops in the tube, nitrate was reduced by the zinc metal, indicating a negative test for nitrate reduction. If no color develops, a nongaseous end product may have been formed, which means your unknown reduced nitrate.

3. Record your results in the Laboratory Report 41–43.

Physiological Characteristics:
Hydrolytic and Degradative Reactions

Because bacteria have a rigid cell wall, they are unable to surround and engulf their food by the process of phagocytosis, which is characteristic of higher cells. To acquire nutrients, bacteria excrete a variety of hydrolytic and degradative exoenzymes that degrade large macromolecules into smaller units that can be transported into the cell for metabolic purposes. For example, amylases and cellulases degrade starch and cellulose, respectively, into simple sugars that are then transported into the cell where they are metabolized by fermentation or oxidation. A variety of proteases degrade proteins, such as casein and gelatin, and polypeptides into amino acids. Triglycerides are degraded into fatty acids and glycerol by various lipases. Sometimes bacteria also hydrolyze small molecules because they can thereby acquire carbon compounds for metabolic purposes. For example, tryptophane is split into pyruvate and indole by the enzyme tryptophanase. The pyruvate is metabolized, but the indole ring accumulates in the growth medium because it cannot be broken down. The accumulation of indole is the basis for a biochemical test that differentiates bacteria that produce trytophanase from those that do not produce the enzyme. Some bacteria hydrolyze urea to produce carbon dioxide and ammonia, thereby causing the pH to become alkaline. The change in pH is detected by a color change in a pH indicator. *Proteus* and other bacteria can oxidatively deaminate phenylalanine to produce phenylpyruvic acid. The latter can be detected with ferric chloride.

The presence of various hydrolytic and degradative enzymes can be used as a basis for identifying bacteria. In this exercise, you will perform biochemical tests for detecting hydrolytic and degradative reactions carried out by bacteria. In each case, you will compare your unknown to reactions carried out by test control organisms.

First Period

(Inoculations)

If each student is working with only one unknown organism, students can work in pairs to share petri plates. Note in figure 42.1 how each plate can serve for two unknowns and a test control organism streaked down the middle of the plate. If each student is working with two unknowns, the plate should not be shared. Whether or not materials will be shared will depend on the availability of materials.

Materials

per pair of students with one unknown each or for one student with two unknowns:

- 1 starch agar plate
- 1 skim milk agar plate
- 1 spirit blue agar plate
- 3 urea slants or broths
- 3 tryptone broths
- phenylalanine agar
- nutrient broth cultures: *Bacillus subtilis, Escherichia coli, Staphylococcus aureus,* and *Proteus vulgaris*

1. Label and streak the three different agar plates as shown in figure 42.1. Note that straight line streaks are made on each plate. Indicate, also, the type of medium in each plate.
2. Label a urea slant or urea broth tube with *P. vulgaris* and a tryptone broth tube with *E.coli.* These are the test control tubes for urea and tryptophan hydrolysis, respectively. Inoculate each with the respective organism.
3. For each unknown, label one urea slant or broth and one tryptone tube with the code number of your unknown. Inoculate each tube with your unknown.
4. Label one phenylalanine agar slant with your unknown number and a second slant with *P. vulgaris.* Inoculate each slant with the respective organism.
5. Incubate the test control cultures at 37° C. Incubate the unknowns at the optimum temperatures that you determined for them.

Second Period

(Evaluation of Tests)

After 24 to 48 hours incubation, compare your unknown and test controls, recording all data on the descriptive charts on page 257.

Starch Hydrolysis

The starch macromolecule consists of two constituents: (1) amylose, a straight chain polymer of 200 to 300 glucose molecules and (2) amylopectin, a larger branched polymer of glucose. Bacteria that hydrolyze starch produce *amylases* that degrade the starch molecule into molecules of maltose, glucose, and dextrins.

Unknown inoculations:
Starch, skim milk and spirit blue agar plates, tryptone broth urea, and phenylalanine agar slants are inoculated with the unknown. Note that straight–line streaks are used on the plates and that test control organisms and unknown organisms are inoculated on the same plate for comparison.

Incubation:
The plates and the tube media should be incubated at the optimum temperature determined for your unknown for 48 hours.

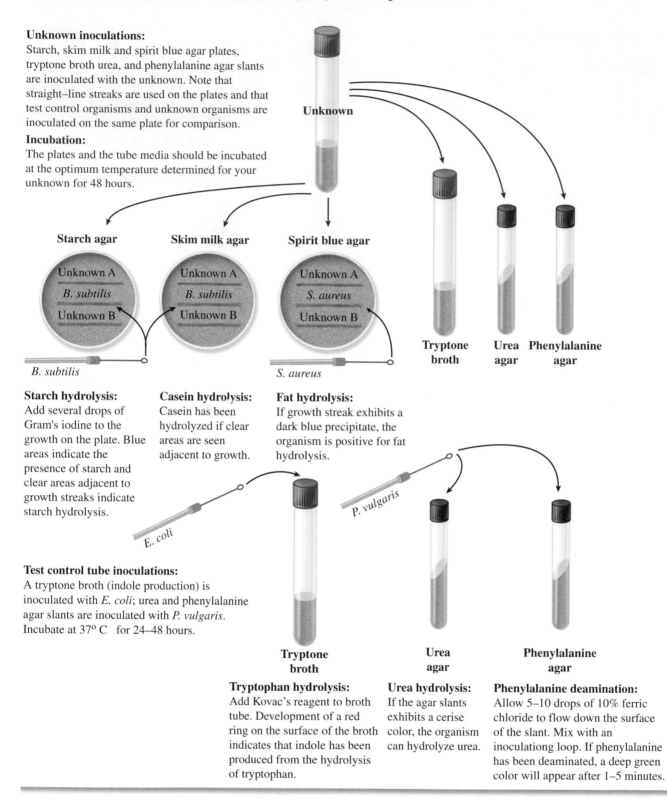

Starch hydrolysis:
Add several drops of Gram's iodine to the growth on the plate. Blue areas indicate the presence of starch and clear areas adjacent to growth streaks indicate starch hydrolysis.

Casein hydrolysis:
Casein has been hydrolyzed if clear areas are seen adjacent to growth.

Fat hydrolysis:
If growth streak exhibits a dark blue precipitate, the organism is positive for fat hydrolysis.

Test control tube inoculations:
A tryptone broth (indole production) is inoculated with *E. coli*; urea and phenylalanine agar slants are inoculated with *P. vulgaris*. Incubate at 37° C for 24–48 hours.

Tryptophan hydrolysis:
Add Kovac's reagent to broth tube. Development of a red ring on the surface of the broth indicates that indole has been produced from the hydrolysis of tryptophan.

Urea hydrolysis:
If the agar slants exhibits a cerise color, the organism can hydrolyze urea.

Phenylalanine deamination:
Allow 5–10 drops of 10% ferric chloride to flow down the surface of the slant. Mix with an inoculationg loop. If phenylalanine has been deaminated, a deep green color will appear after 1–5 minutes.

Figure 42.1 Procedure for doing hydrolysis tests on unknowns

Starch hydrolysis is detected by adding iodine to starch medium. Iodine complexes with the starch macromolecule and causes the medium to turn blue. However, if the starch has been degraded, the medium adjacent to the bacterial growth will be clear after the addition of the iodine.

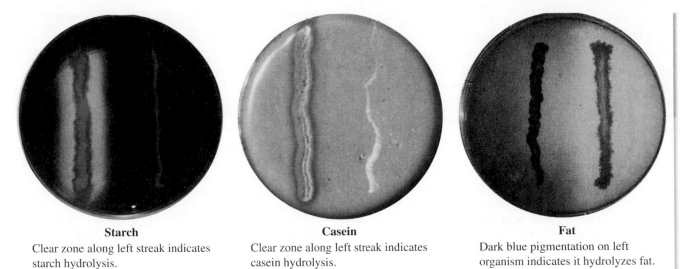

Starch	**Casein**	**Fat**
Clear zone along left streak indicates starch hydrolysis.	Clear zone along left streak indicates casein hydrolysis.	Dark blue pigmentation on left organism indicates it hydrolyzes fat.

Figure 42.2 Hydrolysis test plates: Starch, casein, fat.

Materials

- Gram's iodine
- starch agar plates with test control organism and your unknown

1. Pour enough Gram's iodine over the surface of the starch plate to cover the entire plate. Rotate and tip the plate to spread the iodine. *Bacillus subtilis* produces amylases that degrade starch. If starch is degraded, a clear area will occur next to the growth of the organism. Areas on the plate where no starch hydrolysis has occurred will be blue.
2. Compare your unknown on the same plate to *B. subtilis*. Figure 42.2 illustrates a positive starch hydrolysis result.
3. Record your results.

Casein Hydrolysis

Casein is the predominant protein in milk, and its presence causes milk to have its characteristic white color. Many bacteria produce *proteases,* which are enzymes that degrade protein molecules such as casein into peptides and amino acids. This process is referred to as *proteolysis.*

Examine the growth on the skim milk agar. Note the **clear zone** surrounding *B. subtilis* where proteolysis of the casein has occurred. Figure 42.2 illustrates casein hydrolysis. Compare your unknown to *B. subtilis*. Record the results on the descriptive chart on page 263.

Fat Hydrolysis

Fats or triglycerides are composed of a glycerol molecule to which fatty acid molecules are covalently bonded through ester bonds. Triglycerides are primarily fat storage products in higher organisms such as animals. Some bacteria produce enzymes called *lipases* that cleave the fatty acids from glycerol. The fatty acids and glycerol can then be used for metabolic purposes such as synthesizing phospholipids for membrane construction or for catabolism to produce energy. The decomposition of triglycerides and the breakdown of the fatty acids into short chain volatile organic acids is the reason why butter or margarine becomes rancid.

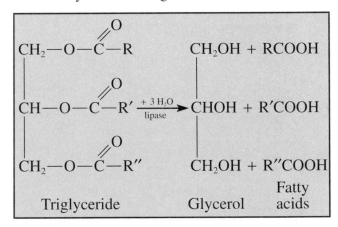

Spirit blue agar contains peptone as a source of carbon, nitrogen, and vitamins. It also contains tributyrin, a simple, natural animal triglyceride that serves as a substrate for lipases. Release of the fatty acids from tributyrin via lipase activity results in the lowering of the pH of the agar to produce a **dark blue precipitate.** However, some bacteria do not completely hydrolyze all the fatty acids from the tributyrin, and as a result, the pH is not sufficiently lowered to give the dark blue precipitate. In this case, all you notice may be simply the depletion of fat or oil droplets in the agar to indicate lipase activity.

Examine the growth of *S. aureus* on the plate. You should be able to see the dark blue reaction as shown in figure 42.2. Compare this to your unknown. *If your unknown appears negative, hold the plate up toward the light and look for a region near the growth where oil droplets are depleted.* If you see the depletion of oil droplets, record this as a positive test in the descriptive chart.

Tryptophan Degradation

Some bacteria have the ability to degrade the amino acid tryptophan producing indole, ammonia, and pyruvic acid. The pyruvic acid can then be used by an organism for various metabolic purposes. The enzyme responsible for the cleavage of tryptophan is *tryptophanase*. The degradation of tryptophan by the enzyme can be detected with Kovac's reagent, which forms a deep red color if indole is present. Tryptone broth (1%) is used for the test because it contains high amounts of tryptophan. Tryptone is derived from casein by a pancreatic digestion of the protein.

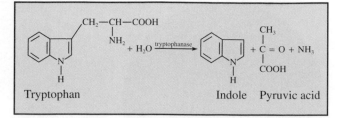

Materials

- Kovac's reagent
- tryptone broth cultures of *E. coli* and your unknown

To test for indole and therefore the activity of tryptophanase, add 10 to 12 drops of Kovac's reagent to the tryptone broth culture of *E. coli*. A red organic layer should form on top of the culture as shown in figure 42.3. Repeat the test for your unknown culture and record the results on the descriptive chart.

Urea Hydrolysis

Urea is a waste product of animal metabolism that is broken down by a number of bacteria. The enzyme responsible for urea hydrolysis is *urease,* which splits the molecule into carbon dioxide and ammonia. Urease is produced by some of the gram-negative enteric bacteria such as *Proteus, Providencia,* and *Morganella,* which can be differentiated from other gram-negative enteric bacteria by this test. Refer to the separation outline in figure 44.2.

Urea medium contains yeast extract, urea, a buffer, and the pH indicator phenol red. Urea is unstable and is broken down by heating under steam pressure at 15 psi. Therefore, the medium is prepared by adding filter-sterilized urea to the base medium after autoclaving it.

When urease is produced by an organism, the resulting ammonia causes the pH to become alkaline. As the pH increases, the phenol red changes from yellow (pH 6.8) to a bright pink or cerise color (pH 8.1 or greater). See figure 42.4.

Figure 42.3 Tryptophan hydrolysis. The left tube shows the presence of indole (red band) at the top of the tube. The right-hand tube is an uninoculated control.
© The McGraw-Hill Companies/Auburn University Photographic Service

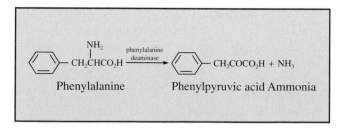

Examine the urea slant inoculated with *Proteus vulgaris* and compare it to your unknown. *If your urea slant is negative, continue the incubation for an additional 7 days to check for slow urease production.* Record your results in the descriptive chart on page 263.

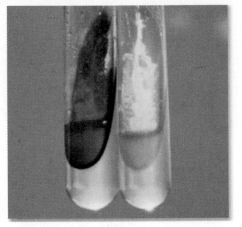

PPA test

Figure 42.5 Left-hand tube exhibits a positive reaction (green). Other tube is negative.

enzyme *phenylalanine deamninase,* a flavoprotein oxidase.

The enzyme can be detected by the addition of 10% ferric chloride, which forms a green colored complex with α-keto acids such as phenylpyruvic acid. The test is useful in differentiating the above bacteria from other Enteriobacteriaceae.

Materials

- dropping bottle of 10% ferric chloride

Allow 5 to 10 drops of 10% ferric chloride to flow down the slant of the test control organism, *P. vulgaris.* To facilitate the reaction, use an inoculating loop to emulsify the culture on the slant with the test reagent. A deep **green color** should appear in 1 to 5 minutes. Refer to figure 42.5. Repeat the test procedure for your unknown. Record your results in the Laboratory Report.

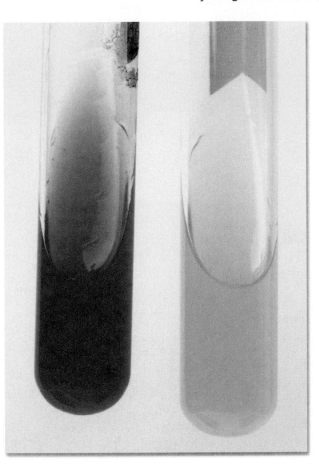

Figure 42.4 Urease test. Tube on the left is positive (*Proteus*); tube on the right is negative.
© The McGraw-Hill Companies/Auburn University Photographic Service

Phenylalanine Deamination

Gram-negative bacteria such as *Proteus, Morganella,* and *Providencia* can oxidatively deaminate the amino acid phenylalanine to produce phenylpyruvic acid and ammonia. The reaction is catalyzed by the

Physiological Characteristics:
Multiple Test Media

Some media are designed to give multiple test results. These include: Kligler's iron agar, which determines fermentation reactions for glucose and lactose and the production of hydrogen sulfide; SIM, which determines hydrogen sulfide and indole production and motility; and litmus milk, which detects fermentation, proteolysis, and other reactions in milk. In addition, the IMViC tests will be discussed; these are an important group of tests used in differentiating some gram-negative enteric bacteria.

First Period

(Inoculations)

As before, test control cultures are included in this exercise. For economy of materials, one set of test control cultures will be made by students working in pairs.

Materials

for test control cultures, per pair of students:

- 1 Kligler's iron agar deep
- 3 SIM deeps
- nutrient broth cultures of *Proteus vulgaris,* *Staphylococcus aureus,* and *Escherichia coli* for each unknown per student
- 1 Kligler's iron agar slant
- 4 SIM deeps
- 1 Litmus milk tube

1. Label one tube of Kligler's iron agar with *P. vulgaris* and additional tubes with your unknown numbers. Inoculate each tube by swabing and then stabbing with an inoculating loop.
2. Label the SIM deeps with *P. vulgaris, S. aureus, E. coli,* and your unknown number.
3. Label one tube of litmus milk with your unknown number. (**Note:** A test control culture for litmus milk will not be made. Interpretation of results will be made based on figure 43.3.)
4. Incubate the test control cultures at 37° C and the unknown cultures at their optimum temperatures for 24 to 48 hours.

Second Period

(Evaluation of Tests)

After the 24 to 48 hours of incubation, examine the tubes and evaluate the results based on the following discussion. Record the test results in the descriptive chart.

Kligler's Iron Agar

Kligler's iron agar is a multiple test medium that will detect the fermentation of glucose and lactose and the production of hydrogen sulfide resulting from the breakdown of the amino acid cysteine. It contains 0.1% glucose, 1% lactose, peptone, ferrous salts, and phenol red as a pH indicator. It is prepared as a slant and is inoculated by streaking the slant and stabbing the butt of the tube. The medium is useful in the differentiation of the gram-negative enteric bacteria.

Fermentation Reactions

The following are the possible results for the fermentation of the carbohydrates in the medium (see figure 43.1 A–C).

1. Alkaline (red) slant/acid (yellow) butt: This means that only glucose was utilized. The organism utilized the low concentration of glucose initially and then degraded the peptone in the medium. The slant is alkaline (red) because glucose was degraded aerobically, and the ammonia released from peptone utilization caused the aerobic slant to become alkaline. However, the butt is yellow (acid) because glucose was fermented anaerobically to produce enough acids to cause the acidic reaction in the butt. If gas is produced, it will be evident by the splitting of the medium and the formation of gas bubbles in the agar slant.

2. Acid (yellow) slant/acid (yellow) butt: The organism has fermented both glucose and lactose, producing acids that cause the pH indicator to turn yellow. Lactose is present in ten times (1%) the concentration of glucose (0.1%), and sufficient acid is produced to cause both the slant and butt to be acidic. However, the tubes must be read at 24 hours because they can revert to alkaline in 48 hours if the lactose becomes depleted and the peptones are utilized, producing ammonia.

3. Alkaline (red) slant/alkaline (red) butt; alkaline (red) slant/no change butt: No fermentation of

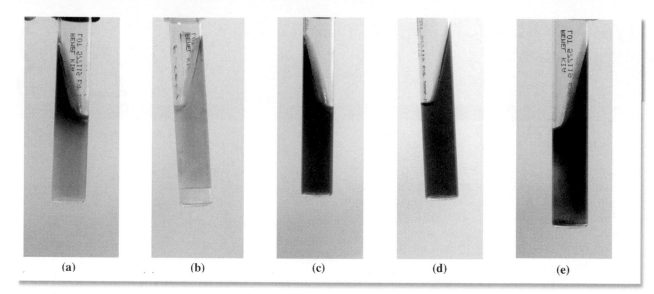

Figure 43.1 Fermentation reactions and hydrogen sulfide production on Kligler's iron agar. (a) Alkaline/alkaline; (b) acid/acid with gas; (c) alkaline/no change; (d) uninoculated; (e) hydrogen sulfide production.
© The McGraw-Hill Companies/Auburn University Photographic Service

either sugar has occurred. Some enteric bacteria can use the peptones both aerobically and anaerobically, causing both the slant and butt to become alkaline. Others can only use the peptone aerobically, producing an alkaline slant but no change in the butt.

Hydrogen Sulfide

Bacteria such as *Proteus vulgaris* can degrade the amino acid cysteine to produce pyruvic acid, ammonia, and hydrogen sulfide. The initial step in the reaction pathway is the removal of sulfide from cysteine, which is catalyzed by *cysteine desulfurase.* This enzyme also requires the coenzyme pyridoxal phosphate for activity.

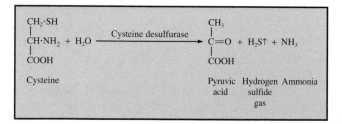

Kligler's iron agar contains ferrous salts that will react with the hydrogen sulfide liberated by cysteine desulfurase to produce an insoluble black precipitate, ferrous sulfide. **Note:** Bacteria such as *Proteus* that produce sulfide can obscure the fermentation reaction in the butt of the tube. If sulfide is produced, an acid reaction has occurred in the butt even if it cannot be observed.

Examine the Kligler's iron agar and record the results for the slant and butt of the tubes for the test organisms. Compare the results to your unknown. *Proteus* will produce hydrogen sulfide and cause the tube butt to turn black (figure 43.1 E). Record your results in the descriptive chart.

SIM Medium

SIM medium is a multiple test medium that detects the production of hydrogen sulfide and indole and that determines if an organism is motile or not. The medium contains hydrolyzed casein, ferrous salts, and agar (0.7%), which makes the medium semisolid. It is inoculated by stabbing. The breakdown of tryptophan in the medium will produce indole, which can be detected by adding Kovac's reagent. If cysteine is degraded, hydrogen sulfide will be released, which will combine with the ferrous salts to produce a black precipitate in the tube.

Because the medium contains a low concentration of agar (0.7%), bacteria that are motile are able to swim in the medium. Motility is determined by diffuse growth out from the line of inoculation or by turbidity throughout the tube. In contrast, non-motile bacteria, such as *Staphylococcus aureus,* will only grow along the line of inoculation (figure 43.2 C, right).

Examine the tubes with control bacteria. *E. coli* is motile and will therefore cause the tube to be turbid (figure 43.2 C, left). Adding Kovac's reagent to the top of the tube results in a red ring, indicating the presence of indole (figure 43.2 B). The SIM tube inoculated with *P. vulgaris* has a black precipitate due to the production of hydrogen sulfide (figure 43.2 A).

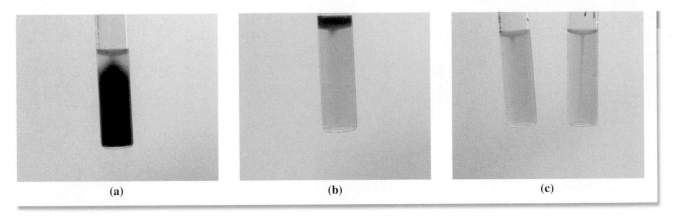

Figure 43.2   Reactions on SIM medium. (a) Hydrogen sulfide production; (b) indole production; (c) motility, left tube, motile organism, versus right tube, non-motile organism.
© The McGraw-Hill Companies/Auburn University Photographic Service

Compare these results with those of your unknown and record them in the descriptive chart.

The IMViC Tests

Sometimes a grouping of tests can be used to differentiate organisms. For example, *E. coli* can be differentiated from *E. aerogenes* by comparing four tests collectively called the IMViC (*I*: indole test; *M*: methyl red test; *V*: Voges-Proskauer test; and *C*: citrate test) tests. The results for the two organisms appear below. The two bacteria give exactly opposite reactions for the tests.

	I	M	V	C
E. coli	+	+	−	−
E. aerogenes	−	−	+	+

These two bacteria are very similar in morphological and physiological characteristics. The IMViC tests can be valuable when testing water for sewage contamination because they can rule out *E. aerogenes,* which is not always associated with sewage.

If your organism is a gram-negative rod and a facultative anaerobe, group these tests and see how your organism fits the combination of tests.

Litmus Milk

Litmus milk contains 10% powdered skim milk and the pH indicator litmus. The medium is adjusted to pH 6.8 and has a purplish blue color before in-oculation. Milk contains the proteins casein, lactalbumin, and lactogloblin as well as the disaccharide lactose, and they provide an excellent growth medium for many microorganisms. Bacteria that ferment the lactose, resulting in acid production, will cause the litmus to turn pink. Other bacteria digest the milk proteins using proteases. This results in the release of ammonia, causing the litmus to turn purple because of the alkaline condition. Some of the proteolytic bacteria can also cause the milk proteins to precipitate and coagulate, thus forming a clot. Clotting can occur because of the production of acid or because of the release of the enzyme *rennin* that converts casein to paracasein.

Certain facultative bacteria can cause the reduction of the litmus dye to a colorless or leuco form. The color change is due to a drop in oxygen levels in the tube that accompanies the production of acids. The reduction of litmus can also occur when bacteria use the dye as an alternative electron acceptor. In these cases, the litmus is acting as a oxidation-reduction indicator as well as a pH indicator.

Figure 43.3 indicates the various color changes and reactions in litmus milk. It should be noted that some of the reactions take 4 to 5 days to fully develop, and therefore cultures should be incubated for this period of time. However, it is important to check the cultures every 24 hours for changes.

Litmus Milk Reactions

Alkaline Reaction   Litmus turns purple or blue because of an alkaline pH produced by the release of ammonia. This can occur in the first 24 hours of incubation and is typical of proteolytic bacteria.

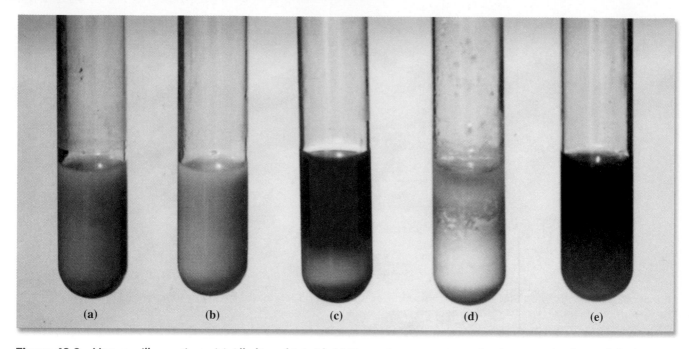

Figure 43.3 Litmus milk reactions: (a) Alkaline. (b) Acid. (c) Upper transparent portion is peptonization; solid white portion in bottom is coagulation and litmus reduction; overall redness is interpreted as acid. (d) Coagulation and litmus reduction in lower half; some peptonization (transparency) and acid in top portion. (e) Litmus indicator is masked by production of soluble pigment (*Pseudomonas*); some peptonization is present but difficult to see in photo.

Acid Reaction Litmus turns pink owing to acidic conditions, which is typical of bacteria that ferments the milk sugar.

Litmus Reduction The culture becomes white because the litmus dye is reduced due to a drop in oxygen levels in the culture or because the dye has been reduced as a result of bacteria using it as an alternative electron acceptor.

Coagulation Curd formation can result from the precipitation and coagulation of proteins. Tilting the tube 45° will confirm whether or not coagulation has occurred.

Peptonization The medium becomes translucent, often turning brown at this stage. This is characteristic of proteolytic bacteria that degrade milk proteins.

Ropiness The formation of a thick slime in the culture that can be demonstrated by inserting a sterile loop and carefully withdrawing it from the culture. The slime will adhere to the loop, forming strings.

Record the results of the litmus milk reactions for your unknown on the descriptive chart on page 263.

Laboratory Report

Complete the Laboratory Report 41–43, which summarizes all the physiological tests performed on your unknown in the last three exercises.

Laboratory Report

41–43 Physiological Characteristics of Bacteria

A. Results

Place the results of the physiological tests in the Descriptive Chart (page 263).

B. Short Answer Questions

1. Why is it important to complete morphological and cultural characterizations before pursuing physiological testing?

   _____

   _____

2. In regards to taxonomic classification of bacteria, what is the relationship between physiological and genetic differentiation of bacteria?

   _____

   _____

3. What is the function of bacterial exoenzymes?

   _____

   _____

4. Differentiate between the following:

   a. anabolism and catabolism

   _____

   b. fermentation and respiration

   _____

5. Why is the catalase test useful for the differentiation of staphylococci from streptococci?

   _____

   _____

6. End products of biochemical reactions are often acids or alkalies.

   a. How are these products typically detected in a culture medium?

   _____

   b. Name two tests for reactions that produce acid end products.

   _____

   c. Name two tests for reactions that produce alkaline end products.

   _____

7. End products of biochemical reactions are sometimes gases.

   a. What types of gases can be produced as a result of sugar fermentation?

   _____

   b. How are these gases detected in fermentation reactions?

   _____

   c. Name a nonfermentative test in which gas production indicates a positive test result.

   _____

8. The tests for the hydrolysis of starch, casein, and triglycerides are similar in terms of setup. How do the methods for the detection of hydrolysis differ between these three tests?

   _____

   _____

9. In addition to the casein hydrolysis test, name two other tests that demonstrate hydrolysis of proteins, or proteolytic digestion.

   _____

   _____

10. For which of the five hydrolysis tests (starch, casein, triglycerides, tryptophan, and urea) would a positive test result be expected for:

    a. the etiologic agent of acne, *Propionibacterium acnes*? Explain.

    _____

    b. the etiologic agent of gastric ulcers, *Helicobacter pylori*? Why would this activity be advantageous in this environment?

    _____

11. Kligler's iron agar and SIM are multiple test media.

    a. What test do these media have in common?

    _____

    b. What components of the media are included for this test?

    _____

    c. Both media are stabbed but for different reasons. Explain.

    _____

12. What difficulties does one encounter when trying to differentiate bacteria on the basis of physiological tests?

    _____

    _____

13. In addition to the morphological, cultural, and physiological tests performed on an unknown, what tests can be conducted to further assist in its identification?

_____

_____

C. Matching Questions

1. *MEDIA*. Match the name of the medium with the physiological test. A media may be used more than once. Tests may require more than one answer.

   a. Kligler's iron agar

   b. MR-VP broth

   c. phenol red lactose

   d. SIM medium

   e. Simmon's citrate agar

   f. skim milk agar

   g. spirit blue agar

   h. tryptone broth

   _____   2,3-butanediol fermentation

   _____   carbohydrate fermentation

   _____   casein hydrolysis

   _____   citrate utilization

   _____   hydrogen sulfide production

   _____   mixed-acid fermentation

   _____   triglyceride hydrolysis

   _____   tryptophan degradation

2. *REAGENTS*. Match the name of the reagent with the physiological test. Tests may require more than one answer.

   a. alpha-naphthol

   b. dimethyl-alpha-naphthylamine

   c. ferric chloride

   d. Gram's iodine

   e. hydrogen peroxide

   f. Kovac's reagent

   g. methyl red

   h. N,N,N′,N′-tetramethyl-p-phenylenediamine dihydrochloride

   i. potassium hydroxide

   j. sulfanilic acid

   _____   2,3-butanediol fermentation

   _____   catalase test

   _____   mixed-acid fermentation

   _____   nitrate reduction

_____ oxidase test

_____ phenylalanine deamination

_____ starch hydrolysis

_____ tryptophan degradation

3. *ENZYMES*. Match the name of the enzyme with the biochemical reaction. Enzymes may be used more than once.

   a. amylase

   b. cysteine desulfurase

   c. lipase

   d. protease

   e. tryptophanase

   f. urease

_____ casein hydrolysis

_____ gelatin liquefaction

_____ hydrogen sulfide production

_____ indole

_____ starch hydrolysis

_____ triglyceride hydrolysis

_____ urea hydrolysis

4. *PRODUCTS*. Match the name of the product with the biochemical reaction. Products may be used more than once. Tests may require more than one answer.

   a. 2,3-butanediol

   b. ammonia

   c. fatty acids

   d. indole

   e. molecular oxygen

   f. phenylpyruvic acid

_____ catalase

_____ phenylalanine deamination

_____ triglyceride hydrolysis

_____ tryptophan degradation

_____ urea hydrolysis

_____ Voges-Proskauer test

Use of Bergey's Manual

Once you have recorded all the data on your descriptive chart pertaining to morphological, cultural, and physiological characteristics of your unknown, you are ready to determine its genus and species. Determination of the genus is relatively easy; species differentiation is more difficult and requires performing more tests.

The most important single source of information we have for identification of bacteria is *Bergey's Manual of Systematic Bacteriology*. The present manual consists of four volumes and lists all the bacteria that have been described and isolated and the various phenotypic and genetic characteristics that describe the organisms. Volumes 1 and 2 will be the primary source for identification of the unknowns in this course.

Bergey's Manual is the collaborative effort of numerous individuals. It is consistently undergoing revision owing to the discovery of new organisms and the refinement of existing information for organisms listed in the manual. The manual is presently under revision. It has an editorial board that oversees the various aspects of the manual and makes sure that proper protocol is followed for listing new isolates or reassigning existing organisms. One of the purposes of this exercise is to help you use *Bergey's Manual* to identify your unknown. Before we approach the mechanics of using the manual, a few comments are in order pertaining to specific problems of bacterial classification.

Classification Problems

The classification of higher organisms utilizes phylogenetic approaches to determine relatedness or phylogeny. These schemes emphasize two major criteria: morphology and an evolutionary record as revealed by fossils deposited in the sedimentary layers of the earth's crust. Organisms are ordered into the **taxons** of kingdoms, phyla, classes, orders/divisions, families, tribes, genera, and species. The species defines the individual organism in the scheme, and for higher organisms, a species is defined as a group of interbreeding organisms that produce fertile offspring.

The application of these phylogenetic schemes to the classification of bacteria is very difficult for several reasons. First, bacteria have very few forms—consisting of rods, cocci, and spirals—compared to the diversity of morphology seen in higher plants and animals. If form were used as a major criterion in bacterial classification, all bacteria would fall into one of three groups. Second, bacteria do not have a fossil record from which family trees can be constructed to show relationships. However, some of the oldest examples of primitive organisms are probably early bacterial cells that have been preserved in fossilized mats called **stromatolites** that occur in tide pools in places such as Australia and that date to 3.5 billion years ago. Lastly, bacteria reproduce primarily by asexual means, using the process of transverse binary fission. Gene transfer does occur in bacteria by transformation, conjugation, and transduction, but this is not the primary means of reproduction. Thus, species as defined for higher organisms does not apply very well to bacteria because they do not reproduce by sexual means to produce fertile offspring. Earlier editions of *Bergey's Manual* attempted to use the hierarchical system of classification being used for higher organisms, but this approach was abandoned when the eighth edition was published owing to the problems cited above. Microbiologists still adhere to the binomial system of classification in which bacteria are named by a genus and species designation. However, in bacteriology, a species is considered to be more of a natural grouping of organisms that share a set of unique characteristics not shared with other organisms. With advances in molecular biology and nucleic acid sequence information, a newer definition of bacterial species will no doubt evolve.

The present system of classification in *Bergey's Manual* is based on a phenetic scheme in which organisms are ordered into groups that share a common set of phenotypic and genetic characteristics not shared by the other groups. A few examples are: the gram-positive cocci; endospore-forming gram-positive rods and cocci; anoxygenic phototrophic bacteria; and mycobacteria. Some of the characteristics listed in *Bergey's Manual* are the Gram stain and other staining reactions, various biochemical tests, serological characteristics, and the mole percent guanine plus cytosine (G+C) content. However, the system is rapidly changing as new information is revealed by genetic information and nucleic acid sequence analysis. In the present edition, each section is described in common terms so that it is easy to understand and can be used for speedy identification of an unknown organism.

Our dependency over the years on *Bergey's Manual* has led many to think that the classification presented is an "official classification." However, as pointed out in the preface to the first volume of the manual, there is no official classification for bacteria. There is however, an official set of rules for naming bacteria. *Bergey's Manual* prevails because it is evolving, and it is a workable system used and agreed upon by the vast majority of bacteriologists. In fact, sequence analysis of nucleic acids from bacteria has resulted in the reordering of bacterial groups that will be reflected in the new edition of *Bergey's Manual*.

Your unknown organism may not match the description stated in *Bergey's Manual*. The groupings are based on the fact that 90% of the isolates conform to the descriptions whereas 10% can vary from the description given. *Remember, you are describing a biological entity, and test results as stated in the manual may not always apply.*

Presumptive Identification

The place to start in identifying your unknown is to determine what genus it fits into. If *Bergey's Manual* is available, scan the tables of contents in volumes 1 and 2 to find the section that seems to describe your unknown. If these books are not immediately available, you can determine the genus by referring to the separation outlines in figures 44.1 and 44.2. Note that

seven groups of gram-positive bacteria are winnowed out in figure 44.1 and four groups of gram-negative bacteria in figure 44.2.

To determine which genus in the group best fits the description of your unknown, compare the genera descriptions provided below. Note that each group has a section designation to identify its position in *Bergey's Manual*.

Group I (section 13, vol. 2) Although there are only three genera listed in this group, section 13 in *Bergey's Manual* lists three additional genera, one of which is *Sporosarcina*, a coccus-shaped organism (see Group V). Most members of Group I are motile and differentiation is based primarily on oxygen needs.

Bacillus Although most of these organisms are aerobic, some are facultative aerobes. Catalase is usually produced. For comparative characteristics of the 34 species in this genus refer to table 13.4.

Clostridium While most of members of this genus are strict anaerobes, some may grow in the presence of oxygen. Catalase is not usually produced. An excellent key for presumptive species identification is provided on pages 1143–1148. Species characterization tables are also provided.

Sporolactobacillus Microaerophilic and catalase-negative. Nitrates are not reduced and indole is not formed. Spore formation occurs very infrequently (1% of cells).

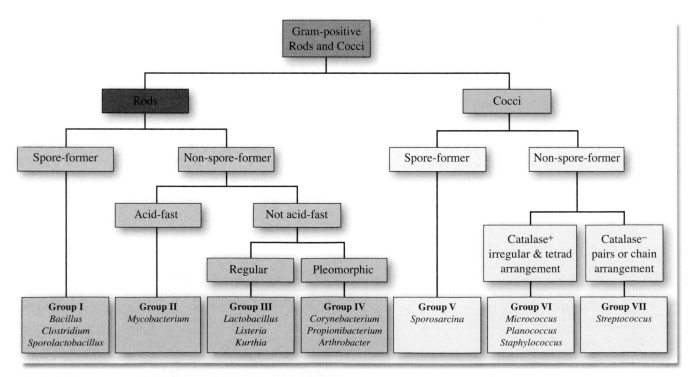

Figure 44.1 Separation outline for gram-positive rods and cocci.

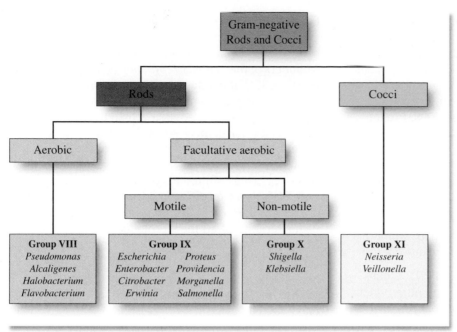

Figure 44.2 Separation outline for Gram-negative rods and cocci.

Since there is only one species in this genus, one needs only to be certain that the unknown is definitely of this genus. Table 13.11 can be used to compare other genera that are similar to this one.

Group II (section 16, vol. 2) This group consists of Family Mycobacteriaceae, with only one genus:

Mycobacterium Fifty-four species are listed in section 16. Differentiation of species within this group depends to some extent on whether the organism is classified as a slow or a fast grower. Tables in this section can be used for comparing the characteristics of the various species.

Group III (section 14, vol. 2) Of the seven diverse genera listed in section 14, only three have been included here in this group.

Lactobacillus Non-spore-forming rods, varying from long and slender to coryneform (club-shaped) coccobacilli. Chain formation is common. Only rarely motile. Facultative anaerobic or microaerophilic. Catalase-negative. Nitrate usually not reduced. Gelatin not liquefied. Indole and H₂S not produced.

Listeria Regular, short rods with rounded ends; occur singly and in short chains. Aerobic and facultative anaerobic. Motile when grown at 20–25° C. Catalase-positive and oxidase-negative. Methyl red positive. Voges-Proskauer positive. Negative for citrate utilization, indole production, urea hydrolysis, gelatinase production, and casein hydrolysis. Table 14.12 provides information pertaining to species differentiation in this genus.

Kurthia Regular rods, 2–4 micrometers long with rounded ends; in chains in young cultures; coccoidal in older cultures. Strictly aerobic. Catalase-positive, oxidase-negative. Also negative for gelatinase production and nitrate reduction. Only two species in this genus.

Group IV (section 15, vol. 2) Although there are 21 genera listed in this section of *Bergey's Manual,* only three genera are described here.

Corynebacterium Straight to slightly curved rods with tapered ends. Sometimes club-shaped. Palisade arrangements common due to snapping division of cells. Metachromatic granules formed. Facultative anaerobic. Catalase-positive. Most species produce acid from glucose and some other sugars. Often produce pellicle in broth. Table 15.3 provides information for species characterization.

Proprionibacterium Pleomorphic rods, often diphtheroid or club-shaped with one end rounded and the other tapered or pointed. Cells may be coccoid, bifid (forked, divided), or even branched. Non-motile. Some produce clumps of cells with "Chinese character" arrangements. Anaerobic to aerotolerant. Generally catalase-positive.

Produce large amounts of proprionic and acetic acids. All produce acid from glucose.

Arthrobacter Gram-positive rod and coccoid forms. Pleomorphic. Growth often starts out as rods, followed by shortening as growth continues, and finally becoming coccoidal. Some V-shaped and angular forms; branching by some. Rods usually non-motile; some motile. Oxidative, never fermentative. Catalase-positive. Little or no gas produced from glucose or other sugars. Type species is *Arthrobacter globiformis*.

Group V   (section 13, vol. 2) This group, which has only one genus in it, is closely related to genus *Bacillus*.

Sporosarcina Cells are spherical or oval when single. Cells may adhere to each other when dividing to produce tetrads or packets of eight or more. Endospores formed (see photomicrographs). Strictly aerobic. Generally motile. Only two species: *S. ureae* and *S. halophila*.

Group VI   (section 12, vol. 2) This section contains two families and 15 genera. Our concern here is with only three genera in this group. Oxygen requirements and cellular arrangement are the principal factors in differentiating the genera. Most of these genera are not closely related.

Micrococcus Spheres, occurring as singles, pairs, irregular clusters, tetrads, or cubical packets. Usually non-motile. Strict aerobes (one species is facultative anaerobic). Catalase- and oxidase-positive. Most species produce carotenoid pigments. All species will grow in media containing 5% NaCl. For species differentiation, see table 12.4.

Planococcus Spheres, occurring singly, in pairs, in groups of three cells, occasionally in tetrads. Although cells are generally gram-positive, they may be gram-variable. Motility is present. Catalase- and gelatinase-positive. Carbohydrates not attacked. Do not hydrolyze starch or reduce nitrate. Refer to table 12.9 for species differentiation.

Staphylococcus Spheres, occurring as singles, pairs, and irregular clusters. Non-motile. Facultative anaerobes. Usually catalase-positive. Most strains grow in media with 10% NaCl. Susceptible to lysis by lysostaphin. Glucose fermentation: acid, no gas. Coagulase production by some. Refer to Exercise 70 for species differentiation, or to table 12.10.

Group VII   (section 12, vol. 2) Note that the single genus of this group is included in the same section of *Bergey's Manual* as the three genera in Group VI. Members of the genus *Streptococcus* have spherical to ovoid cells that occur in pairs or chains when grown in liquid media. Some species, notably, *S. mutans,* will develop short rods when grown under certain circumstances. Facultative anaerobes. Catalase-negative. Carbohydrates are fermented to produce lactic acid without gas production. Many species are commensals or parasites of humans or animals. Refer to Exercise 71 for species differentiation of pathogens. Several tables in *Bergey's Manual* provide differentiation characteristics of all the streptococci.

Group VIII   (section 4, vol. 1) Although there are many genera of gram-negative aerobic rod-shaped bacteria, only four genera are likely to be encountered here.

Pseudomonas Generally motile. Strict aerobes. Catalase-positive. Some species produce soluble fluorescent pigments that diffuse into the agar of a slant. Many tables are available in *Bergey's Manual* for species differentiation.

Alcaligenes Rods, coccal rods, or cocci. Motile. Obligate aerobes with some strains capable of anaerobic respiration in presence of nitrate or nitrite.

Halobacterium Cells may be rod- or disk-shaped. Cells divide by constriction. Most are strict aerobes; a few are facultative anaerobes. Catalase- and oxidase-positive. Colonies are pink, red, or red to orange. Gelatinase not produced. Most species require high NaCl concentrations in media. Cell lysis occurs in hypotonic solutions.

Flavobacterium Gram-negative rods with parallel sides and rounded ends. Non-motile. Oxidative. Catalase-, oxidase-, and phosphatase-positive. Growth on solid media is typically pigmented yellow or orange. Nonpigmented strains do exist. See tables 49.4, 49.5 and 49.6 for differentiation.

Groups IX and X   (section 5, vol. 1) section 5 in *Bergey's Manual* lists three families and 34 genera; of these 34, only 10 genera of Family *Enterobacteriaceae* have been included in these two groups. If your unknown appears to fall into one of these groups, use the separation outline in figure 44.3 to determine the genus. Another useful separation outline is provided in figure 72.1. *Keep in mind, when using these separation outlines, that there are some minor exceptions in the applications of these tests. The diver-*

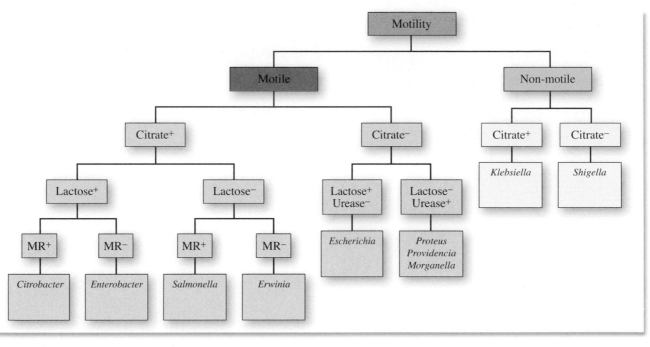

Figure 44.3 Separation outline for Groups IX and X.

sity of species within a particular genus often presents some problematical exceptions to the rule. Your final decision can be made only after checking the species characteristics tables for each genus in *Bergey's Manual.*

Group XI These genera are morphologically quite similar, yet physiologically quite different.

> *Neisseria* (section 4, vol. 1) Cocci, occurring singly, but more often in pairs (diplococci); adjacent sides are flattened. One species (*N. elongata*) consists of short rods. Non-motile. Except for *N. elongata,* all species are oxidase- and catalase-positive. Aerobic.
>
> *Veillonella* (section 8, vol. 1) Cocci, appearing as diplococci, masses, and short chains. Diplococci have flattening at adjacent surfaces. Non-motile. All are oxidase- and catalase-negative. Nitrate is reduced to nitrite. Anaerobic.

Problem Analysis

If you have identified your unknown by following the above procedures, congratulations! Not everyone succeeds at first attempt. If you are having difficulty, consider the following possibilities:

- You may have been given the wrong unknown! Although this is a remote possibility, it does happen at times. Occasionally, clerical errors are made when unknowns are put together.
- Your organism may be giving you a "false-negative" result on a test. This may be due to an

incorrectly prepared medium, faulty test reagents, or improper testing technique.

- Your unknown organisms may not match the description *exactly* as stated in *Bergey's Manual.* By now you are aware that the words *generally, usually,* and *sometimes* are frequently used in the book. It is entirely possible for one of these words to be inadvertently left out in Bergey's assignment of certain test results to a species. *In other words, test results, as stated in the manual, may not always apply!*
- Your culture may be contaminated. If you are not working with a pure culture, all tests are unreliable.
- You may not have performed enough tests. Check the various tables in *Bergey's Manual* to see if there is some other test that will be helpful. In addition, double-check the tables to make sure that you have read them correctly.

Confirmation of Results

There are several ways to confirm your presumptive identification. One method is to apply serological techniques, if your organism is one for which typing serum is available. Another alternative is to use one of the miniature multitest systems that are described in the next section of this manual. Your instructor will indicate which of these alternatives, if any, will be available.

Laboratory Report

There is no Laboratory Report for this exercise.

Miniaturized Multitest Systems

Having run a multitude of tests in Exercises 39 through 44 in an attempt to identify an unknown, you undoubtedly have become aware of the tremendous amount of media, glassware, and preparation time that is involved just to set up the tests. And then, after performing all of the tests and meticulously following all the instructions, you discover that finding the specific organism in "Encyclopedia Bergey" is not exactly the simplest task you have accomplished in this course. The thought must arise occasionally: "There's got to be an easier way!" Fortunately, there are *miniaturized multitest systems*.

Miniaturized systems have the following advantages over the macromethods you have used to study the physiological characteristics of your unknown: (1) minimum media preparation, (2) simplicity of performance, (3) reliability, (4) rapid results, and (5) uniform results. These advantages have resulted in widespread acceptance of these systems by microbiologists.

Since it is not possible to describe all of the systems that are available, only four have been selected here: two by Analytab Products and two by Becton-Dickinson. All four of these products are designed specifically to provide rapid identification of medically important organisms, often within 5 hours. Each method consists of a plastic tube or strip that contains many different media to be inoculated and incubated. To facilitate rapid identification, these systems utilize numerical coding systems that can be applied to charts or computer programs.

The four multitest systems described in this unit have been selected to provide several options. Exercises 45 and 46 pertain to the identification of gram-negative, *oxidase-negative* bacteria (Enterobacteriaceae). Exercise 47 (Oxi/Ferm Tube) is used for identifying gram-negative, *oxidase-positive* bacteria. Exercise 48 (Staph-Ident) is a rapid system for the differentiation of the staphylococci.

As convenient as these systems are, one must not assume that the conventional macromethods of Part 8 are becoming obsolete. Macromethods must still be used for culture studies and confirmatory tests; confirmatory tests by macromethods are often necessary when a particular test on a miniaturized system is in question. Another point to keep in mind is that all of the miniaturized multitest systems have been developed for the identification of *medically important* microorganisms. If one is trying to identify a saprophytic organism of the soil, water, or some other habitat, there is no substitute for the conventional methods.

If these systems are available to you in this laboratory, they may be used to confirm your conclusions that were drawn in Part 8 or they may be used in conjunction with some of the exercises in Part 14. Your instructor will indicate what applications will be made.

Enterobacteriaceae Identification:
The API 20E System

The **API 20E System** is a miniaturized version of conventional tests that is used for the identification of members of the family Enterobacteriaceae and other gram-negative bacteria. It was developed by Analytab Products, of Plainview, New York. This system utilizes a plastic strip (figure 45.1) with 20 separate compartments. Each compartment consists of a depression, or *cupule*, and a small *tube* that contains a specific dehydrated medium (see illustration 4, figure 45.2). The system has a capacity of 23 biochemical tests.

To inoculate each compartment, it is necessary to first make up a saline suspension of the unknown organism; then, with the aid of a Pasteur pipette, fill each compartment with the bacterial suspension. The cupule receives the suspension and allows it to flow into the tube of medium. The dehydrated medium is reconstituted by the saline. To provide anaerobic conditions for some of the compartments, it is necessary to add sterile mineral oil to them.

After incubation for 18–24 hours, the reactions are recorded, test reagents are added to some compartments, and test results are tabulated. Once the test results are tabulated, a *profile number* (7 or 9 digits) is computed. By finding the profile number in a code book, the *Analytical Profile Index*, or on the company website, one is able to determine the name of the organism. If no *Analytical Profile Index* is available, characterization can be done by using chart III in Appendix D.

Although this system is intended for the identification of nonenterics, as well as the Enterobacteriaceae, only the identification of the latter will be pursued in this experiment. Proceed as follows to use the API 20E System to identify your unknown enteric.

First Period

Two things will be accomplished during this period: (1) the oxidase test will be performed if it has not been previously performed, and (2) the API 20E test strip will be inoculated. All steps are illustrated in figure 45.2. Proceed as follows to use this system:

Materials

- agar slant or plate culture of unknown
- test tube of 5 ml 0.85% sterile saline
- API 20E test strip
- API incubation tray and cover
- squeeze bottle of tap water
- test tube of 5 ml sterile mineral oil
- Pasteur pipettes (5 ml size)
- oxidase test reagent
- Whatman no. 2 filter paper
- empty petri dish
- Vortex mixer

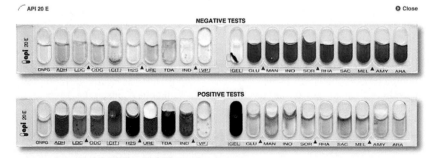

Figure 45.1 Negative and positive test results on API 20E test.
Courtesy of bioMérioux, Inc.

(1) Select one well-isolated colony to make a saline suspension of the unknown organism. Suspension should be well dispersed with a Vortex mixer.

0.85% Saline

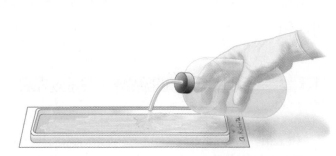

(2) After labeling the end tab of a tray with your name and unknown number, dispense approximately 5 ml of tap water into bottom of tray.

Cupule

Tube

(3) Place an API 20E test strip into the bottom of the moistened tray. Be sure to seal the pouch from which the test strip was removed to prevent contamination of remaining strips.

(4) Dispense saline suspension of organisms into cupules of all twenty compartments. Slightly *underfill* ADH, LDC, ODC, H₂S, and URE. *Completely fill* cupules of CIT, VP, and GEL.

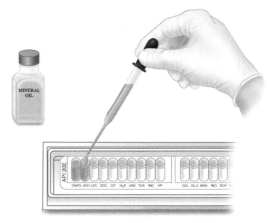

MINERAL OIL

(5) To provide anaerobic conditions for chambers ADH, LDC, ODC, H₂S, and URE, completely fill cupules of these chambers with sterile mineral oil. Use a fresh sterile Pasteur pipette.

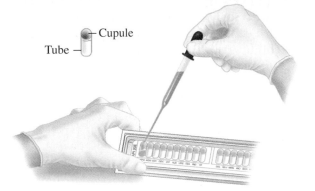

ONPG	ADH	LDC	ODC	CIT	H₂S	URE	TDA	IND	VP		GEL	GLU	MAN
1	2	4	1	2	4	1	2	4	1		2	4	1
+	−	+	+	−	−	−	−	+	−		−	+	+

$\boxed{5}$ $\boxed{1}$ $\square$

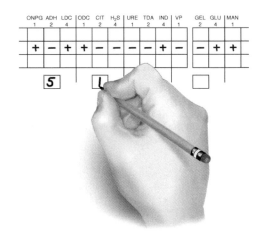

(6) After incubation and after adding test reagents to four compartments, record all results and total numbers to arrive at 7-digit code. Consult the *Analytical Profile Index* to find the unknown.

Figure 45.2 **Procedure for preparing and inoculating the API 20E test strip.**

1. If you haven't already done the **oxidase test** on your unknown, do so at this time. It must be established that your unknown is definitely oxidase-negative before using this system. Use the method that is described on pages 305.

2. Prepare a **saline suspension** of your unknown by transferring organisms from the center of a well-established colony on an agar plate (or from a slant culture) to a tube of 0.85% saline solution. Disperse the organisms well throughout the saline.

3. Label the end strip of the API 20E tray with your name and unknown number. See illustration 2, figure 45.2.
4. Dispense about 5 ml of tap water into the tray with a squeeze bottle. Note that the bottom of the tray has numerous depressions to accept the water.
5. Remove an API 20E test strip from the sealed pouch and place it into the tray (see illustration 3). Be sure to reseal the pouch to protect the remaining strips.
6. Vortex mix the saline suspension to get uniform dispersal, and fill a sterile Pasteur pipette with the suspension. *Take care not to spill any of the organisms on the table or yourself. You may have a pathogen!*
7. Inoculate all the tubes on the test strip with the pipette by depositing the suspension into the cupules as you tilt the API tray (see illustration 4, figure 45.2).
   Important: Slightly *underfill* <u>ADH</u>, <u>LDC</u>, <u>ODC</u>, <u>H$_2$S</u>, and <u>URE</u>. (Note that the labels for these compartments are underlined on the strip.) Underfilling these compartments leaves room for oil to be added and facilitates interpretation of the results.
8. Since the media in |CIT|, |VP|, and |GEL| compartments require oxygen, *completely fill both the cupule and tube* of these compartments. Note that the labels on these three compartments are bracketed as shown here.
9. To provide anaerobic conditions for the <u>ADH</u>, <u>LDC</u>, <u>ODC</u>, <u>H$_2$S</u>, and <u>URE</u> compartments, dispense sterile **mineral oil** to the cupules of these compartments. Use another sterile Pasteur pipette for this step.
10. Place the lid on the incubation tray and incubate at 37° C for 18 to 24 hours. Refrigeration after incubation is not recommended.

Second Period

(Evaluation of Tests)

During this period, all reactions will be recorded on the Laboratory Report, test reagents will be added to four compartments, and the seven-digit profile number will be determined so that the unknown can be looked up in the *API 20E Analytical Profile Index*. Proceed as follows:

Materials

- incubation tray with API 20E test strip
- 10% ferric chloride
- Barritt's reagents A and B
- Kovacs' reagent
- nitrite test reagents A and B
- zinc dust or 20-mesh granular zinc
- hydrogen peroxide (1.5%)
- *API 20E Analytical Profile Index*
- Pasteur pipettes

1. Before any test reagents are added to any of the compartments, consult chart II, Appendix D, to determine the nature of positive reactions of each test, except TDA, VP, and IND.
2. Refer to chart II, Appendix D, for an explanation of the 20 symbols that are used on the plastic test strip.
3. Record the results of these tests on Laboratory Report 45.
4. **If GLU test is negative (blue or blue-green), and there are fewer than three positive reactions** before adding reagents, do not progress any further with this test as outlined here in this experiment. Organisms that are GLU-negative are nonenterics.

   For nonenterics, additional incubation time is required. If you wish to follow through on an organism of this type, consult your instructor for more information.
5. **If GLU test is positive (yellow), or there are more than three positive reactions,** proceed to add reagents as indicated in the following steps.
6. Add one drop of **10% ferric chloride** to the TDA tube. A positive reaction (brown-red), if it occurs, will occur immediately. A negative reaction color is yellow.
7. Add 1 drop each of **Barritt's A** and **B solutions** to the VP tube. Read the VP tube within 10 minutes. The pale pink color that occurs immediately has no significance. A positive reaction is dark pink or red and may take 10 minutes before it appears.
8. Add 1 drop of **Kovacs' reagent** to the IND tube. Look for a positive (red ring) reaction within 2 minutes.

   After several minutes, the acid in the reagent reacts with the plastic cupule to produce a color change from yellow to brownish-red, which is considered negative.
9. Examine the GLU tube closely for evidence of bubbles. Bubbles indicate the reduction of nitrate and the formation of N$_2$ gas. Note on the Laboratory Report that there is a place to record the presence of this gas.
10. Add 2 drops of each **nitrite test reagent** to the GLU tube. A positive (red) reaction should show up within 2 to 3 minutes if nitrates are reduced.

    If this test is negative, confirm negativity with **zinc dust** or 20-mesh granular zinc. A pink-orange color after 10 minutes confirms that nitrate reduction did not occur. A yellow color results if N$_2$ was produced.

11. Add 1 drop of **hydrogen peroxide** to each of the MAN, INO, and SOR cupules. If catalase is produced, gas bubbles will appear within 2 minutes. Best results will be obtained in tubes that have no gas from fermentation.

Final Confirmation

After all test results have been recorded and the seven-digit profile number has been determined, according to the procedures outlined on the Laboratory Report, identify your unknown by looking up the profile number in the *API 20E Analytical Profile Index*.

Cleanup

When finished with the test strip, be sure to place it in a container of disinfectant that has been designated for test strip disposal.

Laboratory Report

Student: _____

Date: _____ Section: _____

45 Enterobacteriaceae Identification: The API 20E System

A. Results

1. Tabulation
   By referring to charts I and II, Appendix D, determine the results of each test and record these results as positive (+) or negative (−) in the table below. Note that the results of the oxidase test must be recorded in the last column on the right side of the table.

ONPG	ADH	LDC	ODC	CIT	H2S	URE	TDA	IND	VP	GEL	GLU	MAN	INO	SOR	RHA	SAC	MEL	AMY	ARA	OXI
1	2	4	1	2	4	1	2	4	1	2	4	1	2	4	1	2	4	1	2	4

☐ ☐ ☐ ☐ ☐ ☐ ☐

NO2	N2 GAS	MOT	MAC	OF-O	OF-F
1	2	4	1	2	4

☐ ☐   Additional Digits

2. Construction of Seven-Digit Profile
   Note in the above table that each test has a value of 1, 2, or 4. To compute the seven-digit profile for your unknown, total up the positive values for each group.

Example:
5 144 572 = *E. coli*

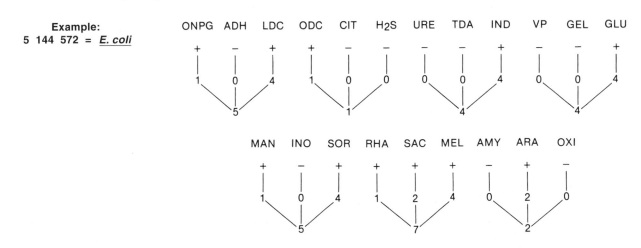

3. Using the *API 20E Analytical Index* or the API Characterization Chart
   If the *API 20E Analytical Index* is available on the demonstration table, use it to identify your unknown, using the seven-digit profile number that has been computed. If no *Analytical Index* is available, use characterization chart III in Appendix D.

   Name of Unknown: _____

4. Additional Tabulation Blank
   If you need another form, use the one below:

api® 20E

Reference Number _____ Patient _____ Date _____

Source/Site _____ Physician _____ Dept./Service _____

	ONPG 1	ADH 2	LDC 4	ODC 1	CIT 2	H₂S 4	URE 1	TDA 2	IND 4	VP 1	GEL 2	GLU 4	MAN 1	INO 2	SOR 4	RHA 1	SAC 2	MEL 4	AMY 1	ARA 2	OXI 4
5 h																					
24 h																					
48 h																					
Profile Number																					

	NO₂ 1	N₂ GAS 2	MOT 4	MAC 1	OF-O 2	OF-F 4
5 h						
24 h						
48 h						
Additional Digits						

Additional Information
_____

Identification
_____

00-42-012   E-3 (7/80)

B. Short Answer Questions

1. What are the advantages and disadvantages of multitest systems for bacterial identification?

   _____

   _____

2. Before using the API 20E System, what test must be performed to confirm the identity of your unknown as a member of the family Enterobacteriaceae? What is the expected result?

   _____

   _____

3. For each of the following aspects, compare and contrast the API 20E System to the Enterotube II System (Exercise 46) for Enterobacteriaceae identification.

   a. time requirement

   _____

   b. specimen preparation

   _____

   c. tests utilized

   _____

   d. anaerobic conditions

   _____

    e. interpretation of results

_____

4. If the seven-digit profile that is tabulated cannot be found in the *API 20E Analytical Profile Index,* what might that indicate about the bacterial culture?

_____

_____

Enterobacteriaceae Identification:
The Enterotube II System

The **Enterotube II** miniaturized multitest system was developed by Becton-Dickinson of Cockeysville, Maryland, for rapid identification of Enterobacteriaceae. It incorporates 12 different conventional media and 15 biochemical tests into a single ready-to-use tube that can be simultaneously inoculated in a moment's time with a minimum of equipment.

If you have an unknown gram-negative rod or coccobacillus that appears to be one of the Enterobacteriaceae, you may wish to try this system on it. Before applying this test, however, *make certain that your unknown is oxidase-negative*, since with only a few exceptions, all Enterobacteriaceae are oxidase-negative. If you have a gram-negative rod that is oxidase-positive, you might try the *Oxi/Ferm Tube II* instead, which is featured in Exercise 47.

Figure 46.1 illustrates an uninoculated tube (upper) and a tube with all positive reactions (lower). Figure 46.2 outlines the entire procedure for utilizing this system.

Each of the 12 compartments of an Enterotube II contains a different agar-based medium. Compartments that require aerobic conditions have openings for access to air. Those compartments that require anaerobic conditions have layers of paraffin wax over the media. Extending through all compartments of the entire tube is an inoculating wire. To inoculate the media, one simply picks up some organisms on the end of the wire and pulls the wire through each of the chambers in a single rotating action.

After incubation, the reactions in all the compartments are noted and the indole test is performed. The Voges-Proskauer test may also be performed as a confirmation test. Positive reactions are given numerical values, which are totaled to arrive at a five-digit code. Identification of the unknown is achieved by consulting a coding manual, the *Enterotube II Interpretation Guide*, which lists these numerical codes for the Enterobacteriaceae. Proceed as follows to use an Enterotube II in the identification of your unknown.

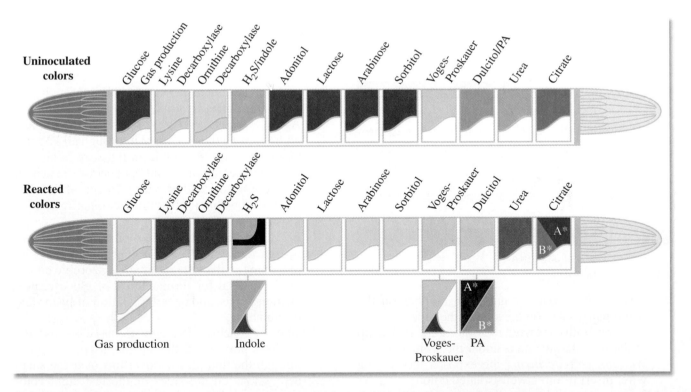

Figure 46.1 Enterotube II color differences between uninoculated and positive tests.

Courtesy and © Becton-Dickinson and Company

(1) Remove organisms from a well-isolated colony. Avoid touching the agar with the wire. To prevent damaging Enterotube II media, do not heat-sterilize the inoculating wire.

(2) Inoculate each compartment by first twisting the wire and then withdrawing it all the way out through the 12 compartments, using a turning movement.

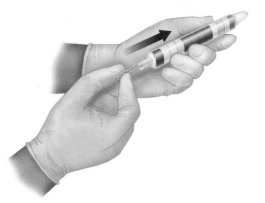

(3) Reinsert the wire (without sterilizing), using a turning motion through all 12 compartments until the notch on the wire is aligned with the opening of the tube.

(4) Break the wire at the notch by bending. The portion of the wire remaining in the tube maintains anaerobic conditions essential for true fermentation.

continued

Figure 46.2   The Enterotube II procedure.

First Period

Inoculation and Incubation

The Enterotube II can be used to identify Enterobacteriaceae from colonies on agar that have been inoculated from urine, blood, sputum, and so on. The culture may be taken from media such as MacConkey, EMB, SS, Hektoen enteric, or trypticase soy agar.

Materials

- culture plate of unknown
- 1 Enterotube II

1. Write your initials or unknown number on the white paper label on the side of the tube.
2. Unscrew both caps from the Enterotube II. The tip of the inoculating end is under the white cap.
3. *Without heat-sterilizing* the exposed inoculating wire, insert it into a well-isolated colony.
4. Inoculate each chamber by first twisting the wire and then withdrawing it through all 12 compartments. Rotate the wire as you pull it through. See illustration 2, figure 46.2.
5. Again, *without sterilizing*, reinsert the wire, and with a turning motion, force it through all 12 compartments until the notch on the wire is aligned with the opening of the tube. (The notch is about $1\frac{5}{8}''$ from handle end of wire.) The tip of the wire should be visible in the citrate compartment. See illustration 3, figure 46.2.
6. Break the wire at the notch by bending, as shown in step 4, figure 46.2. The portion of the wire remaining in the tube maintains anaerobic conditions essential for fermentation of glucose, production of gas, and decarboxylation of lysine and ornithine.
7. Without touching the end of the wire, use the retained portion of the needle to punch holes through the thin plastic coverings over the small depressions on the sides of the last eight compartments (adonitol, lactose, arabinose, sorbitol, Voges-Proskauer, dulcitol/PA, urea, and citrate).

(5) Punch holes with broken-off part of wire through the thin plastic covering over depressions on sides of the last eight compartments (adonitol through citrate). Replace caps and incubate at 35° C for 18–24 hours.

(6) After interpreting and recording positive results on the sides of the tube, perform the indole test by injecting 1 or 2 drops of Kovacs' reagent into the H_2S/indole compartment.

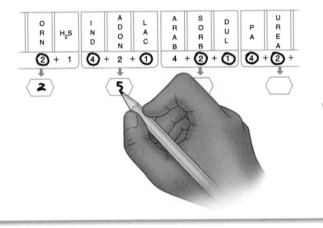

(7) Perform the Voges-Proskauer test, if needed for confirmation, by injecting the reagents into the H_2S/indole compartment.

   After encircling the numbers of the positive tests on the Laboratory Report, total up the numbers of each bracketed series to determine the 5-digit code number. Refer to the *Enterotube II Interpretation Guide* for identification of the unknown by using the code number.

Figure 46.2 (continued)

These holes will enable aerobic growth in these eight compartments.

8. Replace the caps at both ends.
9. Incubate at 35° to 37° C for 18 to 24 hours with the Enterotube II lying on its flat surface. *When incubating several tubes together, allow space between them to allow for air circulation.*

Second Period

Reading Results

Reading the results on the Enterotube may be done in one of two ways: (1) by simply comparing the results with information on chart IV, Appendix D, or (2) by finding the five-digit code number you compute for your unknown in the *Enterotube II Interpretation Guide*. Of the two methods, the latter is much preferred. The chart in the appendix should be used *only* if the *Interpretation Guide* is not available.

   Whether or not the *Interpretation Guide* is available, these three steps will be performed during this

period to complete this experiment: (1) positive test results must *first* be recorded on the Laboratory Report, (2) the indole test, a presumptive test, is performed on compartment 4, and (3) confirmatory tests, if needed, are performed. The Voges-Proskauer test falls in the latter category. Proceed as follows:

Materials

- Enterotube II, inoculated and incubated
- Kovacs' reagent
- 10% KOH with 0.3% creatine solution
- 5% alpha-naphthol in absolute ethyl alcohol
- syringes with needles, or disposable Pasteur pipettes
- test-tube rack
- Enterotube II Results Pad (optional)
- coding manual: *Enterotube II Interpretation Guide*

1. Compare the colors of each compartment of your Enterotube II with the lower tube illustrated in figure 46.1.

2. With a pencil, mark a small plus (+) or minus (−) near each compartment symbol on the white label on the side of the tube.

3. Consult table 46.1 for information as to the significance of each compartment label.

4. Record the results of the tests on the Laboratory Report. *All results must be recorded before doing the indole test.*

5. **Important:** If at this point you discover that your unknown is GLU-negative, proceed no further with the Enterotube II because your unknown is not one of the Enterobacteriaceae. Your unknown may be *Acinetobacter* sp. or *Pseudomonas maltophilia*. If an Oxi/Ferm Tube is available, try it, using the procedure outlined in the next exercise.

6. **Indole Test:** Perform the indole test as follows:
   a. Place the Enterotube II into a test-tube rack with the GLU-GAS compartment pointing upward.

   b. Inject 1 or 2 drops of Kovacs' reagent onto the surface of the medium in the H$_2$S/indole compartment. This may be done with a syringe and needle through the thin Mylar plastic film that covers the flat surface, or with a disposable Pasteur pipette through a small hole made in the Mylar film with a hot inoculating needle.

   c. A positive test is indicated by the development of a **red color** on the surface of the medium or Mylar film within 10 seconds.

7. **Voges-Proskauer Test:** Since this test is used as a confirmatory test, it should be performed *only* when called for in the *Enterotube II Interpretation Guide*. If it is called for, perform the test in the following manner:
   a. Use a syringe or Pasteur pipette to inject 2 drops of potassium hydroxide containing creatine into the VP section.

Table 46.1   Biochemical Reactions of Enterotube II

SYMBOL	UNINOCULATED COLOR	REACTED COLOR	TYPE OF REACTION
GLU-GAS			**Glucose (GLU)** The end products of bacterial fermentation of glucose are either acid or acid and gas. The shift in pH due to the production of acid is indicated by a color change from red (alkaline) to yellow (acidic). Any degree of yellow should be interpreted as a positive reaction; orange should be considered negative.
			Gas Production (GAS) Complete separation of the wax overlay from the surface of the glucose medium occurs when gas is produced. The amount of separation between the medium and overlay will vary with strain of bacteria.
LYS			**Lysine Decarboxylase** Bacterial decarboxylation of lysine, which results in the formation of the alkaline end product cadaverine, is indicated by a change in the color of the indicator from pale yellow (acidic) to purple (alkaline). Any degree of purple should be interpreted as a positive reaction. The medium remains yellow if decarboxylation of lysine does not occur.
ORN			**Ornithine Decarboxylase** Bacterial decarboxylation of ornithine causes the alkaline end product putrescine to be produced. The acidic (yellow) nature of the medium is converted to purple as alkalinity occurs. Any degree of purple should be interpreted as a positive reaction. The medium remains yellow if decarboxylation of ornithine does not occur.
H$_2$S/IND			**H$_2$S Production** Hydrogen sulfide, liberated by bacteria that reduce sulfur-containing compounds such as peptones and sodium thiosulfate, reacts with the iron salts in the medium to form a black precipitate of ferric sulfide usually along the line of inoculation. Some *Proteus* and *Providencia* strains may produce a diffuse brown coloration in this medium, which should not be confused with true H$_2$S production.
			Indole Formation The production of indole from the metabolism of tryptophan by the bacterial enzyme tryptophanase is detected by the development of a pink to red color after the addition of Kovac's reagent.

Table 46.1 Biochemical Reactions of Enterotube II (continued)

SYMBOL	UNINOCULATED COLOR	REACTED COLOR	TYPE OF REACTION
ADON			**Adonitol** Bacterial fermentation of adonitol, which results in the formation of acidic end products, is indicated by a change in color of the indicator present in the medium from red (alkaline) to yellow (acidic). Any sign of yellow should be interpreted as a positive reaction; orange should be considered negative.
LAC			**Lactose** Bacterial fermentation of lactose, which results in the formation of acidic end products, is indicated by a change in color of the indicator present in the medium from red (alkaline) to yellow (acidic). Any sign of yellow should be interpreted as a positive reaction; orange should be considered negative.
ARAB			**Arabinose** Bacterial fermentation of arabinose, which results in the formation of acidic end products, is indicated by a change in color from red (alkaline) to yellow (acidic). Any sign of yellow should be interpreted as a positive reaction; orange should be considered negative.
SORB			**Sorbitol** Bacterial fermentation of sorbitol, which results in the formation of acidic end products, is indicated by a change in color from red (alkaline) to yellow (acidic). Any sign of yellow should be interpreted as a positive reaction; orange should be considered negative.
V.P.			**Voges-Proskauer** Acetylmethylcarbinol (acetoin) is an intermediate in the production of 2,3 butanediol from glucose fermentation. The presence of acetoin is indicated by the development of a red color within 20 minutes. Most positive reactions are evident within 10 minutes.
DUL-PA			**Dulcitol** Bacterial fermentation of dulcitol, which results in the formation of acidic end products, is indicated by a change in color of the indicator present in the medium from green (alkaline) to yellow or pale yellow (acidic).
			Phenylalanine Deaminase This test detects the formation of pyruvic acid from the deamination of phenylalanine. The pyruvic acid formed reacts with a ferric salt in the medium to produce a characteristic black to smoky gray color.
UREA			**Urea** The production of urease by some bacteria hydrolyzes urea in this medium to produce ammonia, which causes a shift in pH from yellow (acidic) to reddish-purple (alkaline). This test is strongly positive for *Proteus* in 6 hours and weakly positive for *Klebsiella* and some *Enterobacter* species in 24 hours.
CIT			**Citrate** Organisms that are able to utilize the citrate in this medium as their sole source of carbon produce alkaline metabolites that change the color of the indicator from green (acidic) to deep blue (alkaline). Any degree of blue should be considered positive.

Source: Courtesy and © Becton-Dickinson and Company

b. Inject 3 drops of 5% alpha-naphthol.
c. A positive test is indicated by a **red color** within 10 minutes.
8. Record the indole and V.P. results on the Laboratory Report.

Laboratory Report

Determine the name of your unknown by following the instructions in Laboratory Report 46. Note that two methods of making the final determination are given.

Laboratory Report

Student: _____

Date: _____ Section: _____

46   Enterobacteriaceae Identification: The Enterotube II System

A. Results

1. Tabulation

   Record the results of each test in the following table with a plus (+)or minus (−).

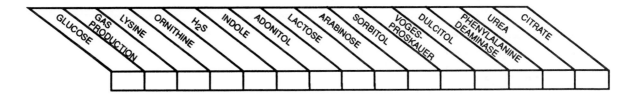

2. Identification by Chart Method

   If no *Interpretation Guide* is available, apply the above results to chart IV, Appendix D, to find the name of your unknown. Note that the spacing of the above table matches the size of the spaces on chart IV. If this page is removed from the manual, folded, and placed on chart IV, the results on the above table can be moved down the chart to make a quick comparison of your results with the expected results for each organism.

3. Using the *Enterotube II Interpretation Guide*

   If the *Interpretation Guide* is available, determine the five-digit code number by circling the numbers (4, 2, or 1) under each test that is positive, and then totaling these numbers within each group to form a digit for that group. Note that there are two tally charts in this Laboratory Report for your use.

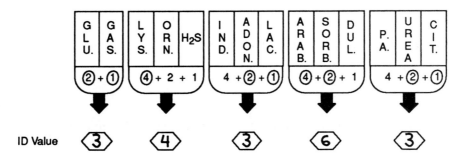

The "ID Value" 34363 can be found by thumbing the pages of the *Interpretation Guide*. The listing is as follows:

ID Value	Organism	Atypical Test Results
34363	*Klebsiella pneumoniae*	None

Conclusion: Organism was correctly identified as *Klebsiella pneumoniae*. In this case, the identification was made independent of the V.P. test.

4. Additional Tabulation Blanks

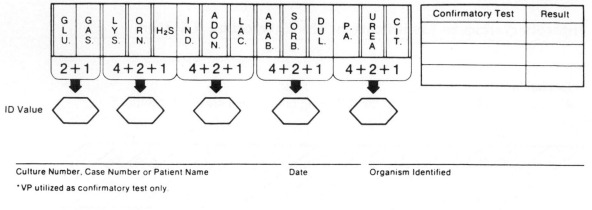

ENTEROTUBE® II*

G L U.	G A S.	L Y S.	O R N.	H₂S	I N D.	A D O N.	L A C.	A R A B.	S O R B.	D U L.	P. A.	U R E A	C I T.
2 + 1		4 + 2 + 1			4 + 2 + 1			4 + 2 + 1			4 + 2 + 1		

Confirmatory Test	Result

ID Value

Culture Number, Case Number or Patient Name   Date   Organism Identified

*VP utilized as confirmatory test only.

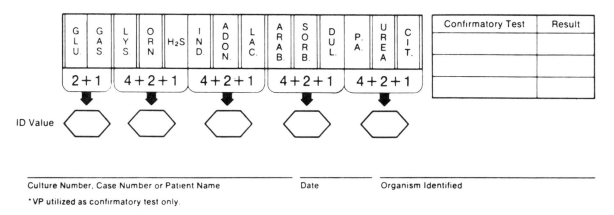

ENTEROTUBE® II*

G L U.	G A S.	L Y S.	O R N	H₂S	I N D.	A D O N.	L A C.	A R A B.	S O R B.	D U L.	P. A.	U R E A	C I T.
2 + 1		4 + 2 + 1			4 + 2 + 1			4 + 2 + 1			4 + 2 + 1		

Confirmatory Test	Result

ID Value

Culture Number, Case Number or Patient Name   Date   Organism Identified

*VP utilized as confirmatory test only.

B. Short Answer Questions

1. What are the advantages and disadvantages of multitest systems for bacterial identification?

   _____

   _____

2. Before using the Enterotube II System, what test must be performed to confirm the identity of your unknown as a member of the Family Enterobacteriaceae? What is the expected result?

   _____

   _____

3. For each of the following aspects, compare and contrast the Enterotube II System to the API 20E System (Exercise 45) for Enterobacteriaceae identification.

   a. time requirement

   _____

   _____

   b. specimen preparation

   _____

   _____

c. tests utilized

_____

_____

d. anaerobic conditions

_____

_____

e. interpretation of results

_____

_____

4. The five-digit code for all members of the Family Enterobacteriaceae starts with the number 2 or 3. What does this indicate about their common biochemistry?

_____

_____

5. If the first number of the five-digit code is 0, what does this indicate? What bacterial genera are likely to give this result? What should you do next?

_____

_____

6. The V.P. test is a confirmatory test. In what situations would this test be utilized?

_____

_____

7. If the five-digit code that is tabulated cannot be found in the *Enterotube II Interpretation Guide,* what might that indicate about the bacterial culture?

_____

_____

O/F Gram-Negative Rods Identification:
The Oxi/Ferm Tube II System

The Oxi/Ferm Tube II, produced by Becton-Dickinson, takes care of the identification of the oxidase-positive, gram-negative bacteria that cannot be identified by using the Enterotube II system. The two multitest systems were developed to work together. If an unknown gram-negative rod is oxidase-negative, the Enterotube II is used. If the organism is oxidase-positive, the Oxi/Ferm Tube II must be used. Whenever an oxidase-negative gram-negative rod turns out to be glucose-negative on the Enterotube II test, one must move on to use the Oxi/Ferm Tube II.

The Oxi/Ferm Tube II System is intended for the identification of nonfastidious species of oxidative-fermentative gram-negative rods from clinical specimens. This includes the following genera: *Aeromonas, Plesiomonas, Vibrio, Achromobacter, Alcaligenes, Bordetella, Moraxella*, and *Pasteurella*. Some other gram-negative bacteria can also be identified with additional biochemical tests. The system incorporates 12 different conventional media that

can be inoculated simultaneously with a minimum of equipment. A total of 14 physiological tests are performed.

Like the Enterotube II system, the Oxi/Ferm Tube II has an inoculating wire that extends through all 12 compartments of the entire tube. To inoculate the media, one simply picks up some organisms on the end of the wire and pulls the wire through each of the chambers in a rotating action.

After incubation, the results are recorded and Kovacs' reagent is injected into one of the compartments to perform the indole test. Positive reactions are given numerical values that are totaled to arrive at a five-digit code. By looking up the code in an *Oxi/Ferm Tube II Biocode Manual*, one can quickly determine the name of the unknown and any tests that might be needed to confirm the identification.

Figure 47.1 illustrates an uninoculated tube and a tube with all positive reactions. Figure 47.2 illustrates the entire procedure for utilizing this system.

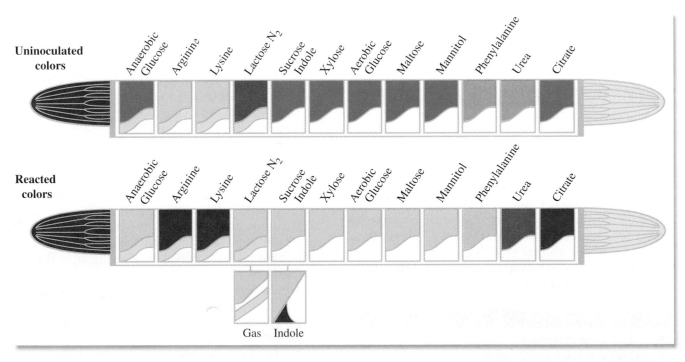

Figure 47.1 Oxi/Ferm Tube II color differences between uninoculated and positive tests.
Courtesy and © Becton-Dickinson and Company

(1) Remove organisms from a well-isolated colony. Avoid puncturing the agar with the wire. To prevent damaging Enterotube II media, do not heat-sterilize the inoculating wire.

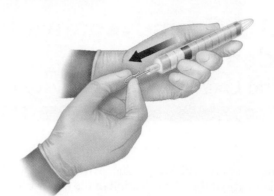

(2) Inoculate each compartment by first twisting the wire and then withdrawing it all the way out through the 12 compartments, using a turning movement.

(3) Reinsert the wire (without sterilizing), using a turning motion through all 12 compartments until the notch on the wire is aligned with the opening of the tube.

(4) Break the wire at the notch by bending. The portion of the wire remaining in the tube maintains anaerobic conditions essential for true fermentation.

Figure 47.2   The Oxi/Ferm Tube II procedure.

continued

A minimum of two periods is required to use this system. Proceed as follows:

First Period

Inoculation and Incubation

The Oxi/Ferm Tube II must be inoculated with a large inoculum from a well-isolated colony. Culture purity is of paramount importance. If there is any doubt of purity, a TSA plate should be inoculated and incubated at 35° C for 24 hours, followed by 24 hours incubation at room temperature. If no growth occurs on TSA, but growth does occur on blood agar, the organism has special growth requirements. *Such organisms are too fastidious and cannot be identified with the Oxi/Ferm Tube II.*

Materials

- culture plate of unknown
- 1 Oxi/Ferm Tube II
- 1 plate of trypticase soy agar (TSA) (for purity check, if needed)

1. Write your initials or unknown number on the side of the tube.
2. Unscrew both caps from the Oxi/Ferm Tube II. The tip of the inoculating end is under the white cap.
3. *Without heat-sterilizing* the exposed inoculating wire, insert it into a well-isolated colony. Do not puncture the agar.
4. Inoculate each chamber by first twisting the wire and then withdrawing it through all 12 compartments. Rotate the wire as you pull it through. See illustration 2, figure 47.2.
5. If a purity check of the culture is necessary, streak a petri plate of TSA with the inoculating wire that has just been pulled through the tube. **Do not flame.**
6. Again, *without sterilizing*, reinsert the wire, and with a turning motion, force it through all 12 compartments until the notch on the wire is aligned with the opening of the tube. (The notch is about $1\frac{5}{8}''$ from the handle end of the wire.) The tip of the wire should be visible in the citrate compartment. See illustration 3, figure 47.2.

(5) Punch holes with broken-off part of wire through the thin plastic covering over depressions on sides of the last eight compartments (sucrose/indole through citrate). Replace caps and incubate at 35° C for 18–24 hours.

(6) After interpreting and recording positive results on the sides of the tube, perform the indole test by injecting 1 or 2 drops of Kovacs' reagent into the sucrose/indole compartment.

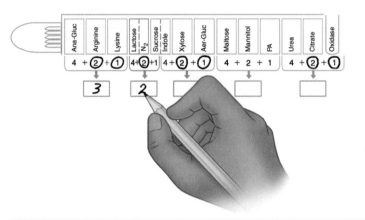

(7) After encircling the numbers of the positive tests on the Laboratory Report, total up the numbers of each bracketed series to determine the 5-digit code number. Refer to the *Biocode Manual* for identification of the unknown by using the code number.

Figure 47.2 (continued)

7. Break the wire at the notch by bending, as noted in step 4, figure 47.2. The portion of the wire remaining in the tube maintains anaerobic conditions essential for true fermentation.
8. With the retained portion of the needle, punch holes through the thin plastic coverings over the small depressions on the sides of the last eight compartments (sucrose/indole, xylose, aerobic glucose, maltose, mannitol, phenylalanine, urea, and citrate). These holes will enable aerobic growth in these eight compartments.
9. Replace both caps on the tube.
10. Incubate at 35° to 37° C for 24 hours, with the tube lying on its flat surface or upright. At the end of 24 hours, inspect the tube to check results and continue incubation for another 24 hours. The 24-hour check may be needed for doing confirmatory tests as required in the *Biocode Manual*. Occasionally, an Oxi/Ferm Tube II should be incubated longer than 48 hours.

Second Preiod

Evaluation of Tests

During this period, you will record the results of the various tests on your Oxi/Ferm Tube II, do an indole test, tabulate your results, use the *Biocode Manual*, and perform any confirmatory tests called for. Proceed as follows:

Materials

- Oxi/Ferm Tube II, inoculated and incubated
- Kovacs' reagent
- syringes with needles, or disposable Pasteur pipettes
- Becton-Dickinson *Biocode Manual* (a booklet)

1. Compare the colors of each compartment of your Oxi/Ferm Tube II with the lower tube illustrated in figure 47.1.

2. With a pencil, mark a small plus (+) or minus (−) near each compartment symbol on the white label on the side of the tube.

3. Consult table 47.1 for information as to the significance of each compartment label.

4. Record the results of all the tests on the Laboratory Report. *All results must be recorded before doing the indole test.*

5. **Indole Test** (illustration 6, figure 47.2): Do an indole test by injecting 2 or 3 drops of Kovacs'

Table 47.1 Biochemical Reactions of the Oxi/Ferm Tube II

REACTION	NEGATIVE	POSITIVE	SPECIAL REMARKS
Anaerobic Glucose			Positive fermentation is shown by change in color from green (neutral) to yellow (acid). Most oxidative-fermentative, Gram-negative rods are negative.
Arginine Dihydrolase			Decarboxylation of arginine results in the formation of alkaline end products that changes bromcresol purple from yellow (acid) to purple (alkaline). Grey is negative.
Lysine			Decarboxylation of lysine results in the formation of alkaline end products that changes bromcresol purple from yellow (acid) to purple (alkaline). Grey is negative.
Lactose			Fermentation of lactose changes the color of the medium from red (neutral) to yellow (acid). Most O/F Gram-negative rods are negative.
N₂Gas-production			Gas production causes separation of wax overlay from medium. Occasionally, the gas will also cause separation of the agar from the compartment wall.
Sucrose			Bacterial oxidation of sucrose causes a change in color from green (neutral) to yellow (acid).
Indole			The bacterial enzyme tryptophanase metabolizes tryptophan to produce indole. Detection is by adding Kovacs' reagent to the compartment 48 hours after incubation.
Xylose			Bacterial oxidation of xylose causes a change in color from green (neutral) to yellow (acid).
Aerobic Glucose			Bacterial oxidation of glucose causes a change in color from green (neutral) to yellow (acid).
Maltose			Bacterial oxidation of maltose causes a change in color from green (neutral) to yellow (acid).
Mannitol			Bacterial oxidation of this carbohydrate is evidenced by a change in color from green (neutral) to yellow (acid).
Phenylalanine			Pyruvic acid is formed by deamination of phenylalanine. The pyruvic acid reacts with a ferric salt to produce a brownish tinge.
Urea			The production of ammonia by the action of urease on urea increases the alkalinity of the medium. The phenol red in this medium changes from beige (acid) to pink or purple. Pale pink should be considered negative.
Citrate			Organisms that grow on this medium are able to utilize citrate as their sole source of carbon. Utilization of citrate raises the alkalinity of the medium. The color changes from green (neutral) to blue (alkaline).

reagent through the flat, plastic surface into the sucrose/indole compartment. Release the reagent onto the inside flat surface and allow it to drop down onto the agar.

If a Pasteur pipette is used instead of a syringe needle, it will be necessary to form a small hole in the Mylar film with a hot inoculating needle to admit the tip of the Pasteur pipette.

A positive test is indicated by the development of a **red color** on the surface of the medium or Mylar film within 10 seconds.

6. Record the results of the indole test on the Laboratory Report.

Laboratory Report

Follow the instructions in Laboratory Report 47 for determining the five-digit code. Use the *Biocode Manual* booklet for identifying your unknown.

Laboratory Report

Student: _____

Date: _____ Section: _____

47 O/F Gram-Negative Rods Identification: The Oxi/Ferm Tube II System

A. Results

1. Tabulation of Results and Code Determination

   Once you have marked the positive reactions on the side of the tube and circled the numbers that are assigned to each of the positive chambers, as indicated in the example below, add the numbers in each bracketed group to get the five-digit code.

   The final step is to look up the code number in the *Oxi/Ferm Tube II Biocode Manual* to determine the genus and species. If confirmatory tests are necessary, the manual will tell you which ones to perform.

   In the example below, the code number is 32303. If you look up this number in the *Biocode Manual* you will find on page 25 that the organism is *Pseudomonas aeruginosa*.

   Use this procedure to identify your unknown by applying your results to the blank diagrams provided.

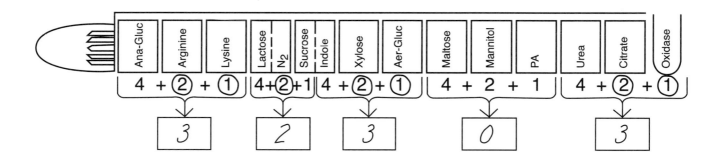

2. Additional Tabulation Blanks

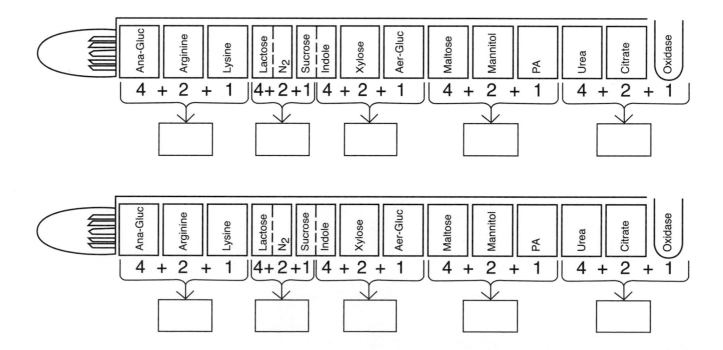

B. Short Answer Questions

1. What are the advantages and disadvantages of multitest systems for bacterial identification?

   _____

   _____

2. Under what circumstances would you choose the Oxi/Ferm Tube II System over the Enterotube II System (Exercise 46) for the identification of gram-negative bacilli?

   _____

   _____

3. If the five-digit code that is tabulated cannot be found in the *Oxi/Ferm Tube II Biocode Manual,* what might that indicate about the bacterial culture?

   _____

   _____

Staphylococcus Identification:
The API Staph System

The API Staph System, produced by bioMérieux of Raleigh, North Carolina is a reliable method for identifying 23 species of gram-positive cocci, including 20 clinically important species of staphylococci. This system consists of 19 microampules that contain dehydrated substrates and/or nutrient media. Except for the coagulase test, all the tests are important in the differentiation of *Staphylococcus, Kocuria* and *Micrococcus*.

Figure 48.1 illustrates two inoculated strips: the upper one just after inoculation and the lower one with positive reactions. Note that the appearance of each microcupule undergoes a pronounced color change when a positive reaction occurs.

Figure 48.2 illustrates the overall procedure. The first step is to make a saline suspension of the organism from an isolated colony. A Staph strip is then placed in a tray that has a small amount of water added to it to provide humidity during incubation. Next, a sterile Pasteur pipette is used to dispense 2 to 3 drops of the bacterial suspension to each microcupule. The inoculated tray is covered and incubated aerobically for 18–24 hours at 35–37° C. Finally a seven-digit profile number is obtained and used to determine the identity of the organism in Appendix D.

As simple as this system might seem, there are a few limitations that one must keep in mind. Final species determination by a competent microbiologist must take into consideration other factors such as the source of the specimen, the catalase reaction, colony characteristics, and antimicrobial susceptibility pattern.

Very often there are confirmatory tests that must also be made.

If you have been working with an unknown that appears to be one of the staphylococci, use this system to confirm your conclusions. If you have already done the coagulase test and have learned that your organism is coagulase-negative, this system will enable you to identify one of the numerous coagulase-negative species that are not identifiable by the procedures in Exercise 70.

First Period

Inoculations and Coagulase Test

Materials

- API Staph test strip
- API incubation tray and cover
- blood agar plate culture of unknown (must not have been incubated over 30 hours)
- blood agar plate (if needed for purity check)
- serological tube of 2 ml sterile saline
- test-tube rack
- sterile swabs (optional in step 2 on next page)
- squeeze bottle of tap water
- tubes containing McFarland No. 3 (BaSO$_4$) standard (see Appendix B)
- sterile Pasteur pipette (5 ml size)

1. If the coagulase test has not been performed, refer to Exercise 70, page 450, for the procedure and perform it on your unknown.

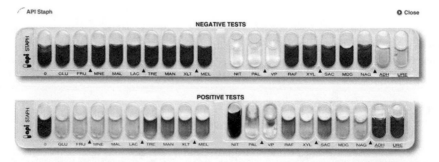

Figure 48.1 **Negative and positive results on API Staph test strips.**
Courtesy of bioMérieux, Inc.

(1) Use several loopfuls of organisms to make saline suspension of unknown. Turbidity of suspension should match McFarland No. 3 barium sulfate standard.

0.85% saline

(2) After labeling the end tab of a tray with your name and unknown number, dispense approximately 5 ml of tap water into bottom of tray.

(3) Place a STAPH test strip into the bottom of the moistened tray. Take care not to contaminate the microcupules with fingers when handling test strip.

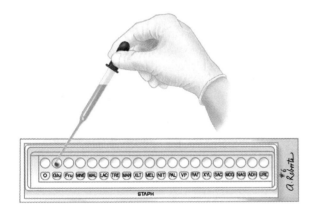

(4) With a Pasteur pipette dispense 2 to 3 drops of the bacterial suspension into each of the 19 microcupules. Cover the tray with the lid and incubate at 35–37° C for 18-24 hours.

(5) Once all results are recorded on Laboratory Report, total up positive values in each group to determine 7-digit profile. Consult chart VII, appendix D, to find unknown.

Figure 48.2 The API Staph procedure.

2. Prepare a saline suspension of your unknown by transferring organisms to a tube of sterile saline from one or more colonies with a loop or sterile swab. Turbidity of the suspension should match a tube of No. 3 McFarland barium sulfate standard.

Important: Do not allow the bacterial suspension to go unused for any great length of time. Suspensions older than 15 minutes become less effective.

3. Label the end strip of the tray with your name and unknown number. See illustration 2, figure 48.2.

4. Dispense about 5 ml of tap water into the bottom of the tray with a squeeze bottle. Note that the bottom of the tray has numerous depressions to accept the water.

5. Remove the API test strip from its sealed envelope and place the strip in the bottom of the tray.

6. After shaking the saline suspension to disperse the organisms, fill a sterile Pasteur pipette with the bacterial suspension.

7. Inoculate each of the microcupules with 2 or 3 drops of the suspension. If a purity check is necessary, use the excess suspension to inoculate another blood agar plate.

8. Place the plastic lid on the tray and incubate the strip aerobically for 18–24 hours at 35° to 37° C.

Second Period

(Five Hours Later)

During this period, the results will be recorded on the Laboratory Report, the profile number will be determined, and the unknown will be identified by looking up the number on the *Staph Profile Register* (or chart VII, Appendix D).

Materials

- API Staph test strip (incubated 18–24 hours)
- *Staph Profile Register*

1. After 18–24 hours incubation, read and interpret the test results (figure 48.3).
   a. For the Voges-Proskauer test, add 1 drop each of Barritt's A and Barritt's B to the V.P. ampule. Incubate 10 minutes. A positive test is indicated by a violet-pink color; a negative test shows no color change.
   b. For the nitrate test, add 1 drop each of nitrate A and nitrate B test reagents to the NIT ampule. Incubate 10 minutes. A positive test is a violet-pink color; a negative test shows no color change.
   c. For the alkaline phosphatase (PAL) test, and 1 drop each of ZYM A and ZYM B to the PAL ampule. Incubate 10 minutes. A positive test is violet in color; a negative test is yellow.
2. Record the results in the Laboratory Report.
3. Construct the profile number according to the instructions on the Laboratory Report and determine the name of your unknown.
   Use chart VII, Appendix D.

Disposal

Once all the information has been recorded be sure to place the entire incubation unit in a receptacle that is to be autoclaved.

Figure 48.3 **Test results of a positive strip.**
Courtesy of bioMérieux, Inc.

Laboratory Report

48 Staphylococcus Identification: The API Staph System

A. Results

1. Tabulation

   By referring to chart VII in Appendix D, determine the results of each test, and record these results as positive (+) or negative (−) in the Profile Determination Table below. Note that two more of these tables have been printed on the next page for tabulation of additional organisms.

	O 1	GLU 2	FRU 4	MNE 1	MAL 2	LAC 4	TRE 1	MAN 2	XLT 4	MEL 1	NIT 2	PAL 4	VP 1	RAF 2	XYL 4	SAC 1	MDG 2	NAG 4	ADH	URE	Lysost*
RESULTS																					

PROFILE NUMBER

GRAM STAIN ☐   COAGULASE ☐

MORPHOLOGY ☐   CATALASE ☐

Additional Information

Identification

*Lysostaphin-lysostatin resistance to be done as a separate test as described by the manufacturer. To be included to determine the seven-digit ID number.

2. Construction of Seven-Digit Profile

   Note in the above table that each test has a value of 1, 2, or 4. To compute the seven-digit profile for your unknown, total up the positive values for each group.

3. Final Determination

   Refer to the *Staph Profile Register* (chart VII, Appendix D) to find the organism that matches your profile number. Write the name of your unknown in the space below and list any additional tests that are needed for final confirmation. If the materials are available for these tests, perform them.

   Name of Unknown: _____

   Additional Tests: _____

	O 1	GLU 2	FRU 4	MNE 1	MAL 2	LAC 4	TRE 1	MAN 2	XLT 4	MEL 1	NIT 2	PAL 4	VP 1	RAF 2	XYL 4	SAC 1	MDG 2	NAG 4	ADH	URE	Lysost*
RESULTS																					

PROFILE NUMBER

GRAM STAIN ☐   COAGULASE ☐

MORPHOLOGY ☐   CATALASE ☐

Additional Information

Identification

*Lysostaphin-lysostatin resistance to be done as a separate test as described by the manufacturer. To be included to determine the seven-digit ID number.

	O 1	GLU 2	FRU 4	MNE 1	MAL 2	LAC 4	TRE 1	MAN 2	XLT 4	MEL 1	NIT 2	PAL 4	VP 1	RAF 2	XYL 4	SAC 1	MDG 2	NAG 4	ADH	URE	Lysost*
RESULTS																					

PROFILE NUMBER

GRAM STAIN ☐   COAGULASE ☐

MORPHOLOGY ☐   CATALASE ☐

Additional Information

Identification

*Lysostaphin-lysostatin resistance to be done as a separate test as described by the manufacturer. To be included to determine the seven-digit ID number.

B. Short Answer Questions

1. What are the advantages and disadvantages of multitest systems for bacterial identification?

2. Before using the API Staph System, what tests should be performed on bacteria to confirm that they are staphylococci?

3. What single test differentiates *Staphylococcus aureus* from other species of staphylococci? What is the expected result?

4. If the seven-digit code that is tabulated cannot be found in the Profile Determination Table (Appendix D: chart VII), what might that indicate about the bacterial culture?

Diversity and Environmental Microbiology

Bacteria are some of the most diverse organisms that inhabit the earth. They occupy just about every environmental niche examined from volcanic effluents in the deepest portions of the sea to the stomachs of animals and insects, whose very existence depends on the metabolic activities of these prokaryotic inhabitants. They produce most of the antibiotics that humans use to treat disease, and they are responsible for the cycling of minerals and elements such as nitrogen and sulfur in the environment. For example, they are the only organisms that can fix nitrogen from the air and return it to the biosphere for use in synthesizing biological molecules. They can exist in unique relationships with one another or they can even be predators on their own kind.

The following exercises reveal some of the diversity of bacteria and the roles that these organisms play in nature. These include the actinomycetes that produce antibiotics, the organisms involved in the fixation of nitrogen, the anaerobic photosynthetic bacteria, the sulfate-reducing bacteria, which will be studied in a model ecosystem called a Winogradsky column, and relationships that bacteria form that can benefit or harm another partner.

Isolation of an Antibiotic Producer:
The *Actinomyces*

The conquest of infectious disease was possible in part because of the discovery of antibiotics. Antibiotics are substances produced by one microorganism that kill or inhibits another microorganism. Most antibiotics that we use to treat infections have been isolated from microorganisms that inhabit the soil. Among these organisms are fungi that produce penicillin, bacteria that produce bacitracin, and the mycelial bacteria, or *Actinomyces,* that produce many of the broad-spectrum antibiotics such as streptomycin and chloramphenicol. The latter organisms are similar to fungi because they produce hyphae, which form an interwoven mass called mycelium. The mycelium comprises the visible colony of these organisms. The mycelium also anchors the colony to its substrate and transports nutrients throughout the colony. Individual hypha in the mycelium specialize to become reproductive structures that produce spores called conidia, or arthrospores, which disseminate to become a new colonies. The organisms belonging to this group are prokaryotes because they have a simple cell structure and contain peptidoglycan in their cell walls. On culture media, they grow as compact "leathery" colonies that penetrate into the agar. These bacteria also produce compounds such as geosmin that are responsible for the distinctive "earthy" odor of soil.

Not all antibiotics produced by microorganisms are useful in treating disease. Many of the antibiotics that have been isolated most certainly kill pathogens, but when tested they proved to be too toxic for use with humans. The search for new antibiotics is a constant endeavor and often begins with isolating soil microorganisms such as the *Actinomyces* and screening individual isolates for their ability to produce substances that inhibit pathogenic bacteria. If antibiotic producers are found, the inhibitory substance can be further characterized and tested to determine if it is a useful clinical compound that can be given to patients.

In the following exercise, you will use glycerol yeast extract agar to plate dilutions of soil. This medium selects for the *Actinomyces*. Isolated colonies will be picked from your plates and tested by cross-streaking a test bacteria to determine if the soil isolate is producing inhibitory compounds.

First Period

(Primary Isolation)

Unless the organisms in a soil sample are diluted sufficiently, the isolation of potential antibiotic producers is nearly impossible. As indicated in figure 49.1, it will be necessary to use a series of six dilution tubes to produce a final soil dilution of 10^{-6}. Proceed as follows:

Materials

per pair of students:
- 6 large test tubes
- 1 bottle of physiological saline solution
- 3 petri plates of glycerol yeast extract agar with 40 µg/ml cycloheximide
- L-shaped glass rod
- beaker of alcohol
- 6 1 ml pipettes
- 1 10 ml pipette

1. Label six test tubes 1 through 6, and with a 10 ml pipette, dispense 9 ml of saline into each tube.
2. Weigh out 1 g of soil and deposit it into tube 1.
3. Vortex mix tube 1 until all soil is well-dispersed throughout the tube.
4. Make a tenfold dilution from tube 1 through tube 6 by transferring 1 ml from tube to tube. Use a fresh pipette for each transfer and be sure to pipette-mix thoroughly before each transfer.
5. Label three petri plates with your initials and the dilutions to be deposited into them.
6. From each of the last three tubes transfer 1 ml to a plate of glycerol yeast extract agar.
7. Spread the organisms over the agar surfaces on each plate with an L-shaped glass rod that has been sterilized each time in alcohol with an open flame. Be sure to cool rod before using.

> **CAUTION:** Keep Bunsen burner flame away from beaker of alcohol. Alcohol fumes are ignitable. Be sure to flame the glass rod when finished.

8. Incubate the plates at 30° C for 7 days.

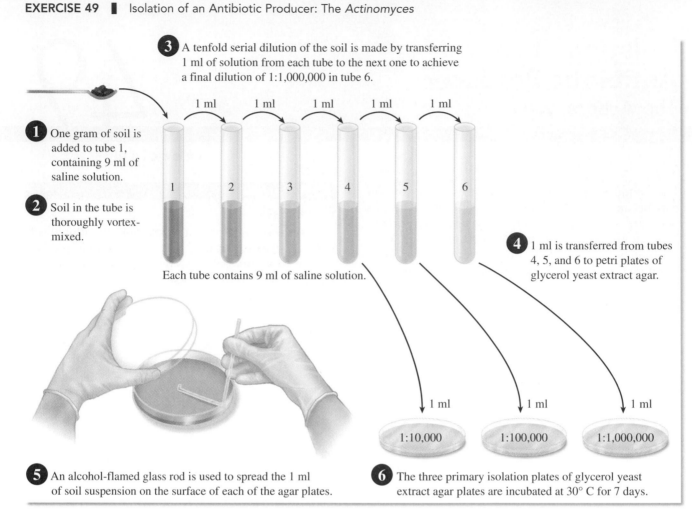

3 A tenfold serial dilution of the soil is made by transferring 1 ml of solution from each tube to the next one to achieve a final dilution of 1:1,000,000 in tube 6.

1 ml   1 ml   1 ml   1 ml   1 ml

1 One gram of soil is added to tube 1, containing 9 ml of saline solution.

2 Soil in the tube is thoroughly vortex-mixed.

Each tube contains 9 ml of saline solution.

4 1 ml is transferred from tubes 4, 5, and 6 to petri plates of glycerol yeast extract agar.

1 ml   1 ml   1 ml

1:10,000   1:100,000   1:1,000,000

5 An alcohol-flamed glass rod is used to spread the 1 ml of soil suspension on the surface of each of the agar plates.

6 The three primary isolation plates of glycerol yeast extract agar plates are incubated at 30° C for 7 days.

Figure 49.1 Primary isolation of antibiotic-producing *Actinomyces*.

Second Period

Colony Selection and Inoculation

The objective in this laboratory period will be to select *Actinomyces*-like colonies that may be antibiotic producers. The organisms will be streaked on trypticase soy agar plates that have been seeded with *Staphylococcus epidermidis*. After incubation, we will look for evidence of antibiosis. Students will continue to work in pairs. Figure 49.2 illustrates the procedure.

Materials

per pair of students:
- 4 trypticase soy agar pours (liquefied)
- 4 sterile petri plates
- TSB culture of *Staphylococcus epidermidis*
- 1 ml pipette
- 3 primary isolate plates from previous period
- water bath at student station (50° C)

1. Place four liquefied agar pours in water bath (50° C) to prevent solidification, and then inoculate each one with 1 ml of *S. epidermidis*.
2. Label the petri plates with your initials and date.
3. Pour the contents of each inoculated tube into petri plates. Allow agar to cool and solidify.
4. Examine the three primary isolation plates for the presence of *Actinomyces*-like colonies. They have a "leathery" appearance due to the presence of spores. They may be white or colored. Your instructor will assist in the selection of colonies.
5. Using a sterile inoculating needle, scrape spores from *Actinomyces*-like colonies on the primary isolation plates to inoculate the seeded TSA plates. Use inoculum from a different colony for each of the four plates.
6. Incubate the plates at 30° C until the next laboratory period.

Third and Fourth Periods

(Evidence of Antibiosis and Confirmation)

Examine the four plates you streaked during the last laboratory period. If you see evidence of antibiosis (inhibition of *S. epidermidis* growth), proceed as follows to confirm results.

Materials

- 1 petri plate of trypticase soy agar
- TSB culture of *S. epidermidis*

If antibiosis is present, make two streaks on the TSA plate as shown in figure 49.2. Make a straight line streak first with spores from the *Actinomyces* colony, using a sterile inoculating needle. Cross-streak with organisms from a culture of *S. epidermidis*. Incubate at 30° C until the next period.

Laboratory Reports

After examining the cross-streaked plate during the fourth period, record your results on Laboratory Report 49 and answer all the questions.

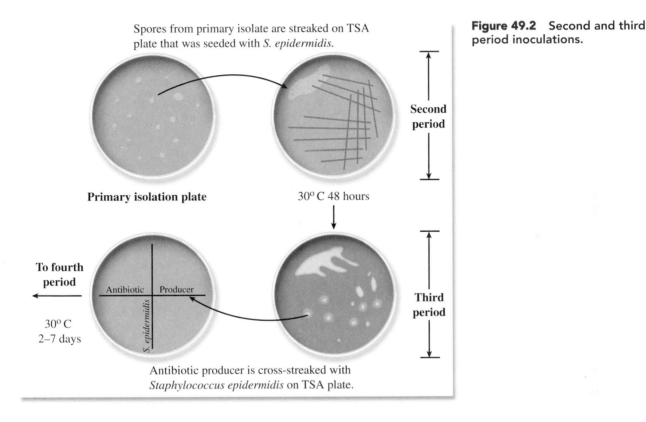

Spores from primary isolate are streaked on TSA plate that was seeded with *S. epidermidis*.

Primary isolation plate

30° C 48 hours

Second period

Third period

To fourth period

30° C 2–7 days

Antibiotic Producer

S. epidermidis

Antibiotic producer is cross-streaked with *Staphylococcus epidermidis* on TSA plate.

Figure 49.2 Second and third period inoculations.

339

Laboratory Report

Student: _____

Date: _____ Section: _____

49 Isolation of an Antibiotic Producer: The *Actinomyces*

A. Results

1. Describe the appearance of *Actinomyces*-like colonies on the agar plate. You may include a drawing.

_____

_____

2. Was there any evidence of antibiosis on your pour plates?

_____

_____

3. Describe what antibiosis looks like on the pour plate. You may include a drawing.

_____

_____

4. From the entire class, how many *Actinomyces*-like isolates demonstrated antibiosis?

_____

_____

5. Did the cross-streak plate(s) confirm your results?

_____

_____

B. Short Answer Questions

1. What is the role of microbial antibiotic production in nature?

   _____

   _____

2. Why does the antibiotic compound not harm the antibiotic-producing microbe?

   _____

   _____

3. What types of microbes are known to produce antibiotic compounds?

   _____

   _____

4. For what likely reason was *S. epidermidis*, but not *S. aureus*, chosen as the test organism?

   _____

   _____

Nitrogen Cycle:
Ammonification

The Nitrogen Cycle

Nitrogen is required by all living organisms for growth. It occurs in amino acids that comprise cell proteins and in nucleotide bases that make up the genetic material of the cell. It also occurs in vitamins that function as coenzymes. Most of the nitrogen present in the biosphere is not accessible to organisms for their metabolic needs because it is part of rock formations that do not turn over readily. Nitrogen is also a thermodynamically stable gas that makes up 78% of the atmospheric gases. Nitrogen can exist in various oxidation states that occur in the soil and water systems. These forms of nitrogen are transformed exclusively by the metabolic activities of microorganisms. The various transformations of nitrogen are part of the **nitrogen cycle** (figure 50.1), which insures that nitrogen in the environment is constantly turned over. These processes are ammonification, nitrification, denitrification, and nitrogen fixation.

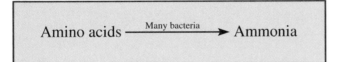

$$\text{Amino acids} \xrightarrow{\text{Many bacteria}} \text{Ammonia}$$

Ammonification   When living organisms die, their remains are decomposed in the soil by a variety of microorganisms. Proteins are degraded by proteases, liberating amino acids in which the amino group is removed by deamination to release ammonia. Nucleic acids are also degraded, releasing ammonia, and animal wastes such as uric acid are degraded to produce ammonia. The process of ammonia production in the soil is called ammonification. Bacteria and plants can assimilate the ammonia to synthesize new proteins and nucleic acids for growth. A variety of bacteria carry out ammonification in the soil, including *Bacillus, Pseudomonas*, and many others.

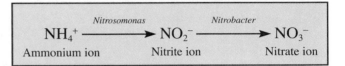

$$\underset{\text{Ammonium ion}}{NH_4^+} \xrightarrow{\textit{Nitrosomonas}} \underset{\text{Nitrite ion}}{NO_2^-} \xrightarrow{\textit{Nitrobacter}} \underset{\text{Nitrate ion}}{NO_3^-}$$

Nitrification   In the soil, ammonia is oxidized to nitrates and nitrites by the nitrifying bacteria in an aerobic process called **nitrification.** These are **chemolithotrophic** bacteria that derive their energy needs from the oxidation of ammonia and nitrite to fix carbon dioxide for their carbon requirements. *Nitrosomonas* oxidizes ammonia to nitrite and the resulting nitrite is then oxidized by *Nitrobacter* to nitrate to complete the process of soil nitrification. Because nitrites and nitrates can be used by plants as a source of nitrogen, these bacteria play a very important role in soil fertility. Farmers select for the activities of the nitrifying bacteria by tilling the soil, creating aerobic conditions that favor nitrification.

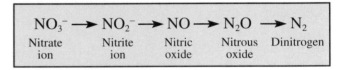

$$\underset{\substack{\text{Nitrate}\\\text{ion}}}{NO_3^-} \rightarrow \underset{\substack{\text{Nitrite}\\\text{ion}}}{NO_2^-} \rightarrow \underset{\substack{\text{Nitric}\\\text{oxide}}}{NO} \rightarrow \underset{\substack{\text{Nitrous}\\\text{oxide}}}{N_2O} \rightarrow \underset{\text{Dinitrogen}}{N_2}$$

Denitrification   There are anaerobic bacteria such as *Paracoccus denitrificans* and *Pseudomonas* that can use nitrate as a terminal electron acceptor in anaerobic or nitrate respiration. In this process called denitrification, nitrate is reduced in a series of steps to form nitrogen gas, which is released into the atmosphere. Denitrification occurs in waterlogged soils, which are anaerobic, and can account for significant losses of the nitrogen fertilizers applied to crops by farmers. The process is purposely promoted in sewage treatment facilities because it prevents the release of nitrates and nitrites into natural water systems where these ions can become toxic to fish and wildlife. Denitrification is an important part of the cycle because it insures that nitrogen in the atmosphere will not be depleted and the cycle will continue to run. Bacteria are the only known organisms to carry out denitrification.

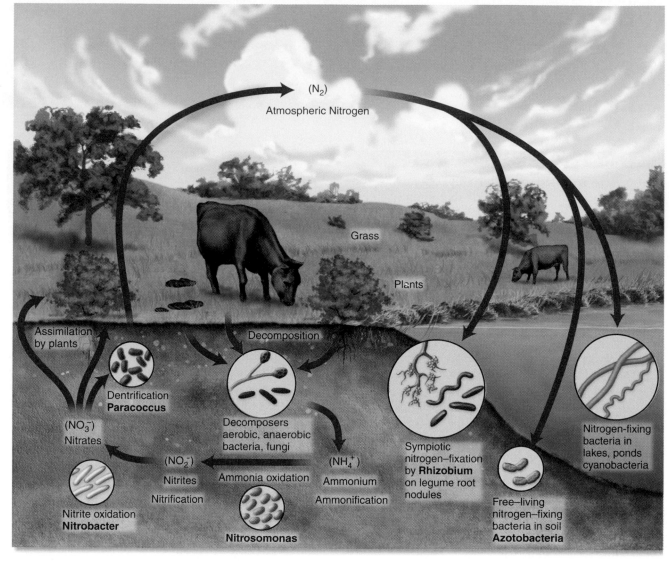

Figure 50.1

Nitrogen Fixation The conversion of nitrogen gas into ammonia is called nitrogen fixation and is restricted to only prokaryotes. This process requires significant energy and reducing power to break the triple bond of nitrogen gas. It is catalyzed by a complex enzyme **nitrogenase** that requires both iron and molybdenum for activity. Two groups of prokaryotes are capable of nitrogen fixation: (1) the free-living, nitrogen-fixing bacteria, exemplified by *Azotobacter*, *Clostridium*, and the filamentous cyanobacteria; and (2) symbiotic, nitrogen-fixing bacteria, represented by *Rhizobium*. The free-living, nitrogen-fixing bacteria occur in the soil and bodies of water where they fix nitrogen that is incorporated into bacterial cell material. *Rhizobium* lives in a **symbiotic** association with legume plants such as alfalfa, clover, and peanuts. The bacteria occur in root nodules on the plant where the bacteria fix nitrogen, which is used by the plant for nutritional purposes. In exchange, the plant provides nutrients to the bacteria and an environment in the nodule that is conducive to nitrogen fixation. Symbiotic nitrogen fixation accounts for most of the nitrogen fixed in the environment.

In the following exercises, you will isolate and study some of the organisms involved in ammonification, nitrogen fixation, and denitrification in the nitrogen cycle. The nitrifying bacteria will not be studied because they require exceedingly long times to grow.

Ammonification

Ammonification is the release of ammonia from biological molecules such as proteins, nucleic acids, and nitrogenous wastes. It occurs primarily in the soil as a result of the decomposition of the remains of dead plants and animals and their waste products. It is carried out by a variety of different soil microorganisms with bacteria such as *Bacillus, Clostridium,* and *Pseudomonas* being some of the primary organisms that participate in this process. These organisms elaborate proteases and nucleases that degrade proteins and nucleic acids into amino acids and nucleotide bases. These compounds are further degraded to release carbon intermediates and ammonia. Nitrogenous waste products such as urea and uric acid are also degraded to release ammonia. Ammonia can be assimilated by most plants and bacteria to synthesize new protein and nucleic acids. *Nitrosomonas*, a common soil bacterium, obtains its energy from the oxidation of ammonia in the process of nitrification.

In this exercise, you will demonstrate the ammonification process by inoculating peptone broth with some soil. After incubation for a few days, the cultures will be tested with Nessler's reagent for the presence of ammonia, and the various bacteria responsible for ammonification will be Gram stained and observed.

First Period

(Inoculation)

Materials

- 2 tubes of peptone broth
- rich garden soil

1. Inoculate one tube of peptone broth with a loopful of soil. Save the other tube for a control.
2. Incubate the tube at room temperature for 3–4 days and 7 days.

Second and Third Periods

(Ammonia Detection)

After 3 or 4 days, test the medium for ammonia with the following procedure. Repeat these tests again after a total of 7 days of incubation.

Materials

- Nessler's reagent
- bromthymol blue pH indicator and spot plate

1. Deposit a drop of **Nessler's reagent** into two separate depressions of a spot plate.
2. Add a loopful of the inoculated peptone broth to one depression and a loopful from the sterile uninoculated tube in the other. Interpretation of ammonia presence is as follows:

   Faint yellow color—small amount of ammonia
   Deep yellow—more ammonia
   Brown precipitate—large amount of ammonia

3. Prepare a Gram stain of the peptone broth culture. What kind of bacterial cells are present?
4. Record your results in Laboratory Report 50.

Laboratory Report

Student: _____

Date: _____ Section: _____

50 Nitrogen Cycle: Ammonification

A. Results

Record the presence or absence of ammonia in the tubes of media:

+	slight ammonia (faint yellow)
+ +	moderate ammonia (deep yellow)
+ + +	much ammonia (brown precipitate)
−	no ammonification

INCUBATION TIME	AMOUNT OF AMMONIA		pH	
	Control	Peptone with Soil	Control	Peptone with Soil
4 days				
7 days				

B. Short Answer Questions

1. Why do all living organisms require a nitrogen source?

_____

_____

2. Differentiate between the nitrogen-containing compounds utilized and produced during the following processes.

a. ammonification

_____

b. nitrification

_____

c. denitrification

_____

d. nitrogen fixation

_____

3. For ammonification, what materials are the sources of the ammonia?

_____

_____

4. What types of microbes carry out ammonification in the soil?

_____

_____

Symbiotic Nitrogen Fixation: *Rhizobium*

One of the most economically important relationships between bacteria and plants is the symbiotic association between the gram-negative bacteria *Rhizobium, Bradyrhizobium, Sinorhizobium, Mesorhizobium,* and *Azorhizobium* with various legumes such as clover, beans, peas, alfalfa, and peanuts. Normally these bacteria are found growing in the soil as part of the soil microflora. However, they are attracted to legumes with which they form a symbiotic relationship. The association of the bacterium with a legume is specific in that only a specific species of bacteria will infect a particular legume, for example, *Rhizobium trifolii* only infects clover. The process of infection begins by the bacterium binding to root hairs on the plant. Binding is facilitated by a protein called **ricadhesin** found on all *Rhizobium* and *Bradyrhizobium*. It is thought that the protein binds to calcium complexes on the root hair. Other binding factors such as carbohydrate-binding **lectins** may also play a role in binding, but their role does not appear to be as important as ricadhesin. Once attached to the root hair, the bacteria invade the deeper root tissue by an **infection thread** in a complicated process that eventually leads to the formation of small tumor-like structures called root nodules. In the nodule, the bacteria become surrounded by a plant membrane, forming a structure called a **symbiosome.** The symbiosome must form before nitrogen fixation can occur.

Inside the root nodule, the bacteria are stimulated to divide forming pleomorphic club-shaped cells called **bacteroids** (figure 51.1) because of the loss of some of their cell wall constituents. The bacteroids begin to elaborate nitrogenase, which fixes nitrogen gas. However, nitrogenase is irreversibly denatured by the presence of oxygen, and *Rhizobium* requires oxygen for the metabolism needed to fix nitrogen. This dilemma is solved by establishing microaerophilic conditions in the nodule via the presence of a pigment called **leghemoglobin,** which is both a plant and bacterial biosynthetic product. Leghemoglobin is an oxygen buffer that protects the oxygen-sensitive nitrogenase but at the same time supplies just enough oxygen to the bacteroids for their metabolism. This pigment also causes the nodule to appear pink in color.

In this symbiotic relationship, the plant supplies carbohydrates that are metabolized by the bacteroid to make ATP and reducing power required to fix nitrogen, and the ammonia resulting from fixation of nitrogen is used by the plant for its biosynthetic purposes. Plants that harbor these bacteria do not require the application of fertilizer. For centuries crops such as alfalfa have been rotated with food crops because it was empirically known that this practice improved soil fertility and increased the yield of the food crops. In fact, life as we know it is obligatorily dependent on this association between plant and bacteria, and life would cease in a short time without the association.

In the following exercise, you will study this fascinating relationship between *Rhizobium* and legumes by observing the bacteroids in crushed nodules.

Materials

- legume plants with nodules, such as clover, peanuts, alfalfa
- scalpels
- alcohol jars
- 1 ml water blanks
- Loeffler's methylene blue

1. Wash off any soil adhering to the roots of the legumes.
2. With a scalpel carefully remove several nodules from the roots of the plant. These can vary in size

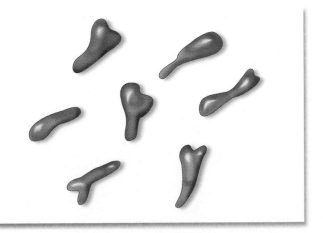

Figure 51.1 **The pleomorphic shape of bacteroids from root nodules.**

and on some plants are very small. Transfer the nodule to a microscope slide.

3. Slice the nodule in half with the scalpel blade. Note the color of the contents of the nodule.

4. Place a drop of water on the nodule and crush the nodule with the scalpel. Continue crushing until the water appears milky and turbid.

5. If any large plant debris is present, try to carefully remove it, leaving only the turbid nodule contents.

6. Allow the contents to air-dry. Heat-fix the smear.

7. Stain the smear with Loeffler's methylene blue for 5 minutes. Wash and blot dry with bibulous paper.

8. Observe the smear under oil immersion. Draw pictures of your observations for Laboratory Report 51.

Note: We will not try to culture bacteroids from nodules because this requires extensive treatment of the nodules with disinfectants prior to crushing the nodules due to the presence of soil bacteria on the surface of the nodule. In most cases, soil bacteria present on nodules overgrow *Rhizobium* on agar plates unless ample sterilization is carried out prior to culturing bacteroids.

Laboratory Report

Student: _____

Date: _____ Section: _____

51 Symbiotic Nitrogen Fixation: *Rhizobium*

A. Results

1. Draw and describe the stained bacteroids seen under oil immersion microscopy.

_____

_____

2. Define pleomorphism. Do you observe any evidence of pleomorphism?

_____

_____

B. Short Answer Questions

1. Define symbiosis using as an example the relationship between *Rhizobium* and legumes.

_____

_____

2. What is responsible for the specificity of the relationship between bacteria and plants?

_____

_____

3. What is responsible for the pink color of the root nodule contents?

_____

_____

Free-living Nitrogen Fixation:
Azotobacter

The first example of a free-living nitrogen-fixing bacterium was described in 1901 by M. W. Beijerinck, the Dutch microbiologist. Beijerinck inoculated soil into an **enrichment culture** composed of glucose and mineral salts but devoid of any nitrogen source except for atmospheric nitrogen. He set up his culture with a thin layer of culture medium covering the bottom of a flask, which selected primarily for aerobic nitrogen-fixing bacteria. In the culture, he observed bacterial cells characteristic of ***Azotobacter.*** Since Beijerinck's initial observations of *Azotobacter,* several other genera of aerobic, free-living, nitrogen-fixing bacteria have also been isolated and characterized. These include *Azomonas, Azospirillum,* and *Beijerinckia.*

These bacteria are interesting because they are strict aerobes but they carry out a process that is poisoned by the presence of oxygen. Nitrogenase is the enzyme that carries out the fixation of nitrogen. The enzyme is composed of two protein components:

Component I—**dinitrogenase** contains molybdenum and iron and is the component that reacts with nitrogen.

Component II—**dinitrogenase reductase** contains iron and is responsible for the reduction of nitrogen gas to form ammonia.

The enzyme is irreversibly denatured by oxygen and therefore it must function under anaerobic conditions. In fact, the nitrogenase from aerobic bacteria such as *Azotobacter* is just as sensitive to oxygen as nitrogenase from anaerobic nitrogen-fixing bacteria. Aerobic bacteria have evolved specialized mechanisms to protect their nitrogenase from oxygen.

There are three proposed mechanisms for how *Azotobacter* is able to protect its sensitive nitrogenase and still fix nitrogen in the presence of air:

Respiratory protection. *Azotobacter* has one of the highest respiratory rates known, even greater than brain tissue, which is exceedingly aerobic. It is thought that the organism's elevated and rapid use of oxygen prevents any "stray" oxygen in the cell from coming in contact with nitrogenase and denaturing it.

Conformational protection. If the nitrogenase in *Azotobacter* accidentally comes in contact with oxygen, the enzyme can combine with a specific protein in the cell that alters the conformation of nitrogenase, thus shielding the oxygen sensitive site and preventing it from interacting with oxygen.

Production of Slime Layers. When *Azotobacter* grows, it produces a copious extracellular polysaccharide slime. It has been postulated that this slime layer prevents too much oxygen from entering the cell where it can interact with the nitrogenase.

In enrichments using nitrogen-free mineral salts and a carbon source such as mannitol, species of *Azotobacter* usually predominate. These are large, gram-negative bacteria that may be rods or coccoid in shape and some may occur in arrowhead-shaped pairs. Many species of *Azotobacter* form resting stages called **cysts.** These cysts are similar to endospores in that they are resistant to drying and to ultraviolet light and ionizing radiation. Also like endospores, their endogenous respiration rate is low. However, they are dissimilar to endospores in that they are not heat resistant and they will actively metabolize some carbon sources if these are available. An additional feature of *Azotobacter* that can be used to differentiate the various species is the production of pigments that can vary from very dark to fluorescent yellow-green.

In the following exercise, you will set up soil enrichments using a nitrogen-free mineral salts supplemented with either 1% mannitol or 1% glucose. Different carbon sources often yield different species of *Azotobacter.* From this medium you will isolate *Azotobacter,* and it will be further characterized by observing isolated colonies, wet mounts, and Gram stains of your isolates. You will also incubate your isolate for a prolonged period of time to determine if it forms cysts and/or pigments.

Materials

- garden soil
- nitrogen-free mineral salts agar supplemented with either 1% mannitol or 1% glucose
- flat-sided bottles
- slides and coverslips
- Gram reagents
- 1 ml sterile saline tubes

1. Inoculate about 0.1 gm of fresh soil into a flat-sided bottle with nitrogen-free salts supplemented with either 1% mannitol or 1% glucose.

Figure 52.1 Enrichment and isolation procedure for *Azotobacter* and *Azomonas.*

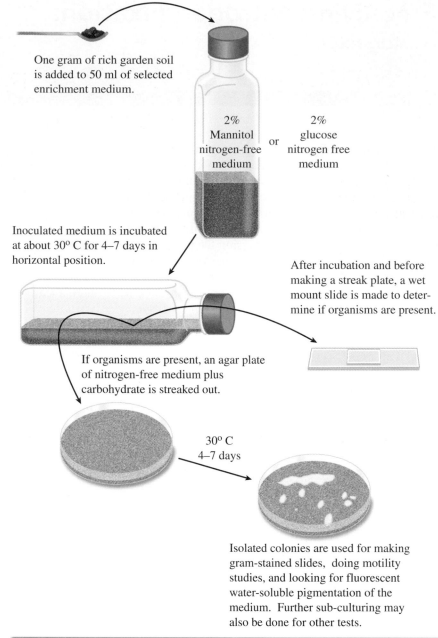

One gram of rich garden soil is added to 50 ml of selected enrichment medium.

2% Mannitol nitrogen-free medium or 2% glucose nitrogen free medium

Inoculated medium is incubated at about 30° C for 4–7 days in horizontal position.

After incubation and before making a streak plate, a wet mount slide is made to determine if organisms are present.

If organisms are present, an agar plate of nitrogen-free medium plus carbohydrate is streaked out.

30° C
4–7 days

Isolated colonies are used for making gram-stained slides, doing motility studies, and looking for fluorescent water-soluble pigmentation of the medium. Further sub-culturing may also be done for other tests.

Shake the bottle to disperse the soil and incubate the bottle on its side for 2–5 days at ambient temperature without shaking. Make sure to loosen the bottle cap so that air is accessible to the culture (figure 52.1).

2. Examine the cultures. Look especially for thin films of growth on the surface of the medium. Prepare wet mounts and examine them under phase-contrast microscopy. Note the presence of any large oval or rod-shaped cells. The cells may also be spherical and occur in pairs. Prepare Gram stains of the culture. Are there other kinds of bacterial cells in the medium?

3. Streak the culture onto Nitrogen-free mineral salts supplemented with the same carbohydrate that was used for the initial enrichment. Incubate the plates for 48 hours at ambient (room) temperature.

4. Examine the colonies on the plate. Look for large, translucent, mucoid colonies. Prepare wet mounts and examine them with a phase-contrast microscope. Compare your results with those of students who used a different carbohydrate. Does the morphology of the cells in the wet mounts appear the same or do they differ? Prepare Gram stains.

5. Transfer a single mucoid colony to a tube containing 1 ml of sterile saline and disperse the

cells. Inoculate a flask of N-free mineral salts supplemented with the carbohydrate used in your initial enrichment. Incubate the culture at ambient temperature for at least two weeks. Also inoculate an agar slant of N-free mineral salts plus appropriate carbohydrate with the culture in the sterile saline.

6. Examine the broth culture periodically during the two weeks. Have any changes in cell morphology occurred in the culture? Has the culture produced cysts? Was there a difference in the cultures grown on different carbohydrates?

7. Did the slant culture produce any pigment? If so, what color is the pigment? Compare the pigment color of your isolate to those in table 52.1. Do you have enough information to make a tentative identification of your organism based on descriptions in the table? Was there a difference in pigment production between cultures grown on different carbon sources?

Note in table 52.1 that two of the water-soluble pigments are fluorescent: one is yellow-green and the other is blue-white. To observe fluorescence, you must expose the cultures to ultraviolet light (wavelength 364 nm) in a darkened room. The characteristics of pigment production in each species may be limited by certain factors, as indicated below:

Brown-black: If the colonies produce this hue of diffusible pigment without becoming red-violet, the organism is *A. nigricans*. Although the table indicates that *A. insignis* can produce the brown-black pigment, it can do so only if the medium contains benzoate.

Brown-black to red-violet: As indicated in the table, *A. nigricans* and *A. armeniacus* are the only genera that produce this type of pigment. Motility is a good way to differentiate these two species.

Red-violet: Although table 52.1 reveals that five species can produce this color of diffusible

Table 52.1 Differential characteristics of the Azotobacteraceae

		Cysts	Motility	Long Filaments	Brown-Black	Brown-Black to Red-Violet	Red-Violet	Green	Yellow-Green Fluorescent	Blue-White Fluorescent
									Water-Soluble Pigments	
Azotobacter	A. chroococcum	+	+	−	−	−	−	−	−	−
	A. vinelandii	+	+	−	−	−	d	d	+	−
	A. beijerinckii	+	−	−	−	−	−	−	−	−
	A. nigricans	+	−	−	d	+	d	−	−	−
	A. armeniacus	+	+	−	−	+	+	−	−	−
	A. paspali	+	+	+	−	−	+	−	+	−
Azomonas	A. agilis	−	+	−	−	−	−	−	+	+
	A. insignis	−	+	−	d¹	−	d	−	d	−
	A. macrocytogenes	−	+	−	−	−	−	−	d	d

d = 11%–89% positive
d¹ = 11%–89% positive on benzoate
Source: From *Bergey's Manual of Systematic Biology*, volume 1, section 4. Reprinted by permission.

pigment, one (*A. insignis*) cannot produce it on the medium we used. A red-violet isolate is unlikely to be *A. paspali* because this organism has been isolated from the rhizosphere of only one species of grass (*Paspalum notatum*). Thus, isolates that produce this pigment are probably one of the other three in the table.

Green: Note that only *A. vinelandii* can produce this water-soluble pigment; however, only 11%–89% of them produce it.

Yellow-green fluorescent: *A. vinelandii*, *A. paspali*, and all species of *Azomonas* are able to produce this pigment on the medium we used. Check for fluorescence with an ultraviolet lamp in a darkened room.

Blue-white fluorescent: Note in table 52.1 that two species of *Azomonas* can produce this type of diffusible pigment; no *Azotobacter* are able to produce it. Check for fluorescence with an ultraviolet lamp in a darkened room.

Laboratory Report

Student: _____

Date: _____ Section: _____

52 Free-Living Nitrogen Fixation: *Azotobacter*

A. Results

Morphology of cells from enrichment I

Description of colonies on nitrogen-free agar plates

Morphology of cells from colonies on nitrogen-free agar plates

Morphology of cells from cultures incubated for 2–3 weeks (cysts)

Description of any pigment produced by the culture

B. Short Answer Questions

1. What causes the denaturation of the nitrogenase enzyme of *Azotobacter*? Why is this unusual?

   _____

   _____

2. Name three ways in which *Azotobacter* protects its nitrogenase against denaturation.

   _____

   _____

3. What happens if nitrogen salts are added to the first enrichment? What organisms might grow?

   _____

   _____

4. *Clostridium,* which is capable of fixing nitrogen, often appears after the growth of *Azotobacter* in these aerobic cultures. How is this possible?

   _____

   _____

Denitrification:
Paracoccus denitrificans

Some bacteria found in the soil and in sewage treatment facilities can use nitrate as an alternative electron acceptor to carry out a type of anaerobic respiration called **nitrate respiration.** This process is called **denitrification** and results in the metabolic reduction of the nitrate ion to nitrogen gas, which is returned to the atmosphere. Nitrate respiration occurs in several enzymatic steps. This involves enzymes that interface with an electron transport system where the nitrate ion serves as the terminal electron acceptor, just as oxygen serves this function in aerobic respiration. These bacteria are facultative because if oxygen is present, nitrate respiration is repressed and these bacteria will grow by normal respiration. Denitrifiers can grow on organic acids, such as succinate, and use nitrate reduction to anaerobically produce ATP for metabolic energy needs. However, denitrification is a **dissimilatory** process because nitrogen is not incorporated into cell material but rather is a metabolic end product.

Denitrification has been identified only in bacteria and is the primary means by which nitrogen gas in the atmosphere is formed biologically. Some organisms such as *Escherichia coli* can partially reduce nitrate to nitrite, which is the basis for the nitrate test used to taxonomically characterize this organism. However, *Paracoccus denitrificans* and *Pseudomonas stutzeri* are true denitrifiers because they completely reduce nitrate to nitrogen gas. These bacteria have been isolated from a variety of habitats worldwide, including soils in the arctic and antarctic regions.

Agriculturally, denitrification can be detrimental because it is a primary means by which nitrogen fertilizers such as ammonium nitrate are lost from soils. This can result in great economic losses to farmers. However, in sewage treatment facilities, denitrification is a beneficial and preferable process because it prevents the release of toxic nitrates and nitrites into natural waters such as lakes and streams where these ions can enhance the overgrowth of algae and be toxic to fish and other organisms in the ecosystem. Furthermore, without the return of nitrogen to the atmosphere, nitrogen in the air would become depleted in a very short time.

In the following exercise, you will inoculate a nitrate-succinate mineral salts medium to enrich for denitrifying bacteria. By inoculating a second enrichment culture with an inoculum from the first culture, an almost pure culture of a denitrifier can be obtained.

Materials

- fresh soil samples
- nitrate-succinate mineral salts broth
- nitrate-succinate mineral salts agar plates
- sterile glass stoppered bottles (60 ml stain bottles)
- GasPaks

First Period

1. Add 1 g of soil to the nitrate-succinate broth and shake vigorously. Allow the contents of the flask to settle (figure 53.1).
2. Carefully decant the supernatant into a glass stoppered bottle, filling the bottle to its total capacity. When the glass stopper is inserted into the bottle, the liquid contents should be expelled from the neck of the bottle. Do not allow any air to become entrapped between the stopper and the medium surface (figure 53.1).
3. Place the bottle on a petri dish lid or bottom and incubate the cultures at 30° C until the next laboratory.

Second Period

1. Examine the culture for the presence of gas. A stream of nitrogen gas bubbles can often be seen streaming from the bottom of the bottle and rising to the top of the bottle (figure 53.1). If enough nitrogen is produced, it can displace the medium, forcing the glass stopper from the bottle.
2. Prepare Gram stains of the culture. What kind of bacterial cells are present? *Paracoccus* is a short, gram-negative rod.

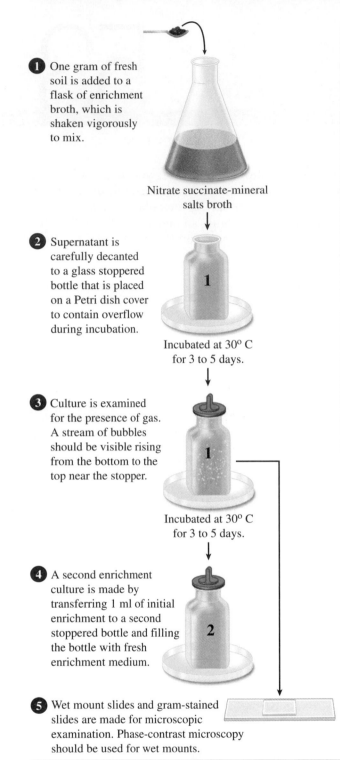

1. One gram of fresh soil is added to a flask of enrichment broth, which is shaken vigorously to mix.

Nitrate succinate-mineral salts broth

2. Supernatant is carefully decanted to a glass stoppered bottle that is placed on a Petri dish cover to contain overflow during incubation.

Incubated at 30° C for 3 to 5 days.

3. Culture is examined for the presence of gas. A stream of bubbles should be visible rising from the bottom to the top near the stopper.

Incubated at 30° C for 3 to 5 days.

4. A second enrichment culture is made by transferring 1 ml of initial enrichment to a second stoppered bottle and filling the bottle with fresh enrichment medium.

5. Wet mount slides and gram-stained slides are made for microscopic examination. Phase-contrast microscopy should be used for wet mounts.

Figure 53.1 Procedure for culturing *Paracoccus denitrificans* from a soil sample

3. Prepare a second enrichment by aseptically transferring 1 ml of culture from the first enrichment to a sterile glass stoppered bottle. Fill the bottle as you did before, not allowing any air to become entrapped between the stopper and medium (figure 53.1).
4. Place the bottle on a petri plate lid or bottom and incubate at 30° C until the next laboratory.

Third Period

1. Has the culture produced any nitrogen gas? Prepare Gram stains and wet mounts. What do you observe? Are the cells gram-negative, short rods? Are the cells motile?
2. Streak a nitrate-succinate mineral salts agar plate from the second enrichment. Incubate the plates in a GasPak at 30° C until next laboratory (figure 53.2).

Fourth Period

1. Describe the colonies on the plate. *Paracoccus* produces small whitish colonies on agar media.
2. Prepare Gram stains. Are the cells still gram-negative short rods?
3. Record all your observations in Laboratory Report 53.

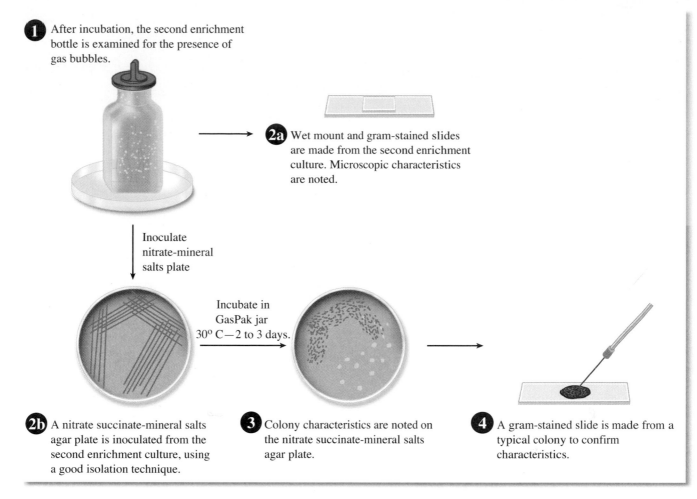

1 After incubation, the second enrichment bottle is examined for the presence of gas bubbles.

2a Wet mount and gram-stained slides are made from the second enrichment culture. Microscopic characteristics are noted.

Inoculate nitrate-mineral salts plate

Incubate in GasPak jar 30° C—2 to 3 days.

2b A nitrate succinate-mineral salts agar plate is inoculated from the second enrichment culture, using a good isolation technique.

3 Colony characteristics are noted on the nitrate succinate-mineral salts agar plate.

4 A gram-stained slide is made from a typical colony to confirm characteristics.

Figure 53.2 Procedure for getting a pure culture out of second enrichment culture.

Laboratory Report

Student: _____

Date: _____ Section: _____

53   Denitrification: *Paracoccus denitrificans*

A. Results

1. **First enrichment culture appearance.** Describe the appearance of the first enrichment culture in the second laboratory period: _____

    _____

    _____

    _____

2. **Microscopic examination:** Describe the appearance of the cells as seen on

    Wet mount: _____

    _____

    Gram-stained slide: _____

    _____

3. **Second enrichment culture appearance.** Describe the appearance of the second enrichment culture after five days incubation: _____

4. **Agar plate colonies.** Describe the characteristics of the colonies on the nitrate-succinate mineral salts agar: _____

    Do these characteristics match those of *Paracoccus denitrificans?* _____

5. **Final Gram-stained slide.** How do the microscopic characteristics of your organism match the description of *Paracoccus denitrificans?* _____

B. Short Answer Questions

1. What is the product of denitrification? _____

    _____

    _____

2. Under what conditions will denitrifying bacteria carry out denitrification? _____

    _____

    _____

3. Why are denitrifying bacteria costly to farmers? _____

    _____

    _____

4. Why are denitrifying bacteria essential to the existence of life on planet earth? _____

_____

_____

5. Of what value is denitrification to the organism? _____

_____

_____

The Winogradsky Column

The Winogradsky column is a model ecosystem that can be used to study a variety of diverse bacteria such as the sulfate reducers, photosynthetic bacteria, and chemolithotrophs. It is named for the Russian microbiologist Sergei Winogradsky who used a similar device to investigate soil bacteria. From his studies, he developed the concept of **chemoautotrophy,** which defined bacteria that fix carbon dioxide for their carbon needs and obtain energy for cell growth by oxidizing inorganic ions. Prior to his studies, autotrophic metabolism was known to occur only in plants.

To set up a Winogradsky column, various salts, a source of sulfate, and mud arc mixed with a fermentable substrate such as cellulose. The mixture is delivered to a glass column that is filled halfway to capacity. The column is then filled with water collected with the mud and illuminated with incandescent light for several weeks. With time, dramatic changes occur in the appearance of the column as various bacterial groups begin to grow (figure 54.1). The various groups of bacteria that develop in the column are listed below.

Algae and Cyanobacteria

In the water on the surface of the column, algae and cyanobacteria initially grow, producing oxygen, which keeps the upper portion of the column aerobic.

Cellulose Digesters

Cellulose is degraded and its end products are fermented by a number of bacteria that occur in the anaerobic mud at the bottom of the column. For example, *Cellulomonas* degrades cellulose and *Clostridium* and other bacteria then ferment the monosaccharides derived from cellulose to produce organic acids such as lactate and acetate.

Sulfate-Reducing Bacteria

Lactate produced by fermentation and the sulfate in the column are utilized by anaerobic sulfate-reducing bacteria, such ***Desulfovibrio desulfuricans.*** These bacteria carry out **sulfate respiration** in which sulfate is reduced to H_2S. In this process, sulfate serves as a terminal electron acceptor and its reduction is interfaced with electron transport to produce ATP, needed for cell growth. Some of the H_2S

produced by these organisms reacts with various metals in the mud to form a black precipitate, which causes the bottom of the column to turn a jet black. H_2S that diffuses upward in the column is used by other bacteria for diverse metabolic needs.

Phototrophic Purple Sulfur Bacteria

The phototrophic purple sulfur bacteria such as *Chromatium* can utilize the hydrogen sulfide for **anoxygenic photosynthesis.** This kind of photosynthesis occurs under anaerobic conditions because oxygen is not an end product. Hydrogen sulfide serves as a source of reducing power and is converted by these organisms to elemental sulfur, which is stored inside of the cell as sulfur granules. The photosynthetic pigments, **bacteriochlorophyll** and **carotenoids,** are produced by

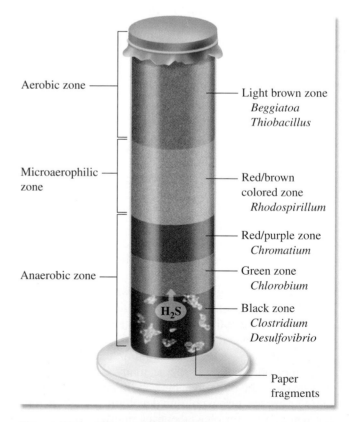

Aerobic zone

Microaerophilic zone

Anaerobic zone

H_2S

Light brown zone
Beggiatoa
Thiobacillus

Red/brown colored zone
Rhodospirillum

Red/purple zone
Chromatium

Green zone
Chlorobium

Black zone
Clostridium
Desulfovibrio

Paper fragments

Figure 54.1 Winogradsky's Column.

these bacteria and these pigments impart spectacular colors, such as purple, red, or brown to areas in the column where these bacteria are growing.

Phototrophic Green Sulfur Bacteria

This group of organisms also grows by anoxygenic photosynthesis. *Chlorobium* uses H_2S as a source of reducing power converting it to elemental sulfur, but unlike the purple sulfur bacteria, the green sulfur bacteria store the sulfur outside the cell. It is interesting that they do not fix carbon dioxide by the Calvin cycle but rather acquire their carbon by fixing carbon dioxide into intermediates in the Krebs cycle. They grow in the anaerobic portion of the column in greenish zones or patches in the mud due to the presence of bacteriochlorophyll.

Purple Nonsulfur Bacteria

The purple nonsulfur bacteria such as *Rhodobacter* and *Rhodospirillum* are the most metabolically diverse of the phototrophic bacteria. Normally they grow photoheterotrophically using various organic compounds (e.g., succinate, glutamate) as a carbon source, and they obtain their energy from anoxygenic photosynthesis. They can also grow photoautotrophically using carbon dioxide for a carbon source and hydrogen gas or low concentrations of sulfide as reducing power. If light becomes very limited, they can completely switch metabolic gears and grow by heterotrophic means on organic acids using respiration. These bacteria can be found in microaerophilic areas in the column, where they grow as brown to reddish cultures in the water or as patches in the mud where it interfaces with the water.

Chemolithotrophic Bacteria

In the aerobic zone of the column, some chemolithotrophic bacteria will begin to grow with time. Bacteria such as *Thiobacillus* and *Beggiatoa* can be found in the aerobic region at the water surface. The chemolithotroph *Thiobacillus* fixes carbon dioxide for its carbon requirements and obtains energy by oxidizing reduced forms of sulfur (e.g., H_2S) to produce sulfate or sulfuric acid.

In the following exercises you will set up a Winogradsky column, and after its development you will subculture some of the bacteria that grow in the column. These will include the purple nonsulfur bacteria and the sulfate-reducing bacteria. The column can also be used to simply observe some of the many diverse bacteria that grow in the column. This can be accomplished by taking samples during column development and preparing wet mounts for phase-contrast microscopy.

First Period

Materials

- graduated cylinder (100 ml size)
- cellulose (cellulose powder, filter paper, or newspaper)
- calcium sulfate, calcium carbonate, dipotassium phosphate
- fresh mud and water collected from various sources
- beakers (250 ml)
- glass stirring rods
- aluminum foil
- incandescent lamps (60–75 watt)

1. Using some source of cellulose, prepare a thick slurry in a beaker by adding water. If you are using paper, tear it into small pieces and macerate it in a small volume of water. If you are using cellulose powder, start with 1–2 g of powder in a small amount of water. A slurry is thick but it is not a paste.
2. Fill the graduated cylinder with the cellulose slurry until it is one-third full.
3. To 200 g of mud, add 1.64 g of calcium sulfate and 1.3 g each of calcium carbonate and dipotassium phosphate. Make a note of the source of your mud.
4. Add some water collected with the mud ("self" water) to your chemical-mud mixture and mix the ingredients well with a stirring rod.
5. Pour the mud into the column on top of the cellulose slurry.
6. Using a glass rod, gently mix the slurry and mud. As you mix, start to pack the mixture into the column. As you pack the column, you may find that you need to add more "self" water to the mixture. The combined slurry/mud/water should fill about two-thirds of the column when you are finished.
7. Top off the column with "self" water until it is 90% full.
8. Cover the column with a piece of aluminum foil to prevent evaporation. Record the initial appearance of the column in Laboratory Report 55.
9. Wrap the sides of the column in aluminum foil to exclude light.
10. Incubate the column at room temperature for two weeks.

Two Weeks Later

1. Remove the foil from the column and note its appearance. The bottom of the column should be turning black due to the presence of the sulfate

reducing bacteria, such as *Desulfovibrio*, that are producing H_2S.

2. With the column uncovered, place a 60–75 watt bulb a few inches from the cylinder. Continue incubation at room temperature for several weeks.

Subsequent Examinations

1. Examine the column periodically. Note the formation of red, brown, purple, or green areas in the mud or water. These will be the anoxygenic photosynthetic bacteria.

2. Prepare wet mounts from the water at the top of the column and examine them by phase-contrast microscopy. The chemolithotrophs, such as *Thiobacillus* and *Beggiatoa*, can be found growing here. Compare your observations to the descriptions of these organisms given below.

Beggiatoa: colorless cells 1 to nearly 25 μm in diameter and 2–10 μm in length, occur in filaments with a constant width. Filaments may contain 50 cells or more. Cells are often wider than they are long. Filaments may occur singly or in cottony masses. Cells may contain sulfur granules when grown in the presence of H_2S.

Thiobacillus: gram-negative, rod-shaped cells (0.5 × 1–4.0 μm); some species are motile. Energy is derived from the oxidation of one or more reduced sulfur compounds. Sulfate is the end product of sulfur compound oxidation. All species fix carbon dioxide by the Calvin-Benson cycle.

3. Save the column and use it as a source of inoculum for Exercise 55, the culturing of the purple nonsulfur bacteria, and Exercise 56, the sulfate reducers.

Laboratory Reports follow Exercise 55 and Exercise 56.

Purple Nonsulfur Photosynthetic Bacteria

The anoxygenic photosynthetic bacteria are a diverse group of bacteria composed of the green sulfur bacteria, the purple sulfur bacteria, and the purple nonsulfur bacteria. Their photosynthetic metabolism is anoxygenic because they do not produce oxygen from splitting water. Reducing power for their photosynthetic metabolism comes from using hydrogen, reduced forms of sulfur, and in some cases organic acids. They grow in anaerobic environments, especially in lakes and ponds, where fermentation is actively occurring and where there is an abundant source of reduced sulfur compounds such as H_2S.

Taxonomically, they are characterized by the photosynthetic pigments that they synthesize, the type of photosynthetic membrane or apparatus that is elaborated in the cell, and other physiological characteristics. All of the organisms in this group synthesize **bacteriochlorophyll** and some also synthesize **carotenoids.** The combination of these pigments imparts rather spectacular colors to cultures of these bacteria, which include brown, purple-reds, red, and green hues.

Bacteriochlorophyll absorbs visible light in the infrared part of the spectrum at 800 to 1,000 nm, whereas chlorophyll absorbs light at 600–700 nm. A consequence of this fact is that the anoxygenic photosynthetic bacteria, which grow in the deeper anaerobic regions of a lake or pond, do not compete for the same wavelengths of light as algae and cyanobacteria, which are growing on the surface of the water. Carotenoids function as accessory pigments in these bacteria to assist in harvesting more light energy. Carotenoids also protect the photosynthetic apparatus from solar damage.

The green sulfur bacteria and the purple sulfur bacteria are primarily photoautotrophs that fix carbon dioxide for their carbon needs and derive reducing power from reduced sulfur or hydrogen. They only grow under strict anaerobic conditions and are difficult to culture without specialized equipment and media that must be purged of oxygen before inoculation. In contrast, the purple nonsulfur bacteria can be cultured under microaerophilic conditions that can be easily established by filling test tubes with culture media filled to full capacity. These organisms usually grow as photoheterotrophs whereby they use organic acids such as succinate for their carbon requirements and gain their energy by photosynthetic reactions. However, some can also grow as strict photoautotrophs, utilizing carbon dioxide and reduced sulfur compounds or hydrogen for growth.

In the following exercise, you will subculture the purple nonsulfur bacteria from the Winogradsky column that you developed in Exercise 54 (figure 54.1). You will collect pigmented material from the column and inoculate it into a succinate mineral salts medium, which will be exposed to incandescent light.

Materials

- 5 screw-cap test tubes (13 × 200 mm size)
- 200 ml of purple nonsulfur enrichment media
- wide-mouth pipettes

1. Label the screw-cap test tubes with the colors of the column areas that you intend to subculture (i.e., brown, purple, red, green). If such areas do not exist, you should collect mud samples from the black areas at the mud-water interface.
2. Fill the screw-cap test tubes to about two-thirds full with the purple nonsulfur enrichment medium.
3. Using a wide-mouth pipette, collect about 1 g of mud or several milliliters of water containing pigmented growth from a Winogradsky column. Deliver the samples to the respective test tubes that you have labeled denoting the specific pigmented areas (figure 55.1).
4. After inoculation, fill the screw-cap test tubes to capacity with additional purple nonsulfur enrichment medium. Tighten each cap and mix the contents of the tube.
5. Place the tubes in a test-tube rack and position the rack a few inches from a 60–75 watt incandescent lamp. Incubate the tubes at room temperature for several days to a week.
6. When the tubes develop coloration, prepare wet mounts and observe them under a phase-contrast microscope. Record your results in Laboratory Report 55. Compare your observations to the following descriptions of purple nonsulfur bacteria that can occur in columns:

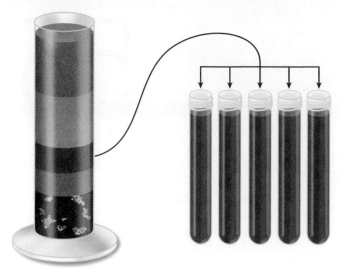

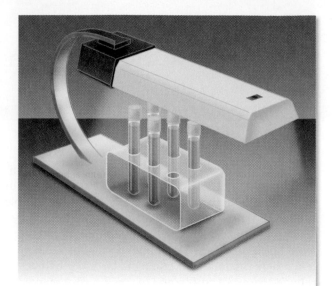

Winogradsky column

1 With a wide mouth pipette deliver approximately 1 gram of mud from each colored layer to a tube containing Rhodospirillaceae enrichment medium.

2 Incubate the inoculated tubes at room temperature while exposed to a 75 watt lamp for 3 to 7 days.

3 Make wet mount slides from each tube and examine with a phase-contrast microscope.

Figure 55.1   **Procedure for subculturing and microscopic examination of the purple nonsulfur photosynthetic bacteria**

Rhodobacter Cells are gram-negative, ovoid to rod-shaped, 0.5–1.2 μm. Some are motile while others are nonmotile. Cells are yellowish-brown when grown photosynthetically.

Rhodocyclus Cells are slender, curved, or straight thin rods, 0.3–1.0 μm. They are non-motile or motile by means of polar flagella, and produce bacteriochlorophyll and carotenoids.

Rhodomicrobium These are ovoid to elongate-ovoid bacteria, 1.0–1.2 μm. Cells reproduce by developing into filaments several times

the length of individual cells. They produce bacteriochlorophyll and carotenoids.

Rhodopseudomonas Cells are rod-shaped, 0.6–2.5 μm wide and 0.6–5.0 μm long, motile or non-motile. Cultures are brown, red, or green in the case of *Rhodopseudomonas viridis*.

Rhodospirillum Cells are spiral, 10.7–1.5 μm wide and motile. Cells produce bacteriochlorophyll and carotenoids. Cultures are red to brown.

Laboratory Report

Student: _____

Date: _____ Section: _____

55 Purple Nonsulfur Photosynthetic Bacteria

A. Observations:

1. **Column Appearance:** Describe in a few words the appearance of the Winogradsky column during the

   First period: _____

   _____

   One week later: _____

   _____

   Two weeks later: _____

   _____

   Subsequent weeks: _____

   _____

2. **Subculture Appearance:** Describe the appearance of your subculture tubes at the end of two weeks:

   _____

   _____

3. **Microscopic Appearance:** Describe the characteristics of the cells you observed on wet mount slides and

   gram-stained slides: _____

   _____

B. Short Answer Questions

1. In terms of photosynthesis, how do anoxygenic photosynthetic bacteria differ from algae and cyanobacteria?

   _____

   _____

2. What roles do the following organisms perform in the Winogradsky column?

   Cellulomonas: _____

   Clostridium: _____

   Desulfovibrio: _____

3. How do *Chromatium* and *Chlorobium* differ as to where they store the sulfur that they produce?

   _____

   _____

4. Give the equations for photosynthesis in the following

   Algae: _____

   Chromatium and *Chlorobium:* _____

C. Tabulation of Results

If different mud samples were used by class members, record your results on the chart below and on a similar chart constructed by your instructor on the chalkboard. Record the presence (+) or absence (−) of the various species identified. In the "Other" column, list any other species that you might have encountered.

Student Initials	Source of Pond Mud	Purple-nonsulfur	Purple Sulfur	Green Sulfur	Chemolithotrophs	Other

D. Conclusions from Above Table:

_____

_____

Sulfate-Reducing Bacteria:
Desulfovibrio

The sulfate-reducing bacteria play an important role in geochemical processes such as petroleum formation and the production of sulfide ores, and as participants in the sulfur cycle where they reduce sulfates in the environment. However, they can also be nuisance organisms because their activities can result in the corrosion of metal pipes due to the formation of iron sulfide. This group of bacteria was first described by Beijerinck in 1895 when he cultured a common isolate, *Desulfovibrio desulfuricans,* which is enriched in media containing lactate and sulfate. This bacterium is a small, curved, gram-negative rod that is motile by polar flagella.

Sulfate-reducing bacteria occur in diverse anaerobic environments, especially marine waters, where sulfate is plentiful and active fermentation is being carried out by other groups of bacteria. The sulfate reducers metabolize a limited number of substrates such as lactate and pyruvate, which they use for carbon requirements. ATP is synthesized by anaerobic sulfate respiration in which the sulfate ion serves as a terminal electron acceptor and is reduced to H_2S, in much the same way that oxygen is converted to water in aerobic respiration. Habitats for the sulfate reducers include anaerobic sewage digesters, waterlogged soils, and paper mill waste ponds. Their presence in these environments is always obvious due to the presence of blackened metal sulfides and the production of H_2S, which imparts a rotten egg smell.

First Period

Materials

- glass stoppered bottles (stain type)
- freshwater enrichment medium
- Winogradsky columns
- wide-mouth pipettes

1. Fill a glass stoppered bottle with freshwater enrichment medium until the liquid just reaches below the neck of the bottle.
2. Inoculate the bottle with 0.1 g of black mud from a Winogradsky column. You can best accomplish this by carefully inserting a wide-mouth pipette into the column and drawing up mud from the blackened areas of the column. Top off the bottle with freshwater enrichment medium and insert the stopper into the neck of the bottle, causing excess medium in the bottle to be expelled. There should be no entrapped air bubbles at the base of the stopper (figure 56.1).
3. Incubate the cultures at 30° C for one week.
4. Observe the cultures for blackening. Carefully remove the stopper and smell the culture to detect the presence of H_2S (rotten eggs).
5. Prepare Gram stains and wet mounts of your culture. Look for small curved rods that are gram-negative. Use a phase-contrast microscope to observe motility in wet mounts.
6. If time permits, transfer 1 ml of the first enrichment culture to a second glass stoppered bottle containing freshwater enrichment medium. Top up the bottle as before and incubate it at 30° C.
7. Prepare Gram stains and wet mounts as before. At this point, you may have an almost pure culture of a sulfate reducer. Record your observations in Laboratory Report 56.

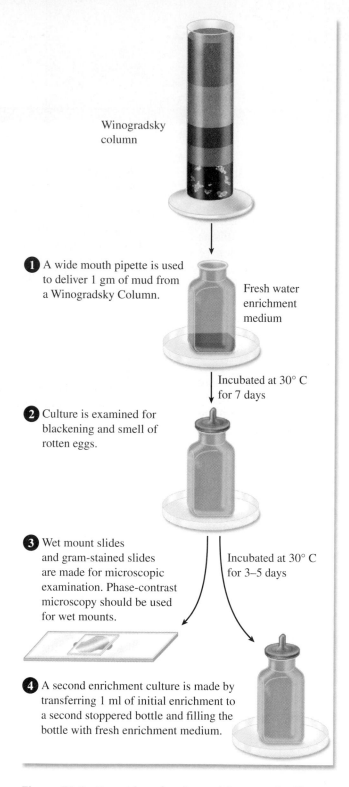

Winogradsky
column

1 A wide mouth pipette is used
to deliver 1 gm of mud from
a Winogradsky Column.

Fresh water
enrichment
medium

Incubated at 30° C
for 7 days

2 Culture is examined for
blackening and smell of
rotten eggs.

3 Wet mount slides
and gram-stained slides
are made for microscopic
examination. Phase-contrast
microscopy should be used
for wet mounts.

Incubated at 30° C
for 3–5 days

4 A second enrichment culture is made by
transferring 1 ml of initial enrichment to
a second stoppered bottle and filling the
bottle with fresh enrichment medium.

Figure 56.1   Procedure for the enrichment of sulfate-reducing bacteria

Laboratory Report

Student: _____

Date: _____   Section: _____

56   Sulfate-Reducing Bacteria: *Desulfovibrio*

A. Microscopic Observations

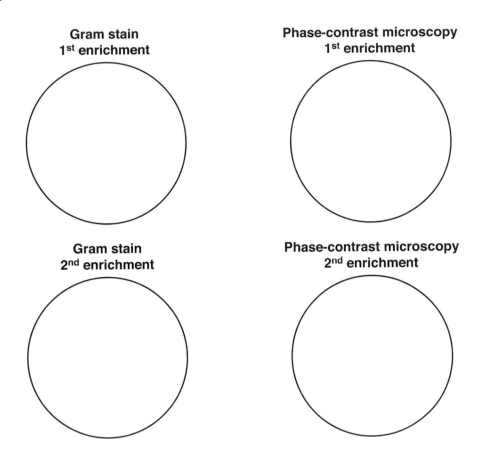

**Gram stain
1st enrichment**

**Phase-contrast microscopy
1st enrichment**

**Gram stain
2nd enrichment**

**Phase-contrast microscopy
2nd enrichment**

B. Short Answer Questions

1. What are the two major products of sulfate reduction by *Desulfovibrio?* What obvious physical characteristics can be used to identify the presence of each?

    _____

    _____

2. Why might these bacteria be more common in marine environments? _____

    _____

    _____

3. How would you isolate and purify the sulfate reducers from the broth enrichment cultures? _____

    _____

    _____

Bacterial Commensalism

Commensal relationships between two organisms occur when one organism benefits from a relationship but the other is unaffected. There are many commensal relationships that exist between organisms in mixed microbial populations. For example, the excretory products of one organism can supply the nutrients for another. Aerobic organisms may utilize the available oxygen in an environment, thus creating anaerobic conditions that allow anaerobes to grow. In every case, the beneficiary of the relationship contributes nothing in the way of either benefit or injury.

In this exercise, you will culture two organisms separately and together to observe an example of commensalism. One of the organisms is *Staphylococcus aureus* and the other is *Clostridium sporogenes*. From your observations of the results, you are to determine which organism profits from the association and what controlling factor is changed when the two are grown together.

Materials

- 3 tubes of nutrient broth
- 1 ml pipette
- nutrient broth culture of *S. aureus*
- fluid thioglycollate medium culture of *C. sporogenes*

First Period

1. Label one tube of nutrient broth *S. aureus*, a second tube *C. sporogenes*, and the third tube *S. aureus* and *C. sporogenes*.
2. Inoculate the first and third tubes with one loopful each of *S. aureus*.
3. Transfer a loopful of *C. sporogenes* to tubes 2 and 3.
4. Incubate the three tubes at 37° C for 48 hours.

Second Period

1. Compare the turbidity in the three tubes, noting which ones are most turbid. Record these results in Laboratory Report 57–59.
2. After shaking the tubes for good dispersion, make a Gram-stained slide of the organisms in each tube and record your observations on combined Laboratory Report 57–59.

Bacterial Synergism

Synergistic relationships occur when two or more microbial populations benefit from the relationship, but the relationship is not obligatory, that is, the populations can survive in their natural environment on their own. For example, synergism would result if two organisms act together to produce a substance that neither of the organisms can produce by themselves. Such relationships are not uncommon among microorganisms. This phenomenon is readily demonstrated in the ability of some bacteria acting synergistically to produce gas by fermenting certain disaccharides.

In this exercise, we will observe the fermentation capabilities of three organisms on two disaccharides. The two sugars, lactose and sucrose, will be inoculated with the individual organisms as well as with various combinations of the organisms to detect which organisms can act synergistically on which sugars. To conserve on media, the class will be divided into three Groups (A, B, and C). Results of inoculations will be shared.

Materials

per pair of students:
- 3 Durham* tubes of lactose broth with bromthymol blue indicator
- 3 Durham* tubes of sucrose broth with bromthymol blue indicator
- 1 nutrient broth culture of *S. aureus*
- 1 nutrient broth culture of *P. vulgaris*
- 1 nutrient broth culture of *E. coli*

  *A *Durham tube* is a fermentation tube of sugar broth that has a small inverted vial in it. See figure 41.6 on page 000.

First Period

Group A

1. Label one tube of each kind of broth *E. coli.*
2. Label one tube of each kind of broth *P. vulgaris.*
3. Label one tube of each kind of broth *E. coli* and *P. vulgaris.*
4. Inoculate each tube with one loopful of the appropriate organisms.
5. Incubate the six tubes at 37°C for 48 hours.

Group B

1. Label one tube of each kind of broth *E. coli.*
2. Label one tube of each kind of broth *S. aureus.*
3. Label one tube of each kind of broth *E. coli* and *S. aureus.*
4. Inoculate each tube with one loopful of the appropriate organisms.
5. Incubate the six tubes at 37°C for 48 hours.

Group C

1. Label one tube of each kind of broth *S. aureus.*
2. Label one tube of each kind of broth *P. vulgaris.*
3. Label one tube of each kind of broth *S. aureus* and *P. vulgaris.*
4. Inoculate each tube with one loopful of the appropriate organisms.
5. Incubate the six tubes at 37°C for 48 hours.

Second Period

1. Look for acid and gas production in each tube, recording your results in Laboratory Report 57–59.
2. Determine which organisms acted synergistically on which disaccharides.
3. Answer the questions for this exercise on combined Laboratory Report 57–59.

Microbial Antagonism

Microbial antagonism, or amensalism, occurs when one population produces a substance or compound that is toxic or inhibitory to other organisms in the environment. The organisms that produce the inhibitory substance are usually unaffected by the substance, and as a result, they may gain a competitive advantage in microenvirnments over others that are also present. Inhibitory substances may be a metabolic end product such as an organic acid that prevents the growth of other bacteria or it may be an antibiotic that kills other organisms by specifically inhibiting some crucial metabolic reaction or function in the affected population. We take advantage of microbial antagonism in certain kinds of food preservation. For example, foods produced by fermentation in which organic acids such as acetic, lactic, or propionic acids are end products (for example, cheeses and vinegar) are not usually subject to microbial spoilage because the acidic pH prevents the growth of other microbes. Some soil bacteria may gain a selective advantage in microenvironments because they produce antibiotics that discourage competition by others for available soil nutrients.

In this exercise, we will attempt to evaluate the antagonistic capabilities of three organisms on two test organisms. The antagonists are *Bacillus cereus* var. *mycoides, Pseudomonas fluorescens,* and *Penicillium notatum.* The test organisms are *Escherichia coli* (gram-negative) and *Staphylococcus aureus* (gram-positive).

Materials

- 6 nutrient agar pours
- 6 sterile petri plates
- nutrient broth cultures of *E. coli, S. aureus, B. cereus* var. *mycoides,* and *P. fluorescens*
- flask culture of *Penicillium notatum* (8–12 day old culture)

First Period

1. Liquefy six nutrient agar pours and cool to 50° C. Hold in 50° C water bath.
2. While the pours are being liquefied, label six plates as follows:

Test Organism	Antagonist
I *S. aureus*	*B. cereus* var. *mycoides*
II *S. aureus*	*P. fluorescens*
III *S. aureus*	*Penicillium notatum*
IV *E. coli*	*B. mycoides*
V *E. coli*	*P. fluorescens*
VI *E. coli*	*Penicillium notatum*

3. Label three liquefied pours *S. aureus,* and label the other three *E. coli.*
4. Inoculate each of the pours with a loopful of the appropriate organisms, flame their necks, and pour into their respective plates.
5. After the nutrient agar in the plates has hardened, streak each plate with the appropriate antagonist. Use a good isolation technique.
6. Invert and incubate the plates for 24 hours at 37° C.

Second Period

1. Examine each plate carefully, looking for evidence of inhibition.
2. Record your results on combined Laboratory Report 57–59 and answer all the questions.

Laboratory Report

Student: _____

Date: _____ Section: _____

57 Bacterial Commensalism

A. Results

Indicate the degree of turbidity (none, + , + + , + + +) in the following table. With colored pencils, draw the appearance of the Gram-stained slides where indicated.

ORGANISMS	TURBIDITY	GRAM STAIN
Staphylococcus aureus		
Clostridium sporogenes		
S. aureus and C. sporogenes		

B. Short Answer Questions

1. What was the reason for the differential growth in nutrient broth of these two species?

   _____

   _____

2. Describe the mechanism of the commensal relationship between these two species.

   _____

   _____

58 Bacterial Synergism

A. Results

Examine the six tubes of media, looking for acid and gas. In the presence of acid, bromthymol blue turns yellow. Record your results in the table below. Consult other students for their results and complete the table.

INDIVIDUAL ORGANISMS	LACTOSE		SUCROSE		COMBINATIONS	LACTOSE		SUCROSE	
	Acid	Gas	Acid	Gas		Acid	Gas	Acid	Gas
E. coli					E. coli and P. vulgaris				
P. vulgaris					E. coli and S. aureus				
S. aureus					S. aureus and P. vulgaris				

B. Short Answer Questions

1. Why is it important to check gas production by organisms individually? Did any of the organisms produce gas in the presence of either lactose or sucrose?

   _____

   _____

2. Why is it important to also take note of acid production during sugar fermentation? How does this help us understand the synergistic relationship between two microbes?

   _____

   _____

3. Based upon the results, name the organism pairs that produce gas synergistically using:

   a. lactose

   _____

   b. sucrose

   _____

59 Microbial Antagonism

A. Short Answer Questions

1. Which organisms are antagonistic to *E. coli?*

   _____

   _____

2. Which organisms are antagonistic to *S. aureus?*

   _____

   _____

3. Why might gram-positive and gram-negative bacteria respond differently to antagonism?

   _____

   _____

4. What role does microbial antagonism play in nature?

   _____

   _____

5. Describe the importance of microbial antagonism for:

   a. controlling food spoilage

   _____

   b. treating infections

   _____

6. Why do antagonistic substances not harm the producer?

   _____

   _____

Applied Microbiology

11

Applied microbiology encompasses many aspects of modern microbiology. We use microorganisms to produce many of the foods we eat such as cheese, yogurt, bread, sauerkraut, and a whole list of fermented beverages. Microorganisms are important in industrial applications where they are involved in producing antibiotics, pharmaceuticals, and even solvents and starting materials for the manufacture of plastics. Their presence and numbers in our foods and drinking water determine if it is safe to consume these substances as they could cause us harm and disease. In the following exercises, you will explore some of the applications of microbiology by determining bacterial numbers and/or kinds in food and water. You will also study the process of alcohol fermentation as an example of food production.

Bacterial Counts of Foods

The presence of microorganisms in food does not necessarily indicate that the food is spoiled or that it has the potential to cause disease. Some foods can have high counts because microorganisms are used in their production. Yogurt, sauerkraut, and summer sausage are examples of foods prepared by microbial fermentation and, therefore, they have hight bacterial counts associated with them during production. However, post-production treatments such as pasteurization or smoking will significantly reduce the numbers of bacteria present. During processing and preparation, food can become contaminated with saprophytic bacteria, which occur in the environment. These bacteria may not be necessarily harmful or pathogenic. Bacteria are naturally associated with some foods when they are harvested. For example, green beans, potatoes, and beets have soil bacteria associated with them when harvested. If these foods are frozen, both the food as well as any bacteria associated with the food, will be preserved. The chalky appearance of grapes is due to yeasts that are naturally associated with grapes and many other fruits. Milk in the udders of healthy cows is sterile but bacteria such as *Streptococcus* and *Lactobacillus* are introduced during milking and processing because they are part of the bacterial flora associated with the animal, especially the outside of the udder. Pasteurization kills many of the bacteria that are introduced during processing, and any pathogens that may be present, but it does not kill all the bacteria present in milk. Some bacteria in milk can survive pasteurization temperatures and eventually cause spoilage and souring of milk. These are called thermoduric bacteria. Hamburger can also have high counts of bacteria that can be introduced during processing and grinding of the meat. Many bacteria in hamburger are harmless saprophytes that come from the environment where processing occurs. For example, endospore-forming bacteria and others can be introduced into ground beef during its preparation.

We must also bear in mind that food can be an important means for the transmission of disease. The Centers for Disease Control estimates that 76 million people per year in the United States become sick; 300,000 are hospitalized; and 5,000 people die from foodborne illnesses. Foodborne illnesses usually result because pathogenic bacteria or their toxins are introduced into food products during processing, handling, or preparation. Food handlers can transmit opportunistic pathogens associated with the human body, like *Staphylococcus aureus* or intestinal bacteria, because of unsanitary practices such as failure to wash their hands before preparing or handling food. Botulism food poisoning results form ingesting a toxin produced by *Clostridium botulinum* when its endospores grow in improperly home canned foods. The endospores occur in the soil and the environment and contaminate the prepared vegetables. *Salmonella* and *Campylobacter* are associated with poultry and eggs and can cause illness if these foods are not properly prepared. *Escherichia coli* O157: H7 is found in the intestines of cattle and can become associated with meat if fecal material from the animal's intestines contaminates meat during the butchering process. This pathogen is then incorporated into hamburger during grinding and processing. Serious illness results form eating improperly cooked hamburger because cooking temperatures are insufficient to kill the organism. Transmission of this pathogen has also occurred when fecal material of cattle contaminated fruits and vegetables such as lettuce and spinach.

Although high bacterial counts in food do not necessarily mean that the food is spoiled or that in harbors disease-causing organisms, it can suggest the potential for more rapid spoilage of the food. Thus, high counts can be important for this reason. One method to ascertain if food is contaminated with fecal bacteria and, therefore, has the potential to spread disease is to perform coliform counts. Coliforms are organisms such *Escherichia coli* that occur in the intestines of humans and warm-blooded animals. Their presence in food or water indicates that fecal contamination has occurred and that there is the high potential for the spread of serious disease such as typhoid fever, bacillary dysentery, cholera, and intestinal viral diseases.

The standard plate count and the coliform count can be used to evaluate foods in much the same manner that they are used on milk and water to determine total bacterial counts and the number of coliforms. However, because most foods are solid in nature, organisms associated with a food sample must be put into suspension by using a food blender before counts can be performed. In this exercise, samples of ground

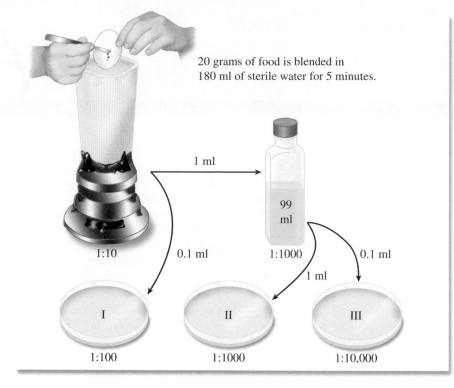

20 grams of food is blended in
180 ml of sterile water for 5 minutes.

1 ml

99
ml

1:10   0.1 ml   1:1000   0.1 ml

1 ml

I   II   III

1:100   1:1000   1:10,000

Figure 60.1   Dilution procedure for bacterial counts of food.

beef, dried fruit, and frozen food will be tested for total numbers of bacteria. This procedure, however, will not determine the numbers of coliforms. Your instructor will indicate the specific kinds of foods to be tested and make individual assignments. Figure 60.1 illustrates the general procedure.

Materials

per student:
- 3 petri plates
- 1 bottle (45 ml) of Plate Count agar or Standard Methods agar
- 1 99-ml sterile water blank
- 2 1.1-ml dilution pipettes

per class:
- food blender
- sterile blender jars (one for each type of food)
- sterile weighing paper
- 180-ml sterile water blanks (one for each type of food)

- samples of ground meat, dried fruit, and frozen vegetables, thawed 2 hours

1. Using aseptic technique, weigh out on sterile weighing paper 20 grams of food to be tested.
2. Add the food and 180 ml of sterile water to a sterile mechanical blender jar. Blend the mixture for 5 minutes. This suspension will provide a 1:10 dilution.
3. With a 1.1-ml dilution pipette dispense from the blender 0.1 ml to plate 1 and 1.0 ml to the water blank. See figure 60.1
4. Shake the water blank 25 times in an arc for 7 seconds with your elbow on the table as done in Exercise 22 (Enumeration of Bacteria).
5. Using a fresh pipette, dispense 0.1 ml to plate III and 1.0 ml to plate II.
6. Pour agar (50°C) into the three plates and incubate them at 35°C for 24 hours.
7. Count the colonies on the best plate and record the results in Laboratory Report 60.

Laboratory Report

Student: _____

Date: _____ Section: _____

60 Bacterial Counts of Foods

A. Results

Record your count and the bacterial counts of various other foods made by other students.

TYPE OF FOOD	PLATE COUNT	DILUTION	ORGANISMS PER ML

B. Short Answer Questions

1. Which type of food had the highest bacterial count? Explain.

   _____

   _____

2. Which type of food had the lowest bacterial count? Explain.

   _____

   _____

3. Coliform bacteria are common contaminants of meats.

   a. Why might one expect to find coliforms in samples of meat?

   _____

   b. What is the danger associated with this type of contamination?

   _____

   c. In terms of food safety, why is it suggested to cook hamburgers medium-well to well-done whereas
      steaks can be cooked rare?

   _____

d. To prevent foodborne illness, what precautions should be followed when preparing a meal that includes ground meat?

_____

4. What considerations should be made to safely thaw frozen foods for later consumption?

_____

_____

5. Why is refrigeration not always an effective means for preventing food spoilage?

_____

_____

6. Why are dried fruits somewhat resistant to spoilage?

_____

_____

Bacteriological Examination of Water:
Qualitative Tests

Prior to the modern age of public health, water was a major means for the spread of infectious diseases such as cholera, dysentery, and typhoid fever. A physician, John Snow, showed in the 1840s that a cholera epidemic in London was the result of cesspool overflow into the Thames River from a tenement where cholera patients lived. When water for drinking was drawn by inhabitants near the cesspool discharge, the contaminated water and pump became the source for the spread of the disease to people in the area. Snow's solution was simply to remove the handle to the pump, and the epidemic abated. Water safety is still a primary concern of municipalities in today's world, and it has become complex. Because of good public health measures, most of us are confident that the water we draw from our faucets is safe and will not cause us disease.

From a microbiological standpoint, it is not the numbers of bacteria that are present in water that is of primary concern to us but rather the kinds of bacteria. Water found in rivers, lakes, and streams can contain a variety of bacteria that may only be harmless saprophytes, which do not cause disease in humans. However, it is important that water not contain the intestinal pathogens that cause typhoid, cholera, and dysentery. In modern cities, treated sewage is discharged into receiving waters of lakes, rivers, and streams, and this constitutes a major sanitary problem because those same bodies of water are the sources of our drinking water. As a result, we have developed methods to treat water to eliminate the potential for disease, and we do microbiological tests to determine if water is potable and safe for consumption.

At first glance, it might seem reasonable to directly examine water for the presence of the pathogens *Vibrio cholerae, Salmonella typhi,* and *Shigella dysenteriae.* However, this not the case because it would be tedious and difficult to specifically test for each of the pathogens. Furthermore, these bacteria are often fastidious, and they might be overgrown by other bacteria in the water if we tried to culture and test for them. It is much easier to demonstrate the presence of some indicator bacterium, such as *Escherichia coli,* which is routinely found in the human intestine but is not found in the soil or water. The presence of these bacteria in water would then indicate the likelihood of fecal contamination and the potential for serious disease.

E. coli is a good indicator of fecal contamination and a good test organism. This is for several reasons: (1) it occurs primarily in the intestines of humans and some warm-blooded animals and it is not found routinely in soil or water; (2) the organism can be easily identified by microbiological tests; (3) it is not as fastidious as the intestinal pathogens, and hence it survives a little longer in water samples. By definition, organisms such as *E. coli* and *Enterobacter aerogenes* are designated as **coliforms,** which are gram-negative, facultative anaerobic, non–endospore forming rods that ferment lactose to produce acid and gas in 48 hours at 35°C. Lactose fermentation with the formation of acid and gas provides the basis for determining the total coliform count of water samples in the United States and therefore designates water purity. The presence of other bacteria, such as *Enterococcus faecalis,* which is a gram-positive coccus that inhabits the human intestine, can also indicate fecal contamination, but testing for this bacterium is not routinely done in the United States.

Three different tests are done to determine the coliform count (figure 61.1): presumptive, confirmed, and completed. Each test is based on one or more of the characteristics of a coliform. A description of each test follows.

Presumptive Test In the presumptive test, 15 tubes of lactose broth are inoculated with measured amounts of water to see if the water contains any lactose-fermenting bacteria that produce gas. If, after incubation, gas is seen in any of the lactose broths, it is *presumed* that coliforms are present in the water sample. This test is also used to determine the most probable number (MPN) of coliforms present per 100 ml of water.

Confirmed Test In this test, plates of Levine EMB agar or Endo agar are inoculated from positive (gas-producing) tubes to see if the organisms that are producing the gas are gram-negative (another coliform characteristic). Both of these media inhibit the growth of gram-positive bacteria and cause colonies of coliforms to be distinguishable from noncoliforms. On EMB agar, coliforms produce small colonies with dark centers (nucleated colonies). On Endo agar,

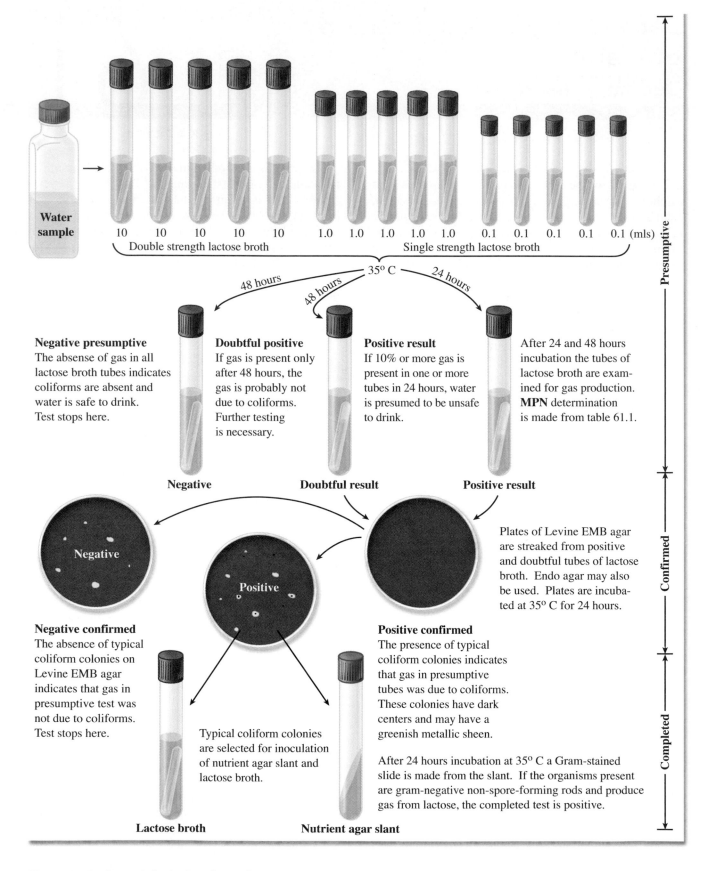

Presumptive

Water sample

10 10 10 10 10 (mls)
Double strength lactose broth

1.0 1.0 1.0 1.0 1.0
Single strength lactose broth

0.1 0.1 0.1 0.1 0.1 (mls)

35° C

48 hours   48 hours   24 hours

Negative presumptive
The absense of gas in all lactose broth tubes indicates coliforms are absent and water is safe to drink. Test stops here.

Doubtful positive
If gas is present only after 48 hours, the gas is probably not due to coliforms. Further testing is necessary.

Positive result
If 10% or more gas is present in one or more tubes in 24 hours, water is presumed to be unsafe to drink.

After 24 and 48 hours incubation the tubes of lactose broth are examined for gas production. **MPN** determination is made from table 61.1.

Negative   **Doubtful result**   **Positive result**

Negative   Positive

Plates of Levine EMB agar are streaked from positive and doubtful tubes of lactose broth. Endo agar may also be used. Plates are incubated at 35° C for 24 hours.

Negative confirmed
The absence of typical coliform colonies on Levine EMB agar indicates that gas in presumptive test was not due to coliforms. Test stops here.

Positive confirmed
The presence of typical coliform colonies indicates that gas in presumptive tubes was due to coliforms. These colonies have dark centers and may have a greenish metallic sheen.

Typical coliform colonies are selected for inoculation of nutrient agar slant and lactose broth.

After 24 hours incubation at 35° C a Gram-stained slide is made from the slant. If the organisms present are gram-negative non-spore-forming rods and produce gas from lactose, the completed test is positive.

Lactose broth   **Nutrient agar slant**

Confirmed

Completed

Figure 61.1   Bacteriological analysis of water.

coliforms produce reddish colonies. The presence of coliform-like colonies confirms the presence of a lactose-fermenting, gram-negative bacterium.

Completed Test In the completed test, our concern is to determine if the isolate from the agar plates truly matches our definition of a coliform. Our media for this test include a nutrient agar slant and a Durham tube of lactose broth. If gas is produced in the lactose tube and a slide from the agar slant reveals that we have a gram-negative, non-spore-forming rod, we can be certain that we have a coliform.

The completion of these three tests with positive results establishes that coliforms are present; however, there is no certainty that *E. coli* is the coliform present. The organism might be *E. aerogenes*. Of the two, *E. coli* is the better sewage indicator since *E. aerogenes* can be of nonsewage origin. To differentiate these two species, one must perform the **IMViC tests,** which are described on page 289 in Exercise 43.

In this exercise, water will be tested from local ponds, streams, swimming pools, and other sources supplied by students and instructor. Enough known positive samples will be evenly distributed throughout the laboratory so that all students will be able to see positive test results. All three tests in figure 61.1 will be performed. If time permits, the IMViC tests may also be performed.

The Presumptive Test

As stated earlier, the presumptive test is used to determine if gas-producing lactose fermenters are present in a water sample. The American Public Health Association's, *Standard Methods for Examination of Water and Waste Water, 21th edition* employs a 15 tube method for the analysis of water samples which is more statistically valid than previous procedures using fewer tubes. This 15 tube procedure is most applicable to clear surface waters but can be used for brackish water and waters with sediments. Figure 61.1 shows the procedure for determining the MPN of a water sample using sample volumes of 10 ml, 1 ml and 0.1 ml each delivered to five tubes of test medium. The figure also describes the subsequent procedures for performing the confirmed and completed tests for tubes in the presumptive test that show gas production. In addition to detecting the presence or absence of coliforms in the presumptive test, the pattern of positive tubes is used to ascertain the **most probable number (MPN)** of coliforms in 100 ml of water. See table 61.1 to determine that value from the number of positive tubes.

Set up the test for your water simple using the procedure outlined in figure 61.1. As stated above, the method will be used for clear surface waters as well as waters with sediments of turbidity.

Materials

- 5 Durham tubes of DSLB
- 10 Durham tubes of SSLB
- 1 10-ml pipette
- 1 1-ml pipette

Note: DSLB designates double-strength lactose broth. It contains twice as much lactose as SSLB (single-strength lactose broth).

1. Set up 5 DSLB and 10 SSLB tubes as illustrated in figure 61.1. Label each tube according to the amount of water that is to be dispensed to it: *10 ml, 1.0 ml,* and *0.1 ml,* respectively.
2. Mix the bottle of water to be tested by shaking it 25 times.
3. With a 10 ml pipette, transfer 10 ml of water to each of the DSLB tubes.
4. With a 1.0 ml pipette, transfer 1 ml of water to each of the middle set of tubes, and 0.1 ml to each of the last five SSLB tubes.
5. Incubate the tubes at 35° C for 24 hours.
6. Examine the tubes and record the number of tubes in each set that have 10% gas or more.
7. Determine the MPN by referring to table 61.1. Consider the following:
   Example: If gas was present in the first five tubes (10 ml samples), and detected in only one tube of the second series (1 ml samples), but was not present in any of the last five tubes (0.1 ml samples), your test results would read as 5-1-0. Table 16.1 indicates that the MPN for this pattern of tubes is 33. This means that this water sample would have approximately 33 organisms per 100 ml with a 95% probability of there being between 10 and 100 organisms. *Keep in mind that the MPN of 33 is only a statistical probability figure.* Please also note that table 61.1 cannot be used to determine the MPN for water samples using less than 15 tubes, for example 9 tubes as was once used by the American Public Health Association. *Standard Methods for Examination of Water and Waste Water* provides tables for interpreting results if other combinations of dilutions are employed. If other dilutions are used for the MPN determination, the MPN can also be estimated using **Thomas' simple formula.**

$$\text{MPN/100 ml} = \frac{\text{no. positive tubes} \times 100}{\sqrt{\frac{\text{ml sample in}}{\text{negative tubes}} \times \frac{\text{ml sample in}}{\text{all tubes}}}}$$

8. Record the data in Laboratory Report 61.

The Confirmed Test

Once it has been established that gas-producing lactose fermenters are present in the water, it is *presumed*

Table 61.1 MPN Index and 95% Confidence Limits for Various Combinations of Positive Results When Five Tubes are Used for Dilution

(10 ML, 1.0 ML, 0.1 ML)*

Combination of Positives	MPN Index/ 100 mL	Confidence Limits		Combination of Positives	MPN Index/ 100 mL	Confidence Limits	
		Low	High			Low	High
0-0-0	<1.8	—	6.8	4-0-3	25	9.8	70
0-0-1	1.8	0.090	6.8	4-1-0	17	6.0	40
0-1-0	1.8	0.090	6.9	4-1-1	21	6.8	42
0-1-1	3.6	0.70	10	4-1-2	26	9.8	70
0-2-0	3.7	0.70	10	4-1-3	31	10	70
0-2-1	5.5	1.8	15	4-2-0	22	6.8	50
0-3-0	5.6	1.8	15	4-2-1	26	9.8	70
1-0-0	2.0	0.10	10	4-2-2	32	10	70
1-0-1	4.0	0.70	10	4-2-3	38	14	100
1-0-2	6.0	1.8	15	4-3-0	27	9.9	70
1-1-0	4.0	0.71	12	4-3-1	33	10	70
1-1-1	6.1	1.8	15	4-3-2	39	14	100
1-1-2	8.1	3.4	22	4-4-0	34	14	100
1-2-0	6.1	1.8	15	4-4-1	40	14	100
1-2-1	8.2	3.4	22	4-4-2	47	15	120
1-3-0	8.3	3.4	22	4-5-0	41	14	100
1-3-1	10	3.5	22	4-5-1	48	15	120
1-4-0	10	3.5	22	5-0-0	23	6.8	70
2-0-0	4.5	0.79	15	5-0-1	31	10	70
2-0-1	6.8	1.8	15	5-0-2	43	14	100
2-0-2	9.1	3.4	22	5-0-3	58	22	150
2-1-0	6.8	1.8	17	5-1-0	33	10	100
2-1-1	9.2	3.4	22	5-1-1	46	14	120
2-1-2	12	4.1	26	5-1-2	63	22	150
2-2-0	9.3	3.4	22	5-1-3	84	34	220
2-2-1	12	4.1	26	5-2-0	49	15	150
2-2-2	14	5.9	36	5-2-1	70	22	170
2-3-0	12	4.1	26	5-2-2	94	34	230
2-3-1	14	5.9	36	5-2-3	120	36	250
2-4-0	15	5.9	36	S-2-4	150	58	400
3-0-0	7.8	2.1	22	5-3-0	79	22	220
3-0-1	11	3.5	23	5-3-1	110	34	250

3-0-2	13	5.6	35		5-3-2	140	52	400
3-1-0	11	3.5	26		5-3-3	170	70	400
3-1-1	14	5.6	36		5-3-4	210	70	400
3-1-2 .	17	6.0	36		5-4-0	130	36	400
3-2-0	14	5.7	36		5-4-1	170	58	400
3-2-1	17	6.8	40		5-4-2	220	70	440
3-2-2	20	6.8	40		5-4-3	280	100	710
3-3-0	17 '	6.8	40		5-4-4	350	100	710
3-3-1	21	6.8	40		5-4-5	430	150	1100
3-3-2	24	9.8	70		5-5-0	240	70	710
3-4-0	21	6.8	40		5-5-1	350	100	1100
3-4-1	24	9.8	70		5-5-2	540	150	1700
3-5-0	25	9.8	70		5-5-3	920	220	2600
4-0-0	13	4.1	35		5-5-4	1600	400	4600
4-0-1	17	5.9	36		5-5-5	>1600	700	—
4-0-2	21	6.8	40					

*Results to two significant figures.

From A.D. Eaton, L.S. Clesceri, E.W. Rice, and A.E. Greenbert, Standard Methods for the Examination of Water and Wastewater, 21st Ed, APHA, 2005: 9–54. Reprinted with permission from the American Public Health Association.

to be unsafe. However, gas formation may be due to noncoliform bacteria. Some of these organisms, such as *Clostridium perfringens*, are gram-positive. To confirm the presence of gram-negative lactose fermenters, the next step is to inoculate media such as Levine eosin-methylene blue (EMB) agar or Endo agar from positive presumptive tubes.

Levine EMB agar contains methylene blue, which inhibits gram-positive bacteria. gram-negative lactose fermenters (coliforms) that grow on this medium will produce "nucleated colonies" (dark centers). Colonies of *E. coli* and *E. aerogenes* can be differentiated on the basis of size and the presence of a greenish metallic sheen. *E. coli* colonies on this medium are small and have this metallic sheen, whereas *E. aerogenes* colonies usually lack the sheen and are larger. Differentiation in this manner is not completely reliable, however. It should be remembered that *E. coli* is the more reliable sewage indicator since it is not normally present in soil, while *E. aerogenes* has been isolated from soil and grains.

Endo agar contains a fuchsin sulfite indicator that makes identification of lactose fermenters relatively easy. Coliform colonies and the surrounding medium appear red on Endo agar. Nonfermenters of lactose, on the other hand, are colorless and do not affect the color of the medium.

In addition to these two media, there are several other media that can be used for the confirmed test. Brilliant green bile lactose broth, Eijkman's medium, and EC medium are just a few examples that can be used.

To demonstrate the confirmation of a positive presumptive in this exercise, the class will use Levine EMB agar and Endo agar. One-half of the class will use one medium; the other half will use the other medium. Plates will be exchanged for comparisons.

Materials

- 1 petri plate of Levine EMB agar (odd-numbered students)
- 1 petri plate of Endo agar (even-numbered students)

1. Select one positive lactose broth tube from the presumptive test and streak a plate of medium according to your assignment. Use a streak method that will produce good isolation of colonies. If all your tubes were negative, borrow a positive tube from another student.
2. Incubate the plate for 24 hours at 35° C.

3. Look for typical coliform colonies on both kinds of media. Record your results on Laboratory Report 61. If no coliform colonies are present, the water is considered bacteriologically safe to drink.

   Note: In actual practice, confirmation of all presumptive tubes would be necessary to ensure accuracy of results.

The Completed Test

A final check of the colonies that appear on the confirmatory media is made by inoculating a nutrient agar slant and lactose broth with a Durham tube. After incubation for 24 hours at 35° C, the lactose broth is examined for gas production. A Gram-stained slide is made from the slant, and the slide is examined under oil immersion optics.

If the organism proves to be a gram-negative, non-spore-forming rod that ferments lactose, we know that coliforms were present in the tested water sample. If time permits, complete these last tests and record the results in Laboratory Report 61.

The IMViC Tests

Review the discussion of the IMViC tests on page 289. The significance of these tests should be much more apparent at this time. Your instructor will indicate whether these tests should be performed if you have a positive completed test.

Laboratory Report

Student: _____

Desk No.: _____ Section: _____

61 Bacteriological Examination of Water: Qualitative Tests

A. Results

1. **Presumptive Test (MPN Determination)**

   Record the number of positive tubes on the chalkboard and on the following table. When all students have recorded their results with the various water samples, complete this tabulation. Determine the MPN according to the instructions on page 389.

WATER SAMPLE (SOURCE)	NUMBER OF POSITIVE TUBES			MPN
	5 Tubes DSLB 10 ml	5 Tubes SSLB 1.0 ml	5 Tubes SSLB 0.1 ml	

2. **Confirmed Test**

   Record the results of the confirmed tests for each water sample that was positive on the presumptive test.

WATER SAMPLE (SOURCE)	POSITIVE	NEGATIVE

3. **Completed Test**

Record the results of completed tests for each water sample that was positive on the confirmed test.

WATER SAMPLE (SOURCE)	LACTOSE FERMENTATION RESULTS	MORPHOLOGY	EVALUATION

B. Short Answer Questions

1. Does a positive presumptive test mean that the water is absolutely unsafe to drink? _____

   Explain: _____

2. What might cause a false positive presumptive test? _____

   _____

3. List three characteristics required of a good sewage indicator:

   a. _____ b. _____ c. _____

4. What enteric **bacterial** diseases are transmitted in polluted water? _____

   _____

5. Name one or more **protozoan** diseases transmitted by polluted water. _____

   _____

6. Why don't health departments routinely test for pathogens instead of using a sewage indicator?

   _____

   _____

7. List five characteristics of coliform bacteria.

   _____

   _____

   _____

8. How is each of the following media used for the detection of coliforms?

    a. lactose broth with Durham tube

    _____

    b. Levine EMB agar

    _____

    c. nutrient agar slant

    _____

9. Once the completed test establishes the presence of coliforms in the water sample, why might you per-form the IMViC tests on these isolates?

    _____

    _____

The Membrane Filter Method

The most probable number method for determining coliform bacteria in water samples is complicated and requires several days to complete. Furthermore, more than one kind of culture medium is needed for each phase of the test to finally establish the presence of coliforms in a water sample. A more rapid method is the **membrane filter method,** also recognized by the United States Public Health Service as a reliable procedure for determining coliforms. In this test, known volumes of a water sample are filtered through membrane filters that have pores 0.45 μm in diameter. Most bacteria, including coliforms, are larger than the pore diameters, and hence bacteria are retained on the membrane filter. Once the water sample has been filtered, the filter disk containing bacterial cells is placed in a petri dish with an absorbent pad saturated with Endo broth. The plate is then incubated at 35° C for 22 to 24 hours during which time individual cells on the filter multiply forming colonies.

Any coliforms that are present on the filter will ferment the lactose in the Endo broth producing acids. The acids produced from fermentation interact with basic fuschin, a dye in the medium, causing coliform colonies to have a characteristic metallic sheen. Noncoliform bacteria will not produce the metallic sheen. Gram-positive bacteria are inhibited from growing because of the presence of bile salts and sodium lauryl sulfate, which inhibit these bacteria. Colonies are easily counted on the filter disk, and the total coliform count is determined based on the volume of water filtered.

Figure 62.1 illustrates the procedure we will use in this experiment.

Materials

- vacuum pump or water faucet aspirators
- membrane filter assemblies (sterile)
- side-arm flask, 1000 ml size, and rubber hose
- sterile graduates (100 ml or 250 ml size)
- sterile, plastic petri dishes, 50 mm dia (Millipore #PD10 047 00)
- sterile membrane filter disks (Millipore #HAWG 047 AO)
- sterile absorbent disks (packed with filters)
- sterile water
- 5 ml pipettes
- bottles of *m* Endo MF broth (50 ml)* water samples

1. Prepare a small plastic petri dish as follows:
   a. With a flamed forceps, transfer a sterile absorbent pad to a sterile plastic petri dish.
   b. Using a 5 ml pipette, transfer 2.0 ml of *m* Endo MF broth to the absorbent pad.
2. Assemble a membrane filtering unit as follows:
   a. *Aseptically* insert the filter holder base into the neck of a 1-liter side-arm flask.
   b. With a flamed forceps, place a sterile membrane filter disk, grid side up, on the filter holder base.
   c. Place the filter funnel on top of the membrane filter disk and secure it to the base with the clamp.
3. Attach the rubber hose to a vacuum source (pump or water aspirator) and pour the appropriate amount of water into the funnel.

   The amount of water used will depend on water quality. No less than 50 ml should be used. Water with few bacteria and low turbidity permit samples of 200 ml or more. Your instructor will advise you as to the amount of water that you should use. Use a sterile graduate for measuring the water.
4. Rinse the inner sides of the funnel with 20 ml of sterile water.
5. Disconnect the vacuum source, remove the funnel, and carefully transfer the filter disk with sterile forceps to the petri dish of *m* Endo MF broth. *Keep grid side up.*
6. Incubate at 35° C for 22 to 24 hours. *Don't invert.*
7. After incubation, remove the filter from the dish and dry for 1 hour on absorbent paper.
8. Count the colonies on the disk with low-power magnification, using reflected light. Ignore all colonies that lack the golden metallic sheen. If desired, the disk may be held flat by mounting between two 2″ × 3″ microscope slides after drying. Record your count on the first portion of Laboratory Report 62.

---

*See Appendix C for special preparation method.

(1) Sterile absorbent pad is aseptically placed in the bottom of a sterile plastic petri dish.

(2) Absorbent pad is saturated with 2.0 ml of *m* Endo MF broth.

(3) Sterile membrane filter disk is placed on filter holder base with grid side up.

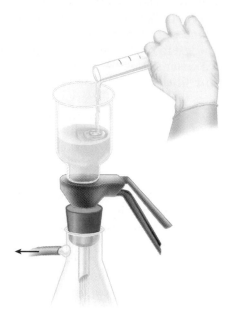

(4) Water sample is poured into assembled funnel, utilizing vacuum. A rinse of 20 ml of sterile water follows.

(5) Filter disk is carefully removed with sterile forceps after disassembling the funnel.

(6) Membrane filter disk is placed on medium-soaked absorbent pad with grid side up. Incubate at 35° C 24 hours.

Figure 62.1 Membrane filter routine.

Laboratory Report

Student: _____

Desk No.: _____ Section: _____

62 The Membrane Filter Method

A. Results

A table similar to the one below will be provided for you, either on the chalkboard or as a photocopy. Record your coliform count on it. Once all data are available, complete this table.

SAMPLE	SOURCE	COLIFORM COUNT	AMOUNT OF WATER FILTERED	MPN*
A				
B				
C				
D				
E				
F				
G				
H				

*MPN = $\dfrac{\text{Coliform Count} \times 100}{\text{Amount of Water Filtered}}$

B. Short Answer Questions

1. In what ways is the membrane filter method for coliform detection superior to the most probable number method (Exercise 61)?

    _____

    _____

2. How is the proper amount of water to be filtered determined? Why is this determination critical to the outcome of the testing?

    _____

    _____

3. Why would the water that passed through the filter not be considered sterile even though bacteria were removed? How can sterility be achieved with the use of filtration?

    _____

    _____

Reductase Test

Milk that contains large numbers of actively growing bacteria will have a lowered oxidation-reduction potential due to the exhaustion of dissolved oxygen by microorganisms. The fact that methylene blue loses its color (becomes reduced) in such an environment is the basis for the **reductase test.** In this test, 1 ml of methylene blue (1:25,000) is added to 10 ml of milk. The tube is sealed with a rubber stopper and slowly inverted three times to mix. It is placed in a water bath at 35° C and examined at intervals up to 6 hours. The time it takes for the methylene blue to become colorless is the **methylene blue reduction time (MBRT).** The shorter the MBRT, the lower the quality of milk. An MBRT of 6 hours is very good. Milk with an MBRT of 30 minutes is of very poor quality (figure 63.1).

The validity of this test is based on the assumption that all bacteria in milk lower the oxidation-reduction potential at 35° C. Large numbers of psychrophiles, thermophiles, and thermodurics (nonpathogenic bacteria that survived pasteurization), which do not grow at this temperature, would not produce a positive test. Raw milk, however, will contain primarily *Streptococcus lactis* and *Escherichia coli,* which are strong reducers; thus, this test is suitable for screening raw milk at receiving stations. Its principal value is that less technical training of personnel is required for its performance.

In this exercise, samples of low- and high-quality raw milk will be tested.

Materials

- 2 sterile test tubes with rubber stoppers for each student
- raw milk samples of low and high quality (samples A and B)
- water bath set at 35° C
- methylene blue (1:25,000)
- 10 ml pipettes
- 1 ml pipettes
- gummed labels

1. Attach gummed labels with your name and type of milk to two test tubes. Each student will test a good-quality as well as a poor-quality milk.
2. Using separate 10 ml pipettes for each type of milk, transfer 10 ml to each test tube. To the milk in the tubes add 1 ml of methylene blue with a 1 ml pipette. Insert rubber stoppers and gently

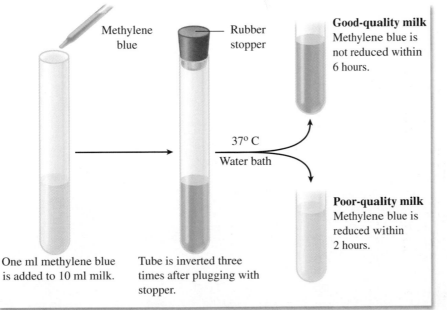

One ml methylene blue is added to 10 ml milk.

Tube is inverted three times after plugging with stopper.

Methylene blue

Rubber stopper

37° C
Water bath

Good-quality milk
Methylene blue is not reduced within 6 hours.

Poor-quality milk
Methylene blue is reduced within 2 hours.

Figure 63.1 Procedure for testing raw milk with the reductase test.

invert three times to mix. Record your name and the time on the labels and place the tubes in the water bath, which is set at 35° C.

3. After 5 minutes incubation, remove the tubes from the bath and invert once to mix. This is the last time they should be mixed.

4. Carefully remove the tubes from the water bath 30 minutes later and every half hour until the end of the laboratory period. *When at least four-fifths of the tube has turned white,* the end point of reduction has taken place. Record this time in Laboratory Report 63. The classification of milk quality is as follows:

Class 1: Excellent, not decolorized in 8 hours

Class 2: Good, decolorized in less than 8 hours but not less than 6 hours

Class 3: Fair, decolorized in less than 6 hours, but not less than 2 hours

Class 4: Poor, decolorized in less than 2 hours

Laboratory Report

Student: _____

Date: _____   Section: _____

63   Reductase Test

A. Results

How would you grade the two samples of milk that you tested? Give the MBRT for each one.

Sample A: _____

Sample B: _____

B. Short Answer Questions

1. Why are the tubes sealed with a rubber stopper rather than left open to the atmosphere?

    _____

    _____

2. Is milk with a short reduction time necessarily unsafe to drink? _____

    Explain: _____

    _____

    _____

3. What other dye can be substituted for methylene blue in this test? _____

    _____

    _____

4. What advantage do you see in this method over the direct count method? _____

    _____

    _____

    _____

    _____

5. What kinds of organisms may be plentiful in a milk sample, yet give a negative reductase test?

    _____

    _____

    _____

    _____

Microbial Spoilage of Canned Food

Spoilage of heat-processed, commercially canned foods is confined almost entirely to the action of bacteria that produce heat-resistant endospores. Canning of foods normally involves heat exposure for long periods of time at temperatures that are adequate to kill bacterial endospores. Particular concern is given to the processing of low-acid foods in which *Clostridium botulinum* can thrive to produce botulism toxin, and thereby cause botulism food poisoning.

Spoilage occurs when the heat processing fails to meet accepted standards. This can occur for several reasons: (1) lack of knowledge on the part of the processor (usually the case in home canning); (2) carelessness in handling the raw materials before canning, resulting in an unacceptably high level of contamination that ordinary heat processing may be inadequate to control; (3) equipment malfunction that results in undetected underprocessing; and (4) defective containers that permit the entrance of organisms after the heat process.

Our concern here will be with the most common types of food spoilage caused by heat-resistant, spore-forming bacteria. There are three types: flat sour, T.A. spoilage, and stinker spoilage.

Flat sour pertains to spoilage in which acids are formed with no gas production; result: sour food in cans that have flat ends. **T.A. spoilage** is caused by thermophilic anaerobes that produce acid and gases (CO_2 and H_2, but not H_2S) in low-acid foods. Cans swell to various degrees, sometimes bursting. **Stinker spoilage** is due to spore-formers that produce hydrogen sulfide and blackening of the can and contents. Blackening is due to the reaction of H_2S with the iron in the can to form iron sulfide.

In this experiment, you will have an opportunity to become familiar with some of the morphological and physiological characteristics of organisms that cause canned food spoilage, including both aerobic and anaerobic endospore formers of *Bacillus* and *Clostridium*, as well as a non-spore-forming bacterium.

Working as a single group, the entire class will inoculate 10 cans of vegetables (corn and peas) with five different organisms. Figure 64.1 illustrates the procedure. Note that the cans will be sealed with solder after inoculation and incubated at different temperatures. After incubation the cans will be opened so that stained microscope slides can be made to determine Gram reaction and presence of endospores. Your instructor will assign individual students or groups of students to inoculate one or more of the 10 cans. One can of corn and one can of peas will be inoculated with each of the organisms. Proceed as follows:

First Period
(Inoculations)

Materials

- 5 small cans of corn
- 5 small cans of peas
- cultures of *Geobacillus stearothermophilus*, *B. coagulans*, *C. sporogenes*, *Thermoanaerobacterium thermosaccharolyticum*, and *E. coli*
- ice picks or awls
- hammer
- solder and soldering iron
- plastic bags
- gummed labels and rubber bands

1. Label the can or cans with the name of the organism that has been assigned to you. Use white gummed labels. In addition, place a similar label on one of the plastic bags to be used after sealing of the cans.
2. With an ice pick or awl, punch a small hole through a flat area in the top of each can. This can be done easily with the heel of your hand or a hammer, if available.
3. Pour off a small amount of the liquid from the can to leave an air space under the lid.
4. Use an inoculating needle to inoculate each can of corn or peas with the organism indicated on the label.
5. Take the cans up to the demonstration table where the instructor will seal the hole with solder.
6. After sealing, place each can in two plastic bags. Each bag must be closed separately with rubber bands, and the outer bag must have a label on it.

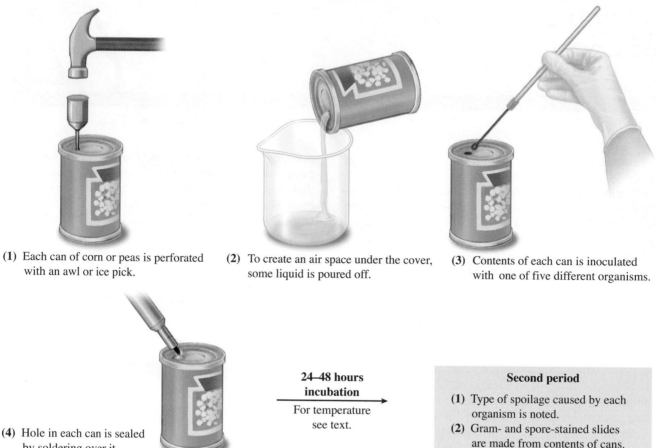

(1) Each can of corn or peas is perforated with an awl or ice pick.

(2) To create an air space under the cover, some liquid is poured off.

(3) Contents of each can is inoculated with one of five different organisms.

(4) Hole in each can is sealed by soldering over it.

24–48 hours incubation
For temperature see text.

Second period
(1) Type of spoilage caused by each organism is noted.
(2) Gram- and spore-stained slides are made from contents of cans.

Figure 64.1 Canned food inoculation procedure.

7. Incubation will be as follows until the next period:
   - **45° C; 72 hours** —T. *thermosaccharolyticum*
   - **37° C; 24–48 hrs.—C.** *sporogenes* and *B. coagulans*
   - **37° C; 24–48 hrs.** —*E. coli*
   Note: If cans begin to swell during incubation, they should be placed in refrigerator.

Second Period

(Interpretation)

After incubation, place the cans under a hood to open them. The odors of some of the cans will be very strong due to H_2S production.

Materials

- can opener, punch-type
- small plastic beakers
- Parafilm
- Gram-staining kit
- spore-staining kit

1. Open each can carefully with a punch-type can opener. If the can is swollen, hold an inverted plastic funnel over the can during perforation to minimize the effects of any explosive release of contents.
2. Remove about 10 ml of the liquid through the opening, pouring it into a small plastic beaker. Cover with Parafilm. This fluid will be used for making stained slides.
3. Return the cans of food to the plastic bags, reclose them, and dispose in the biohazard bin.
4. Prepare Gram-stained and endospore-stained slides from your canned food extract as well as from the extracts of all the other cans. Examine under brightfield oil immersion.
5. Record your observations on the report sheet on the demonstration table. It will be duplicated and a copy will be made available to each student.

Laboratory Report

Complete the first portion of Laboratory Report 64.

Laboratory Report

64   Microbial Spoilage of Canned Food

A. Results

1. Observations
    Record your observations of the effects of each organism on the cans of vegetables. Share results with other students.

ORGANISM	PEAS		CORN	
	Gas Production + or –	Odor	Gas Production + or –	Odor
E. coli				
B. coagulans				
G. stearothermophilus				
C. sporogenes				
T. thermosaccharolyticum				

2. Microscopy
    After making Gram-stained and spore-stained slides of all organisms from the canned food extracts, sketch in representatives of each species:

E. coli	B. coagulans	G. stearothermophilus	C. sporogenes	T. thermosaccharolyticum

B. Short Answer Questions

1. Which organisms, if any, caused flat sour spoilage? _____

    _____

2. Which organisms, if any, caused T.A. spoilage? _____

    _____

3. Which organisms, if any, caused stinker spoilage? _____

    _____

4. Does flat sour cause a health problem? _____
_____

5. Describe how typical spoilage resulting in botulism occurs. _____
_____
_____

6. Why is spoilage more likely to occur for individuals who do home canning than in a canning factory?
_____
_____

Microbiology of Alcohol Fermentation

Fermented food and beverages are as old as civilization. Historical evidence indicates that beer and wine making were well-established as long ago as 2000 B.C. An Assyrian tablet states that Noah took beer aboard the ark.

Beer, wine, vinegar, buttermilk, cottage cheese, sauerkraut, pickles, and yogurt are some of the products of fermentation. Most of these foods and beverages are produced by different strains of yeasts (*Saccharomyces*) or bacteria (*Lactobacillus, Acetobacter,* etc.).

Fermentation is actually a means of food preservation because the acids formed and the reduced environment (anaerobiasis) hold back the growth of many spoilage microbes.

Wine is essentially fermented fruit juice in which alcoholic fermentation is carried out by *Saccharomyces cerevisiae* var. *ellipsoideus*. Although we usually associate wine with fermented grape juice, it may also be made from various berries, dandelions, rhubarb, and so on. Three conditions are necessary: simple sugar, yeast, and anaerobic conditions. The reaction is as follows:

$$C_6H_{12}O_6 \xrightarrow{\text{yeast}} 2C_2H_5OH + 2CO_2$$

Commercially, wine is produced in two forms: red and white. To produce red wines, the distillers use red grapes with the skins left on during the initial stage of the fermentation process. For white wines, either red or white grapes can be used, but the skins are discarded. White and red wines are fermented at 13° C (55° F) and 24° C (75° F), respectively.

In this exercise, we will set up a grape juice fermentation experiment to learn about some of the characteristics of sugar fermentation to alcohol. Note in figure 65.1 that a balloon will be attached over the mouth of the fermentation flask to exclude oxygen uptake and to trap gases that might be produced. To detect the presence of hydrogen sulfide production, we will tape lead acetate test paper inside the neck of the flask. The pH of the substrate will also be monitored before and after the reaction to note any changes that occur.

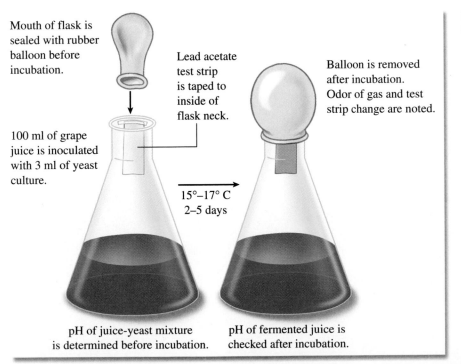

Mouth of flask is sealed with rubber balloon before incubation.

Lead acetate test strip is taped to inside of flask neck.

Balloon is removed after incubation. Odor of gas and test strip change are noted.

100 ml of grape juice is inoculated with 3 ml of yeast culture.

15°–17° C
2–5 days

pH of juice-yeast mixture is determined before incubation.

pH of fermented juice is checked after incubation.

Figure 65.1 Alcohol fermentation setup.

First Period

Materials

- 100 ml grape juice (no preservative)
- bottle of juice culture of wine yeast
- 125 ml Erlenmeyer flask
- 1 10 ml pipette
- balloon
- hydrogen sulfide (lead acetate) test paper
- tape
- pH meter

1. Label an Erlenmeyer flask with your initials and date.
2. Add about 100 ml of grape juice to the flask (fermenter).
3. Determine the pH of the juice with a pH meter and record the pH in Laboratory Report 65.
4. Agitate the container of yeast and juice to suspend the culture, remove 5 ml with a pipette, and add it to the flask.
5. Attach a short strip of tape to a piece of lead-acetate test paper (3 cm long), and attach it to the inside surface of the neck of the flask. Make certain that neither the tape nor the test paper protrudes from the flask.

6. Cover the flask opening with a balloon.
7. Incubate at 15°–17° C for 2–5 days.

Second Period

Materials

- pH meter

1. Remove the balloon and note the aroma of the flask contents. Describe the odor in Laboratory Report 65.
2. Determine the pH and record it in the Laboratory Report.
3. Record any change in color of the lead-acetate-test paper in the Laboratory Report. If any H_2S is produced, the paper will darken due to the formation of lead sulfide as hydrogen sulfide reacts with the lead acetate.
4. Wash out the flask and return it to the drain rack.

Laboratory Report 65

Complete Laboratory Report 65 by answering all the questions.

Laboratory Report

65   Microbiology of Alcohol Fermentation

A. Results

Record here your observations of the fermented product:

Aroma: _____

pH: _____

H$_2$S production: _____

B. Short Answer Questions

1. What compound in the grape juice was fermented? What were the major products of this fermentation?

    _____

    _____

2. Why was the fermentor sealed with a balloon? How would the product be different if the fermentor was sealed with a rubber stopper?

    _____

    _____

3. If the wine were contaminated with *Acetobacter,* what end product would be created? How would this negatively affect the taste? How could this change be measured? Why, for some manufacturers, is this end product actually desired?

    _____

    _____

4. Why is hydrogen sulfide production measured during fermentation?

    _____

    _____

5. Yeast is also used in the production of bread. What is the source of sugar for fermentation? What compounds result from fermentation, and how does each contribute to the final product? What makes sour dough bread sour?

    _____

    _____

6. How does microbial fermentation contribute to food preservation?

    _____

    _____

Bacterial Genetics and Biotechnology

The genetic information encoded in the DNA molecules that make up the genome of the bacterial cell determines its metabolic capabilities and its cellular structure, and thus defines the cell. The information is encoded primarily in the nuclear DNA, but information is also found in small DNA molecules called plasmids, which can replicate and be transferred between cells independent of the nuclear DNA. The information in nuclear DNA and plasmids is arranged into genetic units called genes, which primarily encode for the various enzymes that catalyze the metabolic reactions in a cell and confer upon the cell its unique characteristics that make it different from other cells. The information in nuclear DNA is obligatory for any cell to survive because it defines the cell. However, plasmid DNA supplies accessory information that a bacterial cell might need under special circumstances and its presence in the cell is not essential for viability. Plasmid DNA can encode for characteristics such as antibiotic resistance, specialized metabolic functions, enhanced virulence, or the production of toxins. Such functions can aid a cell in times when the specialized characteristics are advantageous to the cell but are not essential at other times.

It is crucial that the information in the DNA be replicated faithfully when cells divide to insure that the progeny have the same genetic potential as the cells from which they are derived. If changes occur in nucleotide bases during replication because of insertion or deletion of nucelotide bases, this results in changes in the sequence of nucleotide bases, or mutations, which are usually deleterious to the cell. Mutations also result when nucleotide bases are altered by radiation and mutagenic chemicals, such as ultraviolet light and nitrosamines, respectively. Because bacteria only carry a single copy of a particular gene, any mutation is immediate and noticeable. Not every mutation is detrimental, however, and mutation is one way that the genetic information of a bacterial cell can be permanently altered to change the genetic potential of cells.

The genetic potential of a cell can also be changed by the reassortment of genes that occurs during the exchange of genetic material by sexual mechanisms. Bacteria are principally asexual, dividing by transverse binary fission, but they have evolved sexual mechanisms for transferring and exchanging DNA. These include transformation, or the uptake of "naked" DNA; conjugation, involving mating types; and transduction in which genes are transferred by a virus. When genetic information is exchanged, it can recombine, that is, replace existing DNA sequences in a cell. If the changes confer a selective advantage on the progeny, they will become a permanent part of the genetic

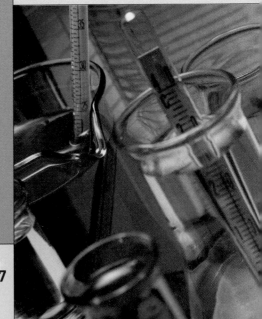

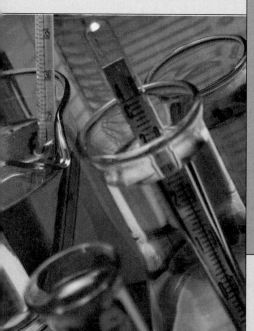

potential, for example, if the new DNA replaces DNA with mutations. The ultimate purpose of sexual exchange is to introduce variability into the gene pool.

In bacteria, variability can also be quickly introduced by the transfer of plasmid DNA. The number of genes carried on a plasmid varies from a few genes to several hundred. Plasmids are exchanged between bacteria primarily by conjugation in which mating strains are involved. As indicated, plasmids encode for specialized functions nonessential to the cell. For example, in a hospital environment, *Staphylococcus aureus* can exchange penicillinase plasmids that confer resistance to penicillin, resulting in resistant strains that can cause serious infections in pediatric and surgical wards.

The knowledge that we have gained from the study of bacterial genetics and extrachromosomal elements such as plasmids has provided new and exciting approaches in molecular biology and spawned the new field of biotechnology. Using molecular biology techniques, we can now introduce "foreign" DNA into bacterial plasmids and cause the expression of desirable products in bacteria. For example, we no longer have to isolate insulin from the pancreases of slaughtered animals to treat diabetics because the genes for human insulin have been genetically engineered into *Escherichia coli*, and strains of this bacterium now produce human insulin or humulin. This is only one example of how pharmaceuticals, such as human growth hormone and interferon, have been genetically engineered into bacterial cells for our purposes.

In the following section, you will perform exercises that involve some genetics and molecular biology techniques. These will include isolation of a mutant, genetic exchange in bacteria, isolation of plasmid DNA, separation of DNA by agarose gel electrophoresis, and amplication of separated DNA by the polymerase chain reaction. These techniques are indispensable and form the basis for many of the procedures in genetic engineering and modern biotechnology.

Mutant Isolation by Replica Plating

When bacteria such as *Escherichia coli* are grown in the presence of an antibiotic such as streptomycin, cells resistant to the antibiotic can grow on the plate. Initially, it was thought that the streptomycin had induced a mutation to produce the resistant cells. We now know that mutations occur randomly and spontaneously in a population of cells and are independent events not related to the presence of a drug. However, to prove this fact was not a trivial task because it was necessary to demonstrate that resistance to the antibiotic occurred spontaneously in the cells growing on a medium that lacks the antibiotic.

In order to show that mutation to streptomycin resistance spontaneously occurs in a population of cells on a plate would involve laboriously picking each individual colony on the plate and transferring the cells in the colony to a medium that contains streptomycin to determine if the cells can grow in its presence. Furthermore, mutations normally occur at a low frequency in a population of cells, and, therefore, hundreds to thousands of transfers would have to be made from a plate containing five hundred to a thousand colonies to demonstrate a single mutant. However, the workload involved in a mutant search can be reduced significantly by using the technique of **replica plating.**

In this technique, sterile velveteen cloth is placed over a wood block that is pressed onto a plate that contains no antibiotic and that has the suspected mutant colonies. Velveteen consists of millions of tiny threads, and each thread acts as a tiny inoculating loop. The exact colony pattern from the medium lacking streptomycin is replicated by the velveteen onto the medium containing the antibiotic. By comparing the two plates, it can be determined which of the colonies on the original plate are spontaneous mutants to streptomycin. Figure 66.1 illustrates the procedure.

Note in figure 66.1 that organisms are first dispersed on nutrient agar with a glass spreading rod. After incubation, all colonies are transferred from the nutrient agar plate to two other plates: first to a nutrient agar plate and second to a streptomycin agar plate. After incubation, streptomycin-resistant strains are looked for on the streptomycin agar.

First Period

Materials

- 1 petri plate of nutrient agar
- 1 bent glass spreading rod
- 1 ml serological pipette
- beaker of 95% ethanol
- Bunsen burner
- broth culture of *E. coli*

1. Pipette 0.1 ml of *E. coli* from broth culture to surface of medium in petri dish.
2. With a sterile bent glass rod, spread the organisms over the plate following the routine shown in figure 66.1.
3. Incubate this plate at 37°C for 24 hours.

(1) Organisms are spread over nutrient agar with a sterile bent glass rod.

(2) After incubation, colonies are picked up with velveteen colony carrier.

(3) Nutrient agar is inoculated by lightly pressing the carrier onto it.

(4) Streptomycin agar is inoculated with same carrier in same manner.

Figure 66.1 Replica plating technique.

Second Period

Materials

- 1 petri plate culture of *E. coli* from previous period
- 1 petri plate of nutrient agar per student
- 1 petri plate of streptomycin agar (100 micrograms of streptomycin per ml of medium)
- 1 sterile colony carrier per student

1. Carefully lower the sterile colony carrier onto the colonies of *E. coli* on the plate from the previous period.
2. Inoculate the plate of nutrient agar by lightly pressing the carrier onto the medium.

3. Now *without* returning the carrier to the original culture plate, inoculate the streptomycin agar in the same manner.
4. Incubate both plates at 37°C for 2 to 4 days in an enclosed cannister.

Third Period

Materials

- Quebec colony counter and hand counter

1. Examine both plates and record the information called for in Laboratory Report 66.
2. Tabulate the results of other members of the class.

Laboratory Report

Student: _____

Date: _____ Section: _____

66 Mutant Isolation by Replica Plating

A. Tabulation of Results

Count the colonies that occur on both plates and record the information on the chalkboard on a table similar to the one below. After all students have recorded their counts, complete this table.

| STUDENT INITIALS | NUMBER OF COLONIES | | STUDENT INITIALS | NUMBER OF COLONIES | |
	Nutrient Agar	Streptomycin Agar		Nutrient Agar	Streptomycin Agar
TOTALS			**TOTALS**		
AVERAGE PER PLATE			**AVERAGE PER PLATE**		

B. Results

1. What was the total number of *E. coli* colonies counted by all groups combined?

   _____

   _____

2. How many *E. coli* isolates were found to be resistant to streptomycin?

   _____

   _____

3. What percentage of the *E. coli* population tested carries spontaneous mutations that confer streptomycin resistance?

   _____

   _____

C. Short Answer Questions

1. What do the results indicate about the natural rate at which bacteria acquire mutations that confer antibiotic resistance? How can the rate be artificially increased? Would doing so be detrimental to the bacteria in any way?

_____

_____

2. Other than by spontaneous mutations, in what ways do bacteria acquire antibiotic resistance?

_____

_____

3. How have humans contributed to the increasing prevalence of antibiotic-resistant bacteria?

_____

_____

4. In what type of gene would a mutation for streptomycin resistance likely be harbored?

_____

_____

Bacterial Transformation

Acquisition of genetic material, in addition to mutation of genes and loss of unnecessary genes, is an important part of bacterial adaptation and evolution. In many cases, acquisition of new genes (i.e., antibiotic resistance) may mean the difference between survival and death of an organism in the environment. Thus, bacteria have developed various mechanisms, including transformation, transduction, and conjugation, to acquire and assimilate external genetic information. Bacterial transformation refers to the uptake of naked DNA molecules by bacterial cells. Transduction involves transfer of genetic material as mediated by bacteriophages, or bacterial viruses. Conjugation involves transfer of genetic material from one bacterium to another and requires close proximity or contact of bacterial cells.

Bacterial transformation was first discovered in the pathogenic bacterium *Streptococcus pneumoniae.* When non-capsule-forming, and thus nonpathogenic, *S. pneumoniae* were coinjected with killed capsule-forming cells, the nonpathogenic cells acquired the ability to make capsules and thus became virulent by acquiring the genes for capsule formation. This ability of a bacterium to take up naked DNA molecules is termed transformation "competence." Some bacteria, including *S. pneumoniae* and *Bacillus subtilis,* are naturally competent and can take up naked DNA molecules from the environment. Many other bacteria, including *Escherichia coli,* are not naturally competent, but they can be made competent via laboratory manipulation. The ability to manipulate *E. coli* to be transformation competent, and hence take up naked DNA, revolutionized the field of molecular biology and facilitated "cloning" of genes in *E. coli.*

There are two major procedures for transforming naturally noncompetent bacteria, chemical and electrical. Chemical transformation does not require special equipment but is not as efficient as electrical transformation. Chemical transformation requires treating bacterial cells with a relatively high concentration of calcium chloride to make them competent. Once a bacterium is suspended in calcium chloride solution and is subjected to "heat shock," DNA molecules can be taken up by the cell. The exact mechanism for the role of heat shock in DNA uptake is not yet understood. Electrical transformation, or electroporation, is more efficient but requires an electroporator or electrical pulser. In this procedure, electrical pulses permeabilize the cell membrane and allow DNA to enter the cell. Electroporation can be used to introduce DNA to a wide variety of cells, including plant and animal cells, in addition to bacterial cells. Most commonly used DNA molecules for transforming bacteria in the laboratory are plasmids.

Plasmids are extrachromosomal DNA molecules that can replicate independent of the chromosome in the cell. Plasmids often carry a gene or genes, such as antibiotic resistance and metabolic genes, which aid in survival of the organism. In nature, plasmids play an important role in adaptation and evolution of microorganisms because they are involved in transfer of genetic information. One of the most common genetic traits to be carried by plasmids is antibiotic resistance. In a bacterial community, a bacterium that carries an antibiotic-resistance plasmid can transfer that genetic information to other bacteria via "horizontal gene transfer," and this can result in a whole community of bacteria that consists of various species of bacteria with resistance to that particular antibiotic. In a microbiology or a molecular biology laboratory, plasmids play an important role as "cloning vectors" in which an investigator inserts a gene of interest in the plasmid to obtain amplification of the gene. Once the recombinant plasmid is transformed into a bacterial cell, amplification of the cloned gene occurs, because plasmids are often present at multiple copies in the cell. A variety of plasmids are commercially available for a molecular biologist. Several things to consider in choosing the right plasmid for a molecular biology experiment include a selectable marker such as antibiotic resistance, restriction endonuclease sites or multiple cloning sites (MCS) for inserting a gene of interest, presence of a screenable marker to facilitate the identification of a recombinant plasmid, and copy number of the plasmid. A commonly used cloning vector pUC19, which carries the gene *bla* for ampicillin resistance, will be used in the next three experiments.

In the next three exercises, students will introduce a new genetic trait and an observable phenotype into a commonly used strain of *E. coli* by transforming the cells with plasmid pUC19 (figure 67.1) and verifying the results using the molecular techniques of polymerase chain reaction (PCR) and plasmid isolation.

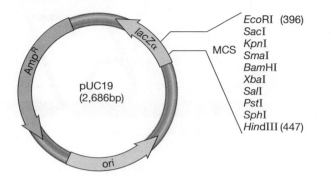

Figure 67.1 Map of plasmid pUC19 showing restriction and multiple cloning sites.

Two methods are presented for transforming bacterial cells. If access to an electroporator is an option, this procedure can be used instead of transforming cells by chemical means.

Method A: Chemical Transformation

Preparation of transformation-competent *E. coli* (to be performed by teaching assistants prior to the laboratory period). See the flow chart in figure 67.2a.

Materials

- *E. coli* strain DH10B (Invitrogen, CA) or equivalent
- 16 × 150 mm test tubes (sterile)
- Luria-Bertani (LB) broth
- L-agar plates
- 500 ml flask (sterile)
- 80 mM $CaCl_2$–50 mM $MgCl_2$ solution (sterile and ice cold)
- 0.1 M $CaCl_2$ (sterile and ice-cold)
- 50% glycerol (sterile and ice-cold)
- 250 ml centrifuge bottles
- 50 ml Oakridge centrifuge bottles
- spectrophotometer
- refrigerated centrifuge (Sorvall RC5 or equivalent)
- Sorvall GSA rotor or equivalent
- Sorvall SS34 rotor or equivalent
- pipettes (5 ml, 10 ml, and 25 ml—sterile) and a pipette aid
- 1.7 ml microcentrifuge tubes (sterile)
- ice and ice bucket
- dry ice or liquid nitrogen (optional)

1. Streak *E. coli* strain DH10B (Invitrogen, CA) on a L-agar plate for single colony isolation and incubate the plate overnight at 37°C.
2. Inoculate 2 ml of Luria-Bertani (LB) broth in a 16 × 150 mm test tube with a single colony of DH10B and grow overnight with aeration at 37°C.
3. On day 2, inoculate 100 ml of LB in a 500 ml flask with 1–2 ml of fresh overnight culture. Grow at 37°C with aeration until the culture reaches OD_{600} of 0.4–0.5.
4. Using aseptic technique, transfer cells to a 250 ml centrifuge bottle and harvest cells by centrifuging in a refrigerated centrifuge (Sorvall, Beckman, or equivalent) at 5000 rpm for 10 minutes at 4° C. From this point on, it is important to keep cells on ice to keep them cold!
5. Decant the supernatant carefully without contaminating the lip of the centrifuge bottle.
6. Gently resuspend the cells in 25 ml of ice-cold 80 mM $CaCl_2$–50 mM $MgCl_2$ solution by pipetting the cells up and down with a 25 ml pipette.
7. Aseptically transfer the cells to a 50 ml centrifuge tube and incubate the cells on ice for 10 minutes.
8. Harvest the cells by centrifuging at 5000 rpm for 10 minutes at 4° C in Sorvall SS34 rotor or equivalent.
9. Decant the supernatant carefully without contaminating the lip of the centrifuge tube.
10. Gently resuspend the cells in 10 ml of ice-cold 80 mM $CaCl_2$–50 mM $MgCl_2$ solution by pipetting the cells up and down with a 10 ml pipette.
11. Harvest the cells by centrifuging at 6000 rpm for 10 minutes at 4° C in Sorvall SS34 rotor or equivalent.
12. Decant the supernatant carefully without contaminating the lip of the centrifuge tube.
13. Gently resuspend the cells in 2.5 ml of ice-cold 0.1 M $CaCl_2$. Add 2.5 ml of ice-cold 50% glycerol and mix gently.

If the transformation-competent cells are to be used within 24 hours, they can be stored at 4° C. However, if they are to be used in the future, they should be aliquoted (200–400 µl/tube) into sterile 1.7 ml microcentrifuge tubes, frozen quickly in dry ice and ethanol bath or liquid nitrogen and stored at −80° C. The cells can be stored for months without appreciable loss of transformation efficiency.

Transformation of *E. coli* strain DH10B with pUC19

See the flow chart in figure 67.2b for steps in the transformation of *E. coli*.

Materials

- pUC19 plasmid DNA at ~ 4 ng/µl (pUC19 DNA can be purchased from various sources including New England Biolabs).
- transformation-competent *E. coli* strain DH10B
- ice and a beaker to hold ice
- shaking water bath or an incubator at 37°C
- heat block or a water bath at 42°C
- SOC medium

A. Preparation of Chemically Competent Cells

B. Transformation with Chemically Competent Cells

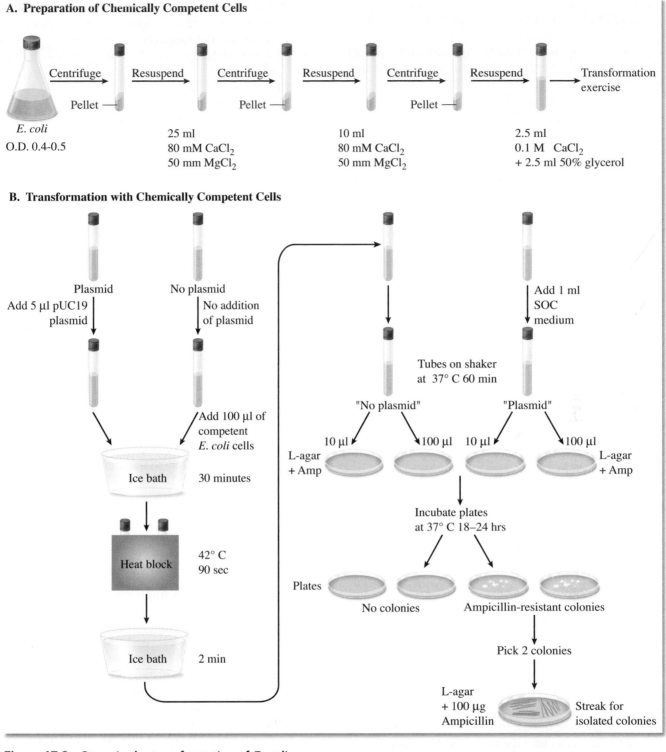

Figure 67.2 Steps in the transformation of *E. coli*.

- L-agar plates with 100 µg/ml of ampicillin (4 plates per group)
- timer
- beaker with 95% alcohol and a spreader
- 13 × 100 mm test tubes (sterile)
- pipetters (P20 and P200) and sterile pipette tips
- sterile 1 ml pipettes and a pipette aid

Remember to practice aseptic technique at each step.

1. Place two sterile 13 × 100 mm test tubes on ice. Mark the first tube as "no plasmid" and the second tube as "plasmid" (figure 67.2b).

2. Use a pipetter (P20) to add 5 µl of pUC19 plasmid DNA (4 ng/µl) to the bottom of the test tube marked as "plasmid."

3. Use a pipetter (P200) to add 100 µl of competent cells to the bottom of each tube and gently mix.

4. Incubate on ice for 30 minutes.

5. Transfer both test tubes to a 42° C heat block or water bath and incubate for 90 seconds.

6. Transfer both test tubes to ice for 2 minutes.

7. Add 1 ml of SOC medium to both test tubes, transfer the test tubes to a shaker, and incubate at 37° C for 60 minutes with gentle shaking. This step is necessary for the plasmid to express the antibiotic-resistance gene (ampicillin resistance, in this case).

8. During incubation, label 4 L-agar plates containing 100 µg/ml of ampicillin as follows: "no plasmid (10 µl)," "no plasmid (100 µl)," "plasmid (10 µl)," and "plasmid (100 µl)."

9. After 60 minutes of incubation, use a pipetter to place cells (either 10 µl or 100 µl) on the appropriate plate. Spread the cells with a cell spreader.

   Note: If you are using a cell spreader made of a glass rod, first dip the spreader in 95% ethanol and burn off the ethanol with flame from the Bunsen burner. After the flame is extinguished, place the spreader on the agar away from the bacterial cells to cool the spreader. Then spread the cells on the agar plate.

10. Incubate the plates overnight at 37° C.

11. The following day, take out the plates and count the number of bacterial colonies present on each plate. If the transformation was successful, one should have no colonies on the "no plasmid" plates but many ampicillin-resistant colonies (Ap^r) on the "plasmid" plates. Calculate the transformation efficiency as number of Ap^r colonies obtained/µg of plasmid DNA used.

12. Pick two ampicillin-resistant colonies and streak them for isolated colonies on fresh L-agar plates containing 100 µg/ml of ampicillin. These colonies will be used for Exercises 68 and 69. Make sure to mark your plates correctly with your names. Place the plates in a 37° C incubator overnight.

13. Next day, take out the plates and store them at 4° C until the day before the next laboratory period.

Method B: Electroporation of DNA

See the flow chart in figure 67.3 for the steps in electroporation of E. coli.

Preparation of electroporation-competent E. coli (to be performed by teaching assistants prior to the laboratory period).

Materials

- *E. coli* strain DH10B (Invitrogen, CA) or equivalent
- 16 × 150 mm test tubes (sterile)
- Luria-Bertani (LB) broth
- L-agar plates
- 500 ml flask (sterile)
- 250 ml centrifuge bottles
- 50 ml Oakridge centrifuge bottles
- spectrophotometer
- refrigerated centrifuge (Sorvall RC5 or equivalent)
- Sorvall GSA rotor or equivalent
- Sorvall SS34 rotor or equivalent
- pipettes (1 ml, 5 ml, 10 ml, and 25 ml—sterile) and a pipette aid
- 1.7 ml microcentrifuge tubes (sterile)
- ice and ice bucket
- 15% glycerol (sterile and ice-cold)

1. Streak *E. coli* strain DH10B on an L-agar plate for single colony isolation and incubate the plate overnight at 37° C.

2. Inoculate 2 ml of Luria-Bertani (LB) broth in a 16 × 150 mm test tube with a single colony of DH10B and grow overnight with aeration at 37° C.

3. The following morning, inoculate 100 ml of LB in a 500 ml flask with 1–2 ml of fresh overnight culture. Grow at 37° C with aeration until the culture reaches OD$_{600}$ of 0.4–0.5.

4. Use aseptic technique to transfer cells to a 250 ml centrifuge bottle and harvest cells by centrifuging in a refrigerated centrifuge (Sorvall, Beckman, or equivalent) at 5000 rpm for 10 minutes at 4° C. From this point on, it is important to keep cells on ice to keep them cold!

5. Decant the supernatant carefully without contaminating the lip of the centrifuge bottle.

6. Resuspend the cells in 10 ml of ice-cold sterile 15% glycerol.

7. Transfer cells to 50 ml centrifuge tubes.

8. Centrifuge to harvest cells at 5000 rpm for 10 minutes at 4° C.

9. Decant the supernatant carefully without contaminating the lip of the centrifuge bottle.

10. Resuspend the cells in 10 ml of ice-cold sterile 15% glycerol.

11. Centrifuge to harvest cells at 5000 rpm for 10 minutes at 4° C.

12. Resuspend cells in 1 ml of ice-cold 15% glycerol.

13. Aliquot cells (200–400 µl/tube) into sterile 1.7 ml microcentrifuge tubes and quick-freeze them in liquid nitrogen or in ethanol and dry-ice bath. The cells can be stored at −80° C for months without appreciable loss of transformation efficiency.

C. Electroporation of Cells

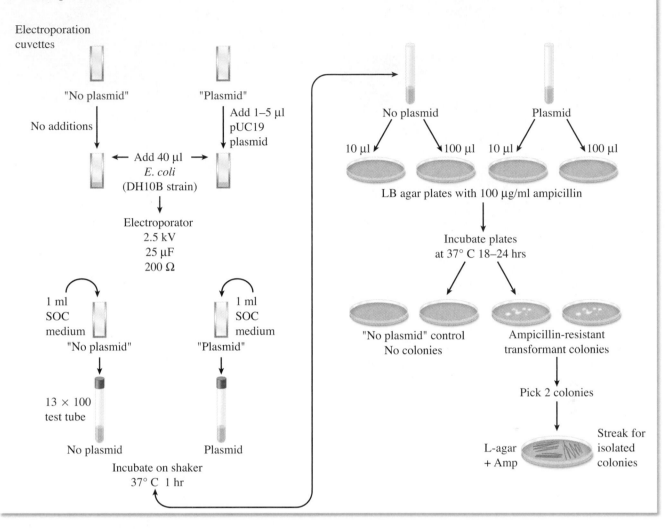

Figure 67.3 Electroporation of *E. coli.*

Electroporation of *E. coli* strain DH10B with pUC19

Materials

- electroporator (Gene Pulser—Bio Rad, CA) or equivalent
- electroporation cuvettes (0.2 cm gap—Bio Rad 165–2086 or equivalent)
- electrocompetent *E. coli* strain DH10B
- pUC19 DNA (4 ng/µl)
- 13 × 100 mm test tubes (sterile)
- Pasteur pipettes with long tips (sterile)
- Pasteur pipette bulb
- shaking water bath or incubator at 37° C
- ice and a beaker to hold ice
- SOC medium (see recipe p. 429)
- L-agar plates with 100 µg/ml of ampicillin (4 plates per group)
- timer
- beaker with 95% alcohol and a spreader

- sterile 13 × 100 mm test tubes (4 tubes per group)
- pipetters (P20 and P200) and sterile pipette tips
- sterile 1 ml pipettes and a pipette aid
- Kim wipes or paper towel

Remember to practice aseptic technique at each step.

1. Place two electroporation cuvettes on ice for 2 minutes to chill. Label one cuvette as "no plasmid" and the other as "plasmid." Keep the cuvettes on ice until you are ready to perform the electroporation.
2. While cuvettes are being chilled, prepare two sterile 13 × 100 mm test tubes by adding 1 ml of SOC medium in each tube. Label one tube as "no plasmid" and the second tube as "plasmid."
3. Use a pipetter (P20) to add 1–5 µl of pUC19 DNA (4 ng/µl) to the "plasmid" cuvette. The volume of DNA added should not exceed 5 µl (figure 67.4).
4. Using a pipetter (P200), add 40 µl of electrocompetent DH10B cells to each cuvette. Maintain

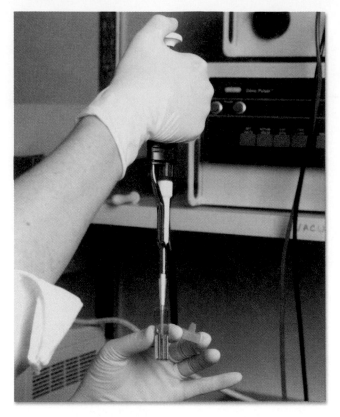

Figure 67.4 Adding DNA to electroporation cuvette.
© The McGraw-Hill Companies/Auburn University Photographic Service

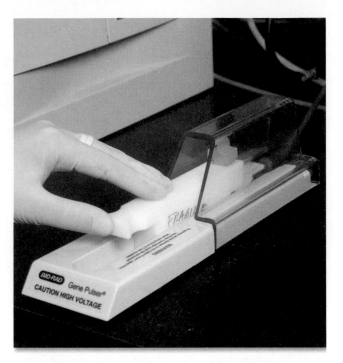

Figure 67.5 Electroporation cuvette.
© The McGraw-Hill Companies/Auburn University Photographic Service

cells and cuvettes on ice. Make certain to mix cells with DNA in the "plasmid" cuvette.

5. Set the electroporator (Bio Rad Gene Pulser or equivalent) to 2.5 kV, 25 μF, and 200 Ω.

6. Remove the "no plasmid" cuvette from ice and quickly wipe off the moisture from the bottom of the cuvette. Check to make sure that cells are at the bottom of the cuvette. Do these steps quickly to keep the cells cold.

7. Place the cuvette in the shocking chamber and apply electrical pulse. Notice the little notch on a cuvette that orients the cuvette in the shocking chamber (figure 67.5).

8. With a sterile pasture pipette, remove 1 ml of SOC from the prepared "no plasmid" test tube and add the medium to the "no plasmid" cuvette that contains the pulsed bacterial cells. This step needs to be done quickly to maintain good transformation efficiency.

9. Transfer cells back to the test tube with a Pasteur pipette.

10. Repeat steps 6 through 9 for the "plasmid" cuvette.

11. Place test tubes in a 37°C shaking incubator and grow with gentle shaking for 1 hour.

12. During incubation, label 4 LB plates containing 100 μg/ml of ampicillin as follows: "no plasmid (10 μl)," "no plasmid (100 μl)," "plasmid (10μl)," and "plasmid (100 μl)."

13. After 60 minutes of incubation, use a pipetter to place cells (either 10 μl or 100 μl) on the appropriate plate. Spread the cells using a cell spreader.

     Note: If you are using a cell spreader made of a glass rod, first dip the spreader in 95% ethanol and burn off the ethanol with flame from the Bunsen burner. After the flame is extinguished, place the spreader on the agar away from the bacterial cells to cool the spreader. Then spread the cells on the agar plate.

14. Incubate the plates overnight at 37° C.

15. The following morning, take out the plates and count the number of bacterial colonies present on each plate. If the transformation was successful, one should have no colonies on the "no plas-

mid" plates but many ampicillin-resistant (Ap^r) colonies on the "plasmid" plates. Calculate the transformation efficiency as number of Ap^r colonies obtained/μg of plasmid DNA used.

16. Take two ampicillin-resistant colonies and streak them on fresh L-agar plates containing 100 μg/ml of ampicillin for isolated colonies. These colonies will be used for Exercises 68 and 69. Make sure to mark your plates correctly with your names. Place the plates at 37° C in the incubator overnight.

17. Next day, take out the plates and store them at 4° C until the day before the next laboratory period.

Use the following directions to make SOC medium.

per liter:

tryptone	20g
yeast extract	5g
NaCl	0.6g
KCl	0.18g
MgCl$_2$	2g
MgSO$_4$	2.5g
glucose	3.6g

Add 950 ml of distilled water and dissolve the solutes. Adjust pH to 7.0 and bring up the volume to 1 L. Autoclave.

Laboratory Report

Student: _____

Date: _____ Section: _____

67   Bacterial Transformation

A. Results

Record the number of ampicillin-resistant colonies obtained on each plate. Calculate the transformation efficiency as # of ampicillin colonies/µg of plasmid DNA used.

Plate	# of Apʳ Colonies	Transformation Efficiency

B. Short Answer Questions

1. Compare the number of Apʳ colonies obtained from plating 10 µl versus 100 µl. Did the 100 µl plating have ten times more Apʳ colonies? Explain.

    _____

    _____

2. What is the purpose of the "no plasmid" sample? What possible explanations are there for getting Apʳ colonies from the "no plasmid" sample?

    _____

    _____

3. What factors are likely to influence the transformation efficiency of E. coli?

    _____

    _____

4. Differentiate between the following types of horizontal gene transfer:

    a. transformation _____

    b. transduction _____

    c. conjugation _____

5. What is competence? Differentiate between natural and artificial competence.

_____

_____

6. How might humans contribute to the development of normal flora bacteria that are antibiotic resistant? Why is this problematic for an individual who is prone to pneumococcal infections?

_____

_____

7. Why is bacterial transformation useful to geneticists and molecular biologists?

_____

_____

Polymerase Chain Reaction for Amplifying DNA

Note: A day before beginning Exercise 68, students need to inoculate 3 ml of Luria broth (LB) in a 16 × 150 mm test tube with a colony of DH10B, and 3 ml of LB + ampicillin (100 μg/ml) in two 16 × 150 mm test tubes each with a colony from two ampicillin-resistant transformants from Exercise 67. Place the test tubes in the shaking 37° C incubator or water bath and grow bacteria overnight with vigorous shaking (220–250 rpm).

Verification of Bacterial Transformation

In this exercise and Exercise 69, presence of pUC19 in the transformed *E. coli* will be verified by two molecular methods: amplification of the ampicillin-resistance gene via the polymerase chain reaction, and isolation of pUC19 from transformed cells. On the first day, students should set up the polymerase chain reaction (PCR) reaction, and while the PCR is being run, start on the plasmid isolation. On the second day, students can finish the plasmid isolation and run both the PCR product and the isolated plasmid on an agarose gel to visualize the DNA.

Polymerase Chain Reaction (PCR)

One of the most important technical developments of the past 20 years in molecular biology is the polymerase chain reaction. The ability to rapidly and easily amplify almost any piece of DNA *in vitro* revolutionized molecular biology and made DNA arrays and the advent of genomics possible. PCR utilizes the heat-stable property of DNA polymerases from organisms that live in extremely warm environments, such as hot springs and thermovents, to rapidly amplify DNA of interest *in vitro*. DNA polymerases are enzymes that replicate DNA. The first thermostable DNA polymerase to be used in this capacity was the *Taq* enzyme, which was isolated from *Thermus aquaticus,* a bacterium found to grow in the hot springs of Yellowstone National Park by Dr. Thomas D. Brock of the University of Wisconsin in 1965. Since then, many other thermostable DNA polymerases have been isolated from other thermophiles and utilized for amplification of DNA *in vitro*. The heat-stable property of *Taq* and other thermostable polymerases

is essential for in vitro amplification of DNA because they can withstand the high temperature (95° C) that is necessary to denature the DNA template between replication cycles.

PCR technology, developed by Kary Mullis in the 1980s, revolutionized molecular biology. The technique involves annealing a pair of oligonucleotide primers to specific sites on the template DNA and using thermostable DNA polymerase to replicate the DNA sequences between two primer sequences. Specifically, the technique involves denaturation of template DNA at 95° C, followed by annealing a pair of oligonucleotide primers at the appropriate temperature, and DNA replication at 72° C, the optimal temperature for activity of many thermostable DNA polymerases. This cycle is repeated 30 to 40 times and, because each newly synthesized DNA molecule, in turn, serves as a template for additional rounds of replication, this results in exponential amplification of the specific DNA sequence between two primers (figure 68.1). This relatively simple technique is used in almost every area of molecular biology, including microbiology, cellular biology, and even forensics. The fact that only one copy of the template DNA is needed to amplify and obtain many copies of the DNA means that one hair follicle or one drop of blood (even dry blood) is enough for an investigator to obtain the necessary genetic information via PCR. Another advantage of PCR is that it is not necessary to isolate the template DNA from the cell for *in vitro* amplification. In a majority of cases, one can PCR amplify a DNA fragment of interest simply by using the whole cell as the source for template DNA. In the first step of the procedure, the sample is heated to 95° C, which lyses and denatures the cellular material including proteins and nucleic acids. Although PCR can be performed manually using three water baths set at appropriate temperatures, due to the labor involved in such an experiment, it is normally performed in a computer-controlled apparatus designated thermocycler. When designing a PCR experiment, several parameters need to be considered: quantity of template DNA, quality of oligonucleotide primers, annealing temperature of primers to the template, and extension time needed to replicate the DNA of interest. Of these considerations, proper design of oligonucleotide primers is the most

2^n = number of copies of desired sequence
n = number of cycles

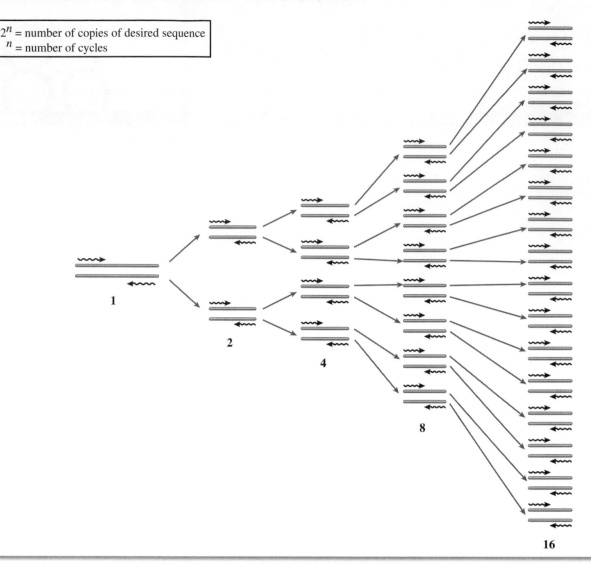

Figure 68.1 Amplification of DNA by polymerase chain reaction.

important factor in achieving success in PCR. A good pair of oligonucleotide primers should hybridize to only one specific site on the template for each primer but should not form extensive secondary structures. Most of the primers are designed using computer programs that take all of these points into account. For this exercise, the oligonucleotide primers have been designed for amplification of the *bla* gene, which encodes for the enzyme β-lactamase on the plasmid pUC19, and gives ampicillin resistance for the plasmid.

In the first part of today's exercise, students will use the power of PCR to verify the presence of the plasmid pUC19 by verifying the presence of the ampicillin-resistance gene, in the transformants obtained from Exercise 67. Although the presence of ampicillin resistance in the transformants suggests the presence of pUC19, it is necessary to verify via molecular methods to be certain.

The second part of today's exercise is to initiate isolation of pUC19 directly from the transformed cells. Plasmid isolation will be completed during Exercise 69 at which time the DNA from PCR and plasmid isolation will be visualized by agarose gel electrophoresis.

Materials

- thermocycler
- 16 × 150 mm test tubes (sterile)
- LB
- LB + ampicillin (100 μg/ml)
- pUC19 DNA (0.01 ng/μl)
- thin-walled PCR tubes (200 μl volume)
- *Taq* polymerase or equivalent
- 10X buffer for thermostable polymerase
- primer 1 (50 μM concentration)
- primer 2 (50 μM concentration)

- dNTP mix (2.5 mM concentration of each dATP, dCTP, dGTP, and dTTP)
- pipetters (P10, P20, P200) and pipette tips (sterile)
- gloves
- ice and a beaker to hold ice
- 1.7 ml microcentrifuge tube (sterile)

Note: For PCR analysis, wear gloves throughout the experiment to avoid contaminating samples with bacteria from your hands or introducing nucleases from your hands that will degrade the template, or inactivating the enzyme.

Refer to the flow chart in figure 68.2.

1. Remove the test tubes from the 37° C shaking incubator. There should be three test tubes with each test tube containing approximately 3 ml of culture. The first test tube should contain the *E. coli* strain DH10B that was used for transformation. The second and the third test tubes contain the two ampicillin-resistant transformants of DH10B from Exercise 67, respectively (save these cultures on the bench top for plasmid isolation).
2. Prepare four thin-walled 200 µl PCR tubes by marking them 1 through 4 and place the tubes on ice.
3. Using a pipetter (P10), add 1 µl of bacterial culture from the test tube containing "plasmid #1"

to the PCR tube #1 and add 1 µl of culture from plasmid #2 to PCR tube #2. To tube #3, add 1 µl of DH10B to serve as the negative control for the experiment. Finally, add 1 µl of purified pUC19 DNA (0.01 ng/µl) to tube #4 to serve as the positive control for the experiment (figure 68.3).

4. In a 1.7 ml microcentrifuge tube on ice, prepare the master mix solution for amplification of the *bla* gene from transformed cells. The master mix should contain everything with the exception of template DNA. To make the master mix for five reactions, add the following:

10X buffer for *Taq*:	25 µl = 1X final concentration
2.5 mM dNTP mix:	20 µl = 0.2 µM final concentration
primer #1:	2.5 µl = 0.5 µM final concentration (stock = 50 µM)
primer #2:	2.5 µl = 0.5 µM final concentration (stock = 50 µM)
deionized water:	195 µl
Taq enzyme:	2.5 units

final volume = 245 µl

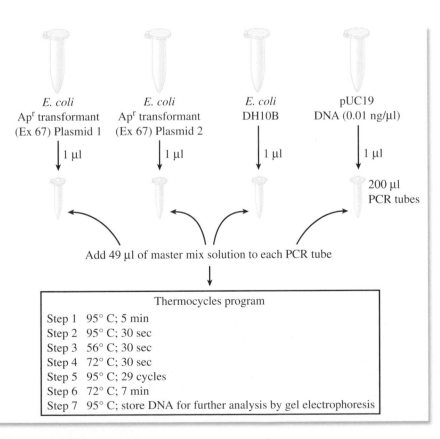

Figure 68.2 Amplification of *E. coli* plasmids by polymerase chain reaction.

E. coli
Ap^r transformant
(Ex 67) Plasmid 1

E. coli
Ap^r transformant
(Ex 67) Plasmid 2

E. coli
DH10B

pUC19
DNA (0.01 ng/µl)

1 µl   1 µl   1 µl   1 µl

200 µl
PCR tubes

Add 49 µl of master mix solution to each PCR tube

Thermocycles program

Step 1 95° C; 5 min
Step 2 95° C; 30 sec
Step 3 56° C; 30 sec
Step 4 72° C; 30 sec
Step 5 95° C; 29 cycles
Step 6 72° C; 7 min
Step 7 95° C; store DNA for further analysis by gel electrophoresis

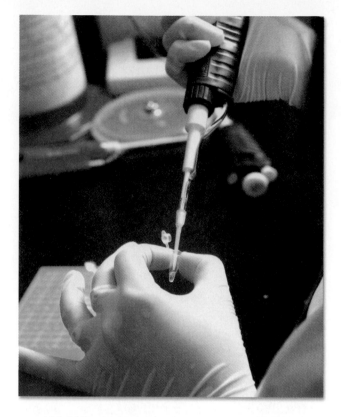

Figure 68.3 Addition of PCR mix to a PCR tube.
© The McGraw-Hill Companies/Auburn University Photographic Services

Figure 68.4 Thermocycler.
© The McGraw-Hill Companies/Auburn University Photographic Services

5. Use a pipetter (P100 or P200) and add 49 µl of the PCR master mix to each of the PCR tubes on ice (figure 68.3). Change the pipetter tips each time to avoid cross-contamination between samples. Close the caps of the microcentrifuge tubes.
6. Place the PCR tubes in a thermocycler and start the program (see figure 68.4).

   Program:
   Step 1:   95° C for 5 minutes to lyse cells and denature DNA
   Step 2:   95° C for 30 seconds to denature
   Step 3:   56° C for 30 seconds to anneal primers
   Step 4:   72° C for 1 minute, 30 seconds to replicate DNA up to 1.5 kb
   Step 5:   Cycle back to Step 2 for additional 29 cycles
   Step 6:   72° C for 7 minutes to complete extensions
   Step 7:   4° C for storage until ready to end the run

7. Once the PCR run is complete, store PCR tubes at 4° C until the next laboratory period.

Primer 1: 5′-GAC GCT CAG TGG AAC GAA AA-3′
Primer 2: 5′-ATG TGC GCG GAA CCC CTA TT-3′

Size of the amplified DNA = ~1.1 kb

Plasmid Isolation

In a molecular biology laboratory, plasmids are important because of their use as **cloning vectors.** Taking advantage of the extrachromosomal replication property of plasmids, molecular biologists developed the use of plasmids to **clone** pieces of DNA fragments to obtain *in vivo* amplification of the DNA for the purpose of genetic manipulation. In order to utilize plasmids as cloning vectors, one has to be able to easily introduce and to isolate plasmids from bacterial cells. In Exercise 67, students learned to introduce plasmids into bacteria via transformation. In this exercise, students will isolate pUC19 from the transformed *E. coli* to verify presence of the plasmid in the bacterial cell to confirm the ampicillin resistance conferred by pUC19.

Over the past 20 years, various methods have been developed for isolating plasmids from bacteria. A common theme in all of the methods for plasmid isolation is the breaking of the cell wall and denaturation of cellular proteins and the chromosomal DNA. The cellular debris are subsequently removed by centrifugation, and plasmid DNA and RNA are recovered. Plasmid DNA can be further isolated away from the contaminating RNA by treating the sample with RNase. In this exercise, students will isolate plasmids via alkaline lysis with SDS, a popular technique developed by Birnboim and Doly in 1979 (*Nucleic Acids Res.* 7: 1513–1523).

Materials

- 1.7 ml microcentrifuge tubes (sterile)
- microcentrifuge
- pipetters (P20, P200, P1000) and pipette tips (sterile)
- ice and a beaker to hold ice
- Lysis Solution 1 (see ingredients p.439)
- Lysis Solution 2 (see ingredients p.439)
- Lysis Solution 3 (see ingredients p.439)
- phenol:chloroform:isoamyl alcohol (25:24:1)
- chloroform:isoamyl alcohol (24:1)
- 100% ethanol (ice-cold)
- 70% ethanol (ice-cold)
- TE (10 mM Tris–1 mM EDTA, pH 8.0)
- deionized water
- RNase A (1 mg/ml in TE, pH 8.0)
- gloves

Part 1: Plasmid Isolation

See the flow chart in figure 69.1.

1. Label two 1.7 ml microcentrifuge tubes as "plasmid 1" and "plasmid 2" (figure 69.1).
2. Transfer 1.5 ml of cultures belonging to the two ampicillin-resistant transformants of DH10B to each of the microcentrifuge tubes.
3. Harvest the bacterial cells by centrifuging at 13,000 rpm for 30 seconds in a microcentrifuge (figure 69.2).
4. Carefully remove the supernatant by pipette, pipetter, or aspiration.
5. Add 100 µl of ice-cold Lysis Solution 1 to each tube and resuspend the cell pellets by pipetting up and down with a pipetter. Make sure the pellet is completely resuspended.
6. Incubate at room temperature for 5 minutes.
7. Add 200 µl of Lysis Solution 2 to each tube. Mix by inverting the tubes 5–10 times. Do not vortex or agitate vigorously!
8. Incubate at room temperature for 5 minutes.
9. Add 150 µl of Lysis Solution 3 to each tube and mix gently by inverting the tubes several times.
10. Incubate on ice for 5 minutes.
11. Centrifuge the lysate at 13,000 rpm for 10 minutes to collect cellular debris. During the spin, prepare two fresh microcentrifuge tubes and label them as "plasmid 1" and "plasmid 2."
12. Carefully transfer the supernatant to a fresh tube using a pipetter (P1000).
13. Add equal volume (~0.4–0.5 ml) of phenol:chloroform:isoamyl alcohol (25:24:1) solution to each tube and mix the contents by vortex. Phenol removes protein contamination from a DNA sample. Wear gloves and exercise caution when handling phenol!
14. Centrifuge at 13,000 rpm for 2 minutes to separate the phases.
15. During the spin, prepare two fresh microcentrifuge tubes and label them as "plasmid 1" and "plasmid 2."
16. Carefully remove the aqueous phase (upper phase) to new tubes. Be careful not to disturb the protein precipitate in the interphase between the aqueous and the organic phases.

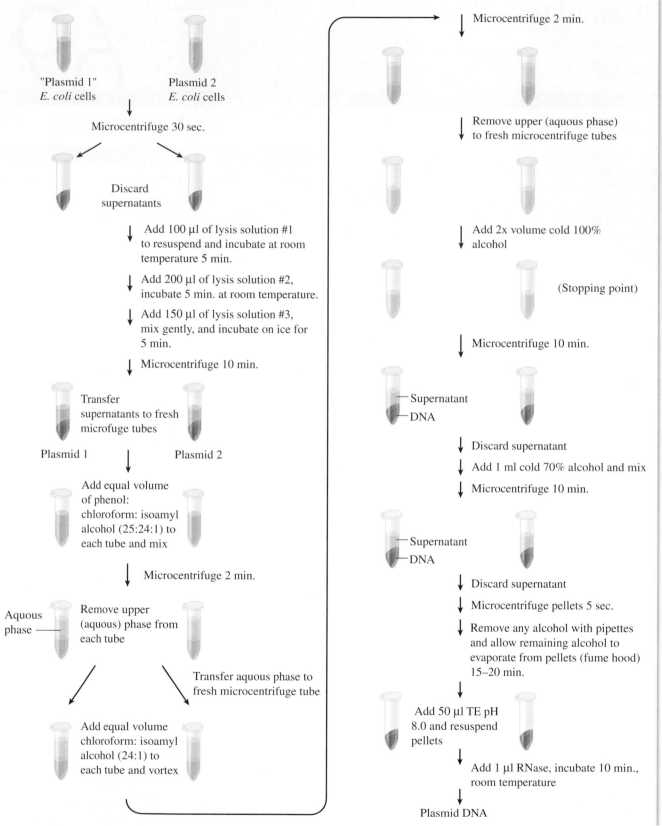

Figure 69.1 Plasmid isolation.

Figure 69.2 Eppendorf Microcentrifuge.
© The McGraw-Hill Companies/Auburn University Photographic Services

17. Add equal volumes of chloroform:isoamyl alcohol (24:1) solution to each tube and vortex. This step removes residual phenol from the sample.
18. Centrifuge at 13,000 rpm for 2 minutes to separate the phases.
19. During the spin, prepare two fresh microcentrifuge tubes and label them as "plasmid 1" and "plasmid 2."
20. Carefully remove the aqueous phase (upper phase) to new tubes.
21. Add 2X volume of ice-cold 100% ethanol to precipitate nucleic acids in the samples.
22. Incubate at room temperature for 5 minutes.
     [This is a good place to stop if time is short. Samples can be stored until the next laboratory period at −20° C.]
23. Centrifuge at 13,000 rpm for 10 minutes to pellet nucleic acids.
24. Carefully discard the supernatant.
25. Add 1 ml of ice-cold 70% ethanol and mix by inverting the tube several times.
26. Centrifuge at 13,000 rpm for 10 minutes to pellet nucleic acids.
27. Carefully discard the supernatant.
28. Centrifuge at 13,000 rpm for 5 seconds to collect remaining ethanol and carefully remove the ethanol with a pipetter.

29. Place tubes with their caps open at room temperature for 15–20 minutes to evaporate away residual ethanol. Placing the tubes in a fume hood will accelerate this step.
30. Add 50 µl of TE (pH 8.0) or deionized water and resuspend the DNA pellets in each tube by pipetting up and down gently with a pipetter.
31. Add 1 µl of RNase A (1 mg/ml) to degrade RNA left in the sample. Incubate at room temperature for 10 minutes.
32. Plasmid DNA can be stored at −20° C.
33. (Optional) Following RNase A digestion, plasmid DNA can be further purified by a second round of phenol extraction and ethanol precipitation.

Lysis Solution 1:
     50 mM glucose
     25 mM Tris, pH 8.0
     10 mM EDTA

Lysis Solution 2:
     0.2 N NaOH
     1% SDS

Lysis Solution 3: 3 M sodium acetate, pH 4.8

Part 2: Agarose Gel Electrophoresis

Nucleic acids can be separated based on size on agarose or polyacrylamide gels. Choice of the matrix depends on the size of the nucleic acids being separated. Polyacrylamide gels are for separation of very small DNA fragments (~10–500 nucleotides) and agarose gels are for separation of DNA fragments that are larger in size (more than 100 nucleotides). By varying the concentration of the matrix, and thus the density of the gel, one can influence the migration of the nucleic acid fragments through the pores of the gel. The higher the density, the more difficulty a larger nucleic acid fragment has in migrating through the pores. Thus, for separation of small fragments, a gel with a high concentration of the matrix is used and, for separation of large fragments, a gel with a low concentration of the matrix is used. Nucleic acids can be visualized on the gel by staining them with intercalating fluorescent dyes. The most commonly used dye for visualization of DNA is ethidium bromide, which intercalates every 10 bases and fluoresces under UV light. Intercalation of ethidium bromide at regular intervals allows detection of DNA fragments on a gel as well as estimation of concentration of the DNA fragment by comparing the intensity to the known DNA standard. Analysis of DNA by agarose gel electrophoresis is one of the most common techniques that is used daily in a molecular biology laboratory. In this exercise, students will run a mini-agarose gel to visualize the DNA molecules

439

from their PCR experiment and plasmid isolation, and learn to appreciate this simple and yet essential technique.

Materials

- agarose
- 1X TBE (see ingredients on p.442)
- 200 ml bottle or a flask
- microwave oven or a boiling water bath
- oven mitt
- ethidium bromide (1 mg/ml)
- mini-gel apparatus (gel tray, a comb with 10 wells, gel chamber)
- power supply
- 1.7 ml microcentrifuge tubes
- 10X DNA sample loading dye (see ingredients on p.442)
- pipetters (P10 or P20) and pipette tips
- 1 kb DNA ladder (Invitrogen, CA, or equivalent)
- UV safety glasses
- UV transilluminator
- photodocument system for UV
- gloves

Refer to the flow chart in figure 69.3.

1. Add 0.6 g of agarose in 50 ml of 1X TBE, pH 8.0 (Tris-Borate-EDTA buffer) in a 200 ml bottle to make a solution of 1.2% agarose (w/v). Bottle cap should be loose.

2. Place the bottle in a microwave oven and heat at 50% power for 5 minutes to melt the agarose. If agarose is not fully melted, continue heating for 2–5 more minutes until it is melted, and allow the agarose to cool on the bench top until it is comfortable to touch (~55° C).

> **CAUTION:** Be careful not to shake the hot liquid because it can boil over and cause severe injury! Wear an oven mitt or equivalent when removing agarose from the microwave oven.

3. Add 25 µl of 1 mg/ml ethidium bromide solution to the gel to achieve final concentration of 0.5 µg/ml of ethidium bromide.

> **CAUTION:** Ethidium bromide is a carcinogen. Please handle the ethidium bromide solution as well as the gel containing ethidium bromide with gloves!

4. Assemble a mini-agarose gel tray comb with 8–10 wells according to the manufacturer's instructions. If using a homemade apparatus, tape the ends of the tray to prevent leakage. Then, place the comb so the bottom of the comb is ~1 mm from the gel tray (figure 69.4).

5. When the agarose has cooled, slowly pour the slurry into the gel tray, preassembled with a comb, until a gel of approximately 5 mm in thickness is obtained.

6. Allow the agarose gel to polymerize at room temperature. As agarose polymerizes, it will turn from clear to opaque gray in color.

7. While the gel is polymerizing, prepare 7 microcentrifuge tubes and label them accordingly as 1 kb DNA size ladder, PCR #1, PCR #2, PCR #3, PCR #4, plasmid #1, and plasmid #2 (figure 69.3).

8. Add 3 µl of the 1 kb DNA ladder, 9 µl of samples from the PCR experiment, and 5 µl of the plasmid samples to appropriate tubes.

9. Add 6 µl of water to the DNA ladder tube and 4 µl of water to the tubes containing the plasmids to bring the total volume of each tube to 9 µl.

10. Add 1 µl of the 10X DNA sample loading dye to each sample and mix.

11. Once the gel is polymerized, place the gel in the running chamber and cover the gel completely with 1X TBE running buffer.

12. Using a pipetter (P20), carefully load 3 µl of the 1 kb DNA ladder (0.1 µg/µl) to the first well. Change the pipette tip and load 10 µl of each sample to the other wells. (To load DNA, carefully break the surface of the buffer, place the pipette tip directly over the well, and slowly dispense the sample. Do not penetrate the well with the pipette tip. The loading dye contains sucrose, and thus the sample will sink into the well if it is dispensed directly above the well.)

13. Place the chamber cover with the negative electrode (black wire) on the side of the chamber that has the top of the gel and the positive electrode (red wire) on the side of the chamber that has the bottom of the gel. Connect the electrodes to a power supply (black to black and red to red), turn on the power supply, and run the gel at 100 V until the blue dye (bromophenol blue) is approximately 1–2 cm from the bottom of the gel. If the gel is 10 cm in length, the run time is approximately 1 hour under these conditions.

14. Turn off the power supply. Remove the gel and place it on a UV transilluminator. Put on protective eyewear to protect your eyes from the UV light, turn on the UV lamp, and visualize the DNA fragments on a UV transilluminator. Take a picture of the gel for analysis (figure 69.5).

WARNING

Dispose of the gel containing ethidium bromide only in a special designated container. Do not dispose of the gel in the regular trash.

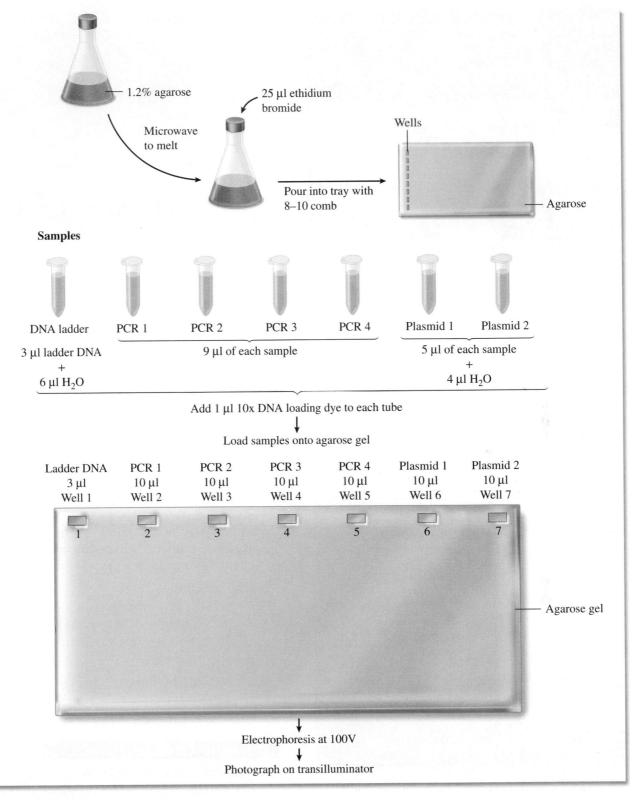

Figure 69.3 **Agarose gel preparation.**

Figure 69.4 Electrophoresis apparatus and a gel assembly.
© The McGraw-Hill Companies/Auburn University Photographic Services

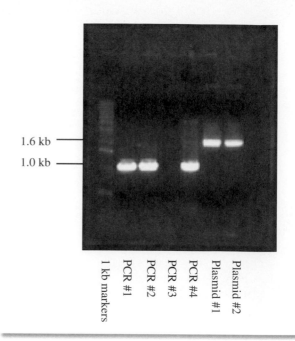

Figure 69.6 Expected results.

Figure 69.5 Transilluminator and a gel document system.
© The McGraw-Hill Companies/Auburn University Photographic Services

15. Analyze the picture of the gel (figure 69.6). The PCR product should be approximately 1.1 kb in size. In the lanes containing plasmid, one may see more than one band. This is because isolated plasmids can exist in three conformations depending on how carefully the isolation was performed. In the cell, plasmids exist mostly in closed circular "supercoiled" form. During the isolation, some plasmids may have been converted to "open circle" or even "linear" forms. These different forms of plasmids migrate differently in a gel and result in multiple bands on a gel.

10X DNA loading buffer:
    0.25% (w/v) bromophenol blue
    0.25% (w/v) xylene cyanol FF
    40% (w/v) sucrose

1X TBE:
    90 mM Tris-Borate
    2 mM EDTA, pH 8.0

Literature Cited

1. Birnboim and Doly. 1979. *Nucleic Acids Res.* 7: 1513–1523.

Laboratory Report

Student: _____

Date: _____ Section: _____

68–69   Polymerase Chain Reaction and Plasmid Isolation

A. Results

Analyze and describe the DNA band patterns present on the gel. Assign an approximate molecular weight to the DNA bands in the test samples by comparing their migration to the molecular weight markers. Are the sizes of the bands appropriate? PCR fragment should be approximately 1.1 kb and pUC19 is approximately 2.7 kb.

B. Short Answer Questions

1. PCR analysis indicated the presence of the bla gene in an Ap^r bacterium but no plasmid DNA could be isolated from the same bacterium. Give possible reasons for this outcome.

    _____

    _____

2. For PCR, what property of DNA does temperature influence? Why do the "denaturation" and "annealing" steps proceed at different temperatures?

    _____

    _____

3. Before the isolation of a thermostable DNA polymerase and the development of thermocyclers, how do you think PCR cycling could have been accomplished? What would have been the drawbacks of such a procedure?

    _____

    _____

4. Why are primers required for DNA amplification? Why is the correct concentration of primers in the reaction critical to successful amplification?

    _____

    _____

5. Why is it important to wear gloves when setting up the PCR tubes?

    _____

    _____

6. Why is PCR so valuable for forensic analyses? Are there any drawbacks to its application?

    _____

    _____

7. What causes the separation of DNA fragments during agarose gel electrophoresis?

8. Following gel electrophoresis, why might more than one band be visualized for a single plasmid?

9. What properties of ethidium bromide make is useful for the detection of DNA in agarose gels? Which of these properties also makes it a carcinogen?

Medical Microbiology and Immunology

Most of the exercises up to this point in this manual pertain to applications that are used in all aspects of microbiology such as staining, aseptic technique, culture methods, and identification of unknown bacteria. The following exercises will focus on bacteria that are of medical importance, on the methods used to isolate and characterize these bacteria, and on immunological tests used to diagnose disease. The techniques and tests used for studying each type of organism will vary, and in some cases, they will be specific for a group.

A complete coverage of medical microbiology is not possible in an introductory course because of the number of organisms involved and because some of the pathogens, for example, *Mycobacterium tuberculosis*, require very specialized techniques for handling owing to their pathogenicity. However, this manual would not be complete if it did not include some of the procedures and tests used to identify some of the common pathogens such as the staphylococci, streptococci, and some of the enteric bacteria.

Exercises 70 through 72 will involve the specific methods used to characterize the pyogenic cocci, *Staphylococcus* and *Streptococcus*, and the gram-negative intestinal bacteria. In Exercises 73 through 75, immunological and serological methods used to diagnose bacterial diseases will be explored. Hematological tests involving the counting of blood cells and blood typing are covered in Exercises 76 and 77; these are often included in a medical microbiology course. Finally, Exercise 78 is a study of how disease is spread from one person to another; this is accomplished by studying a "synthetic epidemic."

445

The Staphylococci:
Isolation and Identification

Collectively, the staphylococci and streptococci are referred to as the **pyogenic** (pus-forming) gram-positive cocci. As a group, they cause abscesses, boils, carbuncles, osteomyelitis, and fatal septicemias. The staphylococci were originally isolated from pus in wounds but were subsequently demonstrated to be part of the normal flora of the nasal membranes, hair follicles, skin, and perineum in healthy individuals. Most of the human population are carriers of staphylococci. This is especially important in hospitals where 90% of the personnel can be carriers and where these bacteria are responsible for many **nosocomial** (hospital-acquired) infections. To further complicate matters, *Staphylococcus aureus* has developed resistance to many antibiotics including methicillin. MRSA, or methicillin-resistant *S. aureus,* is a major epidemiological problem in hospitals where it is responsible for a variety of hospital-acquired infections. Recently, a community form of MRSA has been isolated from infections in individuals who have not been hospitalized.

The staphylococci are gram-positive spherical bacteria that divide in more than one plane to form irregular clusters of cells (figure 70.1). In fact, the name "staphylococcus" is derived from the Greek, meaning "bunch of grapes." In the present edition of *Bergey's Manual,* the staphylococci are grouped in Family I, Micrococcaceae, with three other genera. The streptococci differ significantly from the staphylococci and are placed in a separate family, Deinococcaceae. The staphylococci are a coherent phylogenetic group that have very little in common with the streptococci except that both are gram-positive non-spore-forming cocci.

There are nineteen species of staphylococci listed in *Bergey's Manual,* with the most medically important being *S. aureus, S. epidermidis,* and *S. saprophyticus.* A significant characteristic separates S. aureus from the other two species: *S. aureus* can coagulate plasma, which is often correlated with its pathogenicity. However, the non-coagulase-producing organisms *S. epidermidis* and *S. saprophyticus* do cause infections of the cerebrospinal fluid, prosthetic joints, and vascular grafts.

The focus of this experiment will be to differentiate the three medically important species of staphylococci. If other species are encountered, the student may wish to employ the API-Staph miniaturized test strip system (Exercise 48).

In this experiment, we will attempt to isolate staphylococci from (1) the nose, (2) a **fomite** or inanimate object, and (3) an "unknown control." If the nasal membranes and fomite fail to yield a positive isolate, the unknown control will yield a positive isolate provided all the inoculations and tests are performed correctly.

Since *S. aureus* is the most significant pathogen in the group, it will be the focus of our concern in the exercise. Therefore, only its characteristics will be outlined. However, the non-coagulase-producing organisms do cause disease. *S. saprophyticus,* a causative agent of urinary tract infections in sexually active young women, is responsible for 10–20% of these infections. The non-coagulase-producing staphyloocci are also important pathogens in nosocomial, or hospital-acquired, infections. They have been associated with infections in cerebrospinal fluid, prosthetic joints, and vascular graphs.

S. aureus cells are 0.8 to 1μm in diameter and may occur singly, in pairs, or in clusters. Colonies of the bacterium grown on trypticase soy agar or blood agar are 1 to 3 mm in diameter, and they may be yellow, orange, or white in color. *S. aureus* is

Figure 70.1 Staphylococci.
© Science VU/Charles W. Stratton/Visuals Unlimited

Glycocalyx slime

Catheter surface

Cell cluster

Table 70.1 Differentiation of Three Species of Staphylococci

	S. aureus	S. epidermidis	S. saprophyticus
α-toxin	+	–	–
Mannitol (acid only)	+	–	(+)
Coagulase	+	–	–
Biotin for growth	–	+	NS
Novobiocin	S	S	R

Note: NS = not significant; S = sensitive; R = resistant; (+) = mostly positive

salt-tolerant and will grow well in media containing 10% sodium chloride. In addition, *S. aureus* ferments mannitol to produce acid. Virtually all strains are coagulase positive and will cause serum to form a clot. The organism also produces an α-toxin that causes a wide, clear zone of beta-hemolysis on blood agar. In rabbits, this α-toxin has been shown to cause localized necrosis of tissue and death. The other two species of staphylococci do not produce coagulase or the α-toxin. *S. saprophyticus* will ferment mannitol to produce acid without gas. Table 70.1 lists the principal characteristics of the three species of staphylococci.

To determine the incidence of carriers in our classroom, as well as the incidence of the organism on common fomites, we will follow the procedure illustrated in figure 70.2. Results of class findings will be tabulated on the chalkboard so that all members of the class can record data required in Laboratory Report 70. The characteristics we will look for in our isolates will be (1) beta-type hemolysis (α-toxin), (2) mannitol fermentation, and (3) coagulase production. Organisms found to be positive for these three characteristics wll be *presumed* to be *S. aureus*. Final confirmation will be made with additional tests. Proceed as follows:

First Period

(Specimen Collection)

Note in figure 70.2 that swabs that have been applied to the nasal membranes and fomites will be placed in tubes of enrichment medium containing 10% NaCl (*m-staphylococcus* broth). Since your unknown control will lack a swab, initial inoculations from this culture will have to be done with a loop.

Materials

- 1 tube containing numbered unknown control
- tubes of *m*-staphylococcus broth
- 2 sterile cotton swabs

1. Label the three tubes of *m*-staphylococcus broth NOSE, FOMITE, and the number of your unknown control.
2. Inoculate the appropriate tube of *m*-staphylococcus broth with one or two loopfuls of your unknown control.
3. After moistening one of the swabs by immersing partially into the "nose" tube of broth, swab the nasal membrane just inside your nostril. A small amount of moisture on the swab will enhance the pick-up of organisms. Place this swab into the "nose" tube.
4. Swab the surface of a fomite with the other swab that has been similarly moistened and deposit this swab in the "fomite" tube.

   The fomite you select may be a coin, drinking glass, telephone mouthpiece, or any other item that you might think of.
5. Incubate these tubes of broth for 4 to 24 hours at 37° C.

Second Period

(Primary Isolation Procedure)

Two kinds of media will be streaked for primary isolation: mannitol salt agar and staphylococcus medium 110.

Mannitol salt agar (MSA) contains mannitol, 7.5% sodium chloride, and phenol red indicator. The NaCl inhibits organisms other than staphylococci. If the mannitol is fermented to produce acid, the phenol red in the medium changes color from red to yellow.

Staphylococcus medium 110 (SM110) also contains NaCl and mannitol, but it lacks phenol red. Its advantage over MSA is that it favors colony pigmentation by different strains of *S. aureus*. Since this medium lacks phenol red, no color change takes place as mannitol is fermented.

Materials

- 3 culture tubes from last period
- 2 petri plates of MSA
- 2 petri plates of SM110

1. Label the bottoms of the MSA and SM110 plates as shown in figure 70.2. Note that to minimize the number of plates required, it will be necessary to make half-plate inoculations for the nose and fomite. The unknown control will be inoculated on separate plates.
2. Quadrant streak the MSA and SM110 plates with the unknown control.
3. Inoculate a portion of the nose side of each plate with the swab from the nose tube; then, with a sterile loop, streak out the organisms on the remainder of the agar on that half of each plate. The swabbed areas will provide massive growth; the streaked-out areas should yield good colony isolation.

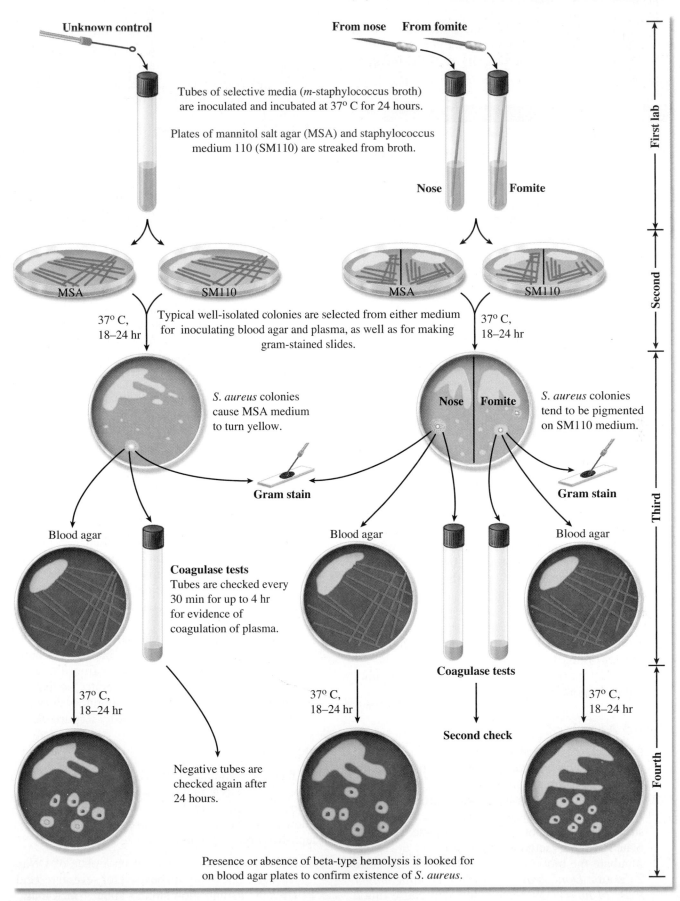

Unknown control

From nose **From fomite**

Tubes of selective media (*m*-staphylococcus broth)
are inoculated and incubated at 37° C for 24 hours.

Plates of mannitol salt agar (MSA) and staphylococcus
medium 110 (SM110) are streaked from broth.

Nose **Fomite**

MSA SM110 MSA SM110

37° C,
18–24 hr

Typical well-isolated colonies are selected from either medium
for inoculating blood agar and plasma, as well as for making
gram-stained slides.

37° C,
18–24 hr

S. aureus colonies
cause MSA medium
to turn yellow.

Nose **Fomite**

S. aureus colonies
tend to be pigmented
on SM110 medium.

Gram stain **Gram stain**

Blood agar **Blood agar** **Blood agar**

Coagulase tests
Tubes are checked every
30 min for up to 4 hr
for evidence of
coagulation of plasma.

Coagulase tests

37° C,
18–24 hr

37° C,
18–24 hr

37° C,
18–24 hr

Second check

Negative tubes are
checked again after
24 hours.

Presence or absence of beta-type hemolysis is looked for
on blood agar plates to confirm existence of *S. aureus*.

First lab **Second** **Third** **Fourth**

Figure 70.2 **Procedure for presumptive identification of staphylococci.**

4. Repeat step 3 to inoculate the other half of each agar plate with the swab from the fomite tube.
5. Incubate the plates aerobically at 37° C for 24 to 36 hours.

Third Period

(Plate Evaluations and Coagulase/DNase Tests)

During this period, we will perform the following tasks: (1) evaluate the plates from the previous period, (2) inoculate blood agar plates, (3) make gram-stained slides, and (4) perform coagulase and/or DNase tests on organisms from selected colonies. Proceed as follows:

Materials

- MSA and SM110 plates from previous period
- 2 blood agar plates
- serological tubes containing 0.5 ml of 1:4 saline dilution of rabbit or human plasma (one tube for each isolate)
- petri plates of DNase agar
- Gram-staining kit

Evaluation of Plates

1. Examine the mannitol salt agar plates. Has the phenol red in the medium surrounding any of the colonies turned yellow?

   If this color change exists, it can be presumed that you have isolated a strain of *S. aureus*. Record your results in Laboratory Report 70 and on the chalkboard. (Your instructor may wish to substitute a copy of the chart from the Laboratory Report to be filled out at the demonstration table.)
2. Examine the plates of SM110. The presence of growth here indicates that the organisms are salt-tolerant. Note color of the colonies (white, yellow, or orange).
3. Record your observations of these plates in Laboratory Report 70 and on the chalkboard.

Blood Agar Inoculations

1. Label the bottom of one blood agar plate with your unknown-control number, and streak out the organisms from a staph-like colony.
2. Select staphylococcus-like colonies from the MSA and SM110 plates from the nose and fomites for streaking out on another blood agar plate. Use half-plate streaking methods, if necessary.
3. Incubate the blood agar plates at 37° C for 18 to 24 hours. *Don't leave plates in incubator longer than 24 hours.* Overincubation will cause blood degeneration.

Coagulase Tests

The fact that 97% of the strains of *S. aureus* have proven to be coagulase positive and that the other two species are *always* coagulase negative makes the coagulase test an excellent definitive test for confirming identification of *S. aureus*.

The procedure is simple. It involves inoculating a small tube of plasma with several loopfuls of the organism and incubating it in a 37° C water bath for several hours. If the plasma coagulates, the organism is coagulase positive. Coagulation may occur in 30 minutes or several hours later. *Any degree of coagulation, from a loose clot suspended in plasma to a solid immovable clot, is considered to be a positive result, even if it takes 24 hours to occur.*

It should be emphasized that this test is valid only for gram-positive, staphylococcus-like bacteria because some gram-negative rods, such as *Pseudomonas,* can cause a false-positive reaction. The mechanism of clotting in such organisms is not due to coagulase. Proceed as follows:

1. Label the plasma tubes NOSE, FOMITE, or UNKNOWN, depending on which of your plates have staph-like colonies.
2. With a wire loop, inoculate the appropriate tube of plasma with organisms from one or more colonies on SM110 or MSA. Use several loopfuls. Success is more rapid with a heavy inoculation. If positive colonies are present on both nose and fomite sides, be sure to inoculate a separate tube for each side.
3. Place the tubes in a 37° C water bath.
4. Check for solidification of the plasma every 30 minutes for the remainder of the period. Note in figure 70.3 that solidification may be complete, as in the lower tube, or be semisolid as seen in the middle tube.

   Any cultures that are negative at the end of the period will be left in the water bath. At 24 hours your instructor will remove them from the water bath and place them in the refrigerator, so that you can evaluate them in the next laboratory period.
5. Record your results in Laboratory Report 70.

DNase Test

The fact that coagulase-positive bacteria are also able to hydrolyze DNA makes the DNase test a reliable means of confirming *S. aureus* identification. The following procedure can be used to determine if a staph-like organism can hydrolyze DNA.

1. Heavily streak the organism on a plate of DNase test agar. One plate can be used for several test cultures by making short streaks about 1 inch long.
2. Incubate for 18–24 hours at 35° C.

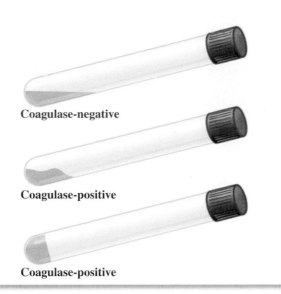

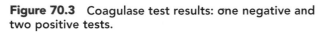

Figure 70.3 Coagulase test results: one negative and two positive tests.

Gram-Stained Slides

While your tubes of plasma are incubating in the water bath, prepare gram-stained slides from the same colonies that were used for the blood agar plates and coagulase tests.

    Examine the slides under oil immersion lens and draw the organisms in the appropriate areas of the Laboratory Report.

Fourth Period

(Confirmation)

During this period we will make final assessment of all tests and perform any other confirmatory tests that might be available to us.

Materials

- coagulase tubes from previous tests
- blood agar plates from previous period
- DNase test agar plates from previous period
- 0.1N HCl

1. Examine any coagulase tubes that were carried over from the last laboratory period that were negative at the end of that period. Record your results in Laboratory Report 70.
2. Examine the colonies on your blood agar plates. Look for clear (beta-type) hemolysis around the colonies. The presence of α-toxin is a definitive characteristic of *S. aureus*. Record your results in the Laboratory Report.
3. Look for zones of clearing near the streaks on the DNase agar plate. If none is seen, develop by flooding the plate with 0.1N HCl. The acid will render the hydrolyzed areas somewhat opaque.
4. Record your results on the chart on the chalkboard or chart on the demonstration table. If an instructor-supplied tabulation chart is used, the instructor will have copies made of it to be supplied to each student.

Further Testing

In addition to using the API Staph miniaturized test strip system (Exercise 48) to confirm your identification of staphylococci, you may wish to use the latex agglutination slide test described in Exercise 74. Your instructor will inform you as to the availability of these materials and the desirability of proceeding further.

Laboratory Report

After recording your results on the chalkboard (or on the chart on the demonstration table), complete the chart in the Laboratory Report and answer all the questions.

Laboratory Report

Student: _____

Date: _____ Section: _____

70 The Staphylococci: Isolation and Identification

A. Results

1. **Tabulation**

   At the beginning of the third laboratory period, the instructor will construct a chart similar to this one on the chalkboard. After examining your mannitol salt agar and staphylococcus 110 medium plates, record the presence (+) or absence (−) of staphylococcus growth in the appropriate columns. After performing coagulase tests on the various isolates, record the results also as (+) or (−) in the appropriate columns.

STUDENT INITIALS	NOSE			FOMITES			
	Staph Colonies		Coagulase	Item	Staph Colonies		Coagulase
	MSA	SM110			MSA	SM110	

2. **Microscopy**

Provide drawings here of the various isolates as seen under oil immersior (Gram-staining).

UNKNOWN CONTROL	NOSE	FOMITE

3. **Percentages**

From the data in the table on the previous page, determine the incidence (percentage) of individuals and fomites that harbor coagulase-positive and coagulase-negative staphylococci in this experiment.

SOURCE	TOTAL TESTED	TOTAL POSITIVE	PERCENTAGE POSITIVE	TOTAL NEGATIVE	PERCENTAGE NEGATIVE
Humans (Nose)					
Fomites					

4. **Record of Test Results**

Record here the results of each test performed in this experiment. Under GRAM STAIN indicate cellular arrangement as well as Gram reaction.

ISOLATE	GRAM STAIN	α TOXIN	MANNITOL (ACID)	COAGULASE
Unknown Control No.____				
Nose Isolate No. 1				
Nose Isolate No. 2				
Fomite Isolate				

5. **Final Determination**

Record here the name of your unknown control. If API Staph miniaturized multitest strips are available, confirm your conclusions by testing each isolate. See Exercise 48.

Name of unknown control: _____

Staph results: _____

B. Short Answer Questions

1. Describe the selective and differential properties of mannitol salt agar (MSA) for the isolation and identification of staphylococci.

_____

_____

2. Why is the coagulase test considered to be the definitive test for *S. aureus*?

_____

_____

3. What is the role of coagulase in the pathogenesis of *S. aureus*?

    _____

    _____

4. What is the role of α-toxin in the pathogenesis of *S. aureus*?

    _____

    _____

5. What are nosocomial infections?

    _____

    _____

6. Why are the staphylococci among the leading causes of nosocomial infections?

    _____

    _____

7. Why are staphylococcal infections becoming increasingly difficult to treat?

    _____

    _____

8. What precautions are taken in nosocomial settings to prevent staphylococcal infections?

    _____

    _____

9. Why might hospital patients be tested for nasal carriage of *S. aureus*?

    _____

    _____

10. Why might hospital staff be tested for nasal carriage of *S. aureus*?

    _____

    _____

The streptococci differ from the staphylococci in two significant characteristics: (1) They occur in chains rather than in clusters (figure 71.1), and (2) they lack the enzyme catalase, which degrades hydrogen peroxide to form water and oxygen. The streptococci oftentimes cause mixed infections with staphylococci, but they independently can cause diseases such as pneumonia, meningitis, endocarditis, pharyngitis, erysipelas, and glomerulonephritis.

Streptococci are normal inhabitants of the human pharynx and mouth where they occur on the surfaces of the teeth, tongue, cheek, throat, and in saliva. They can also be found on the skin and in the vagina. The enterococci have been established as a separate genus in the first edition of *Bergey's Manual of Systematic Bacteriology* and include organisms such as *Enterococcus faecalis* and *E. faecium*. The latter two bacteria occur primarily in the human colon and rectum.

The streptococci of greatest medical importance to humans are *S. pyogenes, S. agalactiae,* and *S. pneumoniae.* Of lesser importance are the enterococci *E. faecalis* and *E. faecium* and a streptococcus species usually associated with animals, *S. bovis.*

Several systems have been used to classify the streptococci. One of the first was proposed by J. H. Brown in 1919 that was based on the ability of these bacteria to cause lysis of red blood cells when grown on blood agar. Today, we still use hemolysis as an important phenotypic characteristic for differentiating the streptococci, and we now better understand the mechanism of hemolysis. Streptococci that completely destroy red cells in blood agar produce a clear zone around the colonies and are termed "beta-hemolytic." These bacteria produce streptolysin O or S, which is responsible for the lysis. Other streptococci produce a greenish or brownish zone around colonies, termed "alpha-hemolytic." The viridans group of streptococci produce alpha-hemolysis, and in fact, the name "viridans" comes from a Latin word, *viridis,* meaning green. Alpha-hemolysis is considered to be a partial hemolysis of the red cells. This type of hemolysis may be associated with the loss of potassium from the red cells, resulting in the discoloration.

In 1933, Rebecca Lancefield classified the streptococci on the basis of immunological groups related to carbohydrate antigens associated with the cells. Her original groups have been expanded to groups A through V. Lancefield groups are still used as an important characteristic in differentiating the streptococci, but the system is not inclusive because some streptococci, such as *S. pneumoniae,* do not possess the carbohydrate antigens.

In 1978, Jones proposed that streptococci be divided into seven groups, designated pyogenic, pneumococci, oral, fecal, lactic anaerobic, and other streptococci. A similar grouping was adopted in the first edition of *Bergey's Manual of Systematic Bacteriology* but with the pneumococci grouped with the pyogenic streptococci. The streptococci associated with human diseases are *S. pyogenes, S. agalactiae, S. pneumoniae, E. faecalis, E. faecium,* and *S. bovis.* A differentiation of these medically important streptococci based on biochemical, immunological, and physiological characteristics is shown in table 71.1. A further description is given in Appendix E.

The purpose of this exercise is twofold: (1) to learn the standard procedures for isolating streptococci and (2) to learn how to differentiate the medically important streptococci.

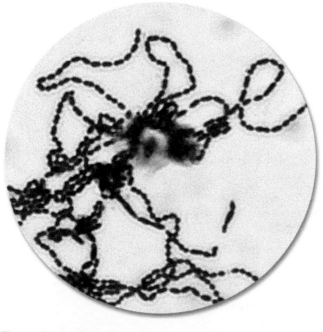

Figure 71.1 Streptococci.

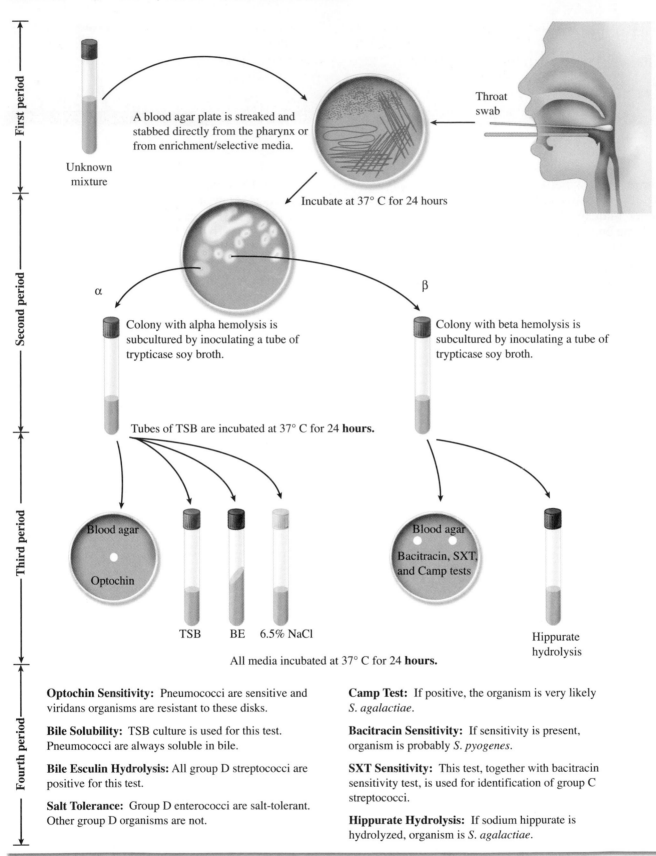

Throat swab

A blood agar plate is streaked and stabbed directly from the pharynx or from enrichment/selective media.

Unknown mixture

Incubate at 37° C for 24 hours

α

Colony with alpha hemolysis is subcultured by inoculating a tube of trypticase soy broth.

β

Colony with beta hemolysis is subcultured by inoculating a tube of trypticase soy broth.

Tubes of TSB are incubated at 37° C for 24 **hours.**

Blood agar

Optochin

TSB   BE   6.5% NaCl

Blood agar

Bacitracin, SXT, and Camp tests

Hippurate hydrolysis

All media incubated at 37° C for 24 **hours.**

Optochin Sensitivity: Pneumococci are sensitive and viridans organisms are resistant to these disks.

Bile Solubility: TSB culture is used for this test. Pneumococci are always soluble in bile.

Bile Esculin Hydrolysis: All group D streptococci are positive for this test.

Salt Tolerance: Group D enterococci are salt-tolerant. Other group D organisms are not.

Camp Test: If positive, the organism is very likely *S. agalactiae*.

Bacitracin Sensitivity: If sensitivity is present, organism is probably *S. pyogenes*.

SXT Sensitivity: This test, together with bacitracin sensitivity test, is used for identification of group C streptococci.

Hippurate Hydrolysis: If sodium hippurate is hydrolyzed, organism is *S. agalactiae*.

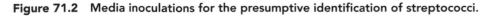

Figure 71.2   Media inoculations for the presumptive identification of streptococci.

Figure 71.2 illustrates the overall procedure to be followed in the pursuit of the two goals on p. 457. Note that blood agar is used to separate the streptococci into two groups on the basis of the type of hemolysis they produce on blood agar. Those organisms that produce alpha-hemolysis on blood agar can be differentiated by four tests. Those that produce beta-type hemolysis can be differentiated with the CAMP test and three other tests. The procedure outlined here is, primarily, designed to achieve *presumptive identification* of seven groups of streptococci. A few extra tests are usually required to confirm identification.

To broaden the application of these tests, your instructor may give you two or three unknown cultures of streptococci to be identified along with the pharyngeal isolates. *If unknowns are to be used, they will not be issued until physiological media are to be inoculated.*

First Period

(Making a Streak-Stab Agar Plate)

During this period, a plate of blood agar is swabbed and streaked in a special way to determine the type of hemolytic bacteria that are present in the pharynx and in an unknown mixture. Before making such a streak plate, however, clinicians prefer to use a tube of enrichment broth (TSB) or a selective medium of TSB with a little crystal violet added to it (TSBCV). Media of this type are usually incubated at 37° C for 24 hours. This is particularly useful if the number of organisms might be low or if the swab cannot be applied to blood agar immediately. *Although this enrichment/selective step has been omitted here, it should be understood that the procedure is routine.*

Since swabbing one's own throat properly can be difficult, it will be necessary for you to work with your laboratory partner to swab each other's throats. Once your throat has been swabbed, you will proceed to use the swab to streak and stab your own agar plate according to a special procedure shown in figure 71.3.

Materials

- 1 tongue depressor
- 1 sterile cotton swab
- inoculating loop
- 1 blood agar plate

1. With the subject's head tilted back and the tongue held down with the tongue depressor, rub the back surface of the pharynx up and down with the sterile swab.

   Also, *look for white patches* in the tonsillar area. Avoid touching the cheeks and tongue.

2. Since streptococcal hemolysis is most accurately analyzed when the colonies develop anaerobically beneath the surface of the agar, it will be necessary to use a streak-stab technique as shown in figure 71.3. The essential steps are as follows:

   - Roll the swab over an area approximating one-fifth of the surface. The entire surface of the swab should contact the agar.
   - With a wire loop, streak out three areas as shown to thin out the organisms.
   - Stab the loop into the agar to the bottom of the plate at an angle perpendicular to the surface to make a clean cut without ragged edges.
   - Be sure to make one set of stabs in an unstreaked area so that streptococcal hemolysis will be easier to interpret with a microscope.

> **CAUTION:** Dispose of swabs and tongue depressors in beaker of disinfectant.

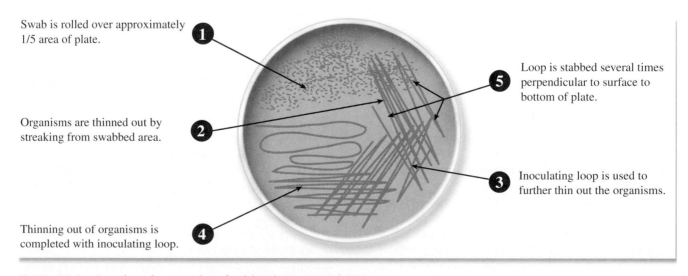

Swab is rolled over approximately 1/5 area of plate. **1**

Organisms are thinned out by streaking from swabbed area. **2**

Thinning out of organisms is completed with inoculating loop. **4**

5 Loop is stabbed several times perpendicular to surface to bottom of plate.

3 Inoculating loop is used to further thin out the organisms.

Figure 71.3 **Streak-stab procedure for blood agar inoculations.**

3. Repeat the inoculation procedure for the unknown mixture.
4. Incubate the plates aerobically at 37°C for 24 hours. Do not incubate the plates longer than 24 hours.

Second Period

(Analysis and Subculturing)

During this period, two things must be accomplished: first, the type of hemolysis must be correctly determined and, second, well-isolated colonies must be selected for making subcultures. The importance of proper subculturing cannot be overemphasized: without a pure culture, future tests are certain to fail. Proceed as follows:

Materials

- blood agar plate from previous period
- tubes of TSB (one for each different type of colony)
- dissecting microscope

1. Look for isolated colonies that have alpha- or beta-hemolysis surrounding them. Streptococcal colonies are characteristically very small.
2. Do any of the stabs appear to exhibit hemolysis? Examine these hemolytic zones near the stabs under 60X magnification with a dissecting microscope.
3. Consult figure 71.4 to analyze the type of hemolysis. Note that the illustrations on the left side indicate what the colonies would look like if they were submerged under a layer of blood agar (two-layer pour plate). The illustrations on the right indicate the nature of hemolysis around stabs on streak-stab plates. Although this illustration is very diagrammatic, it reveals the microscopic differences between three kinds of hemolysis: alpha, alpha-prime, and beta.

    Only those stabs that are completely free of red blood cells in the hemolytic area are considered to be **beta-hemolytic.** The chance of isolating a colony of this type from your own throat is very slim, for the beta-hemolytic streptococci are the most serious pathogens.

    If some red blood cells are seen dispersed throughout the hemolytic zone, the organism is classified as **alpha-prime-hemolytic.** Viridans streptococci often fall in this category.
4. Record your observations in Laboratory Report 71.
5. Select well-isolated colonies that exhibit hemolysis (alpha, beta, or both) for inoculating tubes of TSB. Be sure to label the tubes ALPHA or BETA. Whether or not the organism is alpha or beta is crucial in identification.

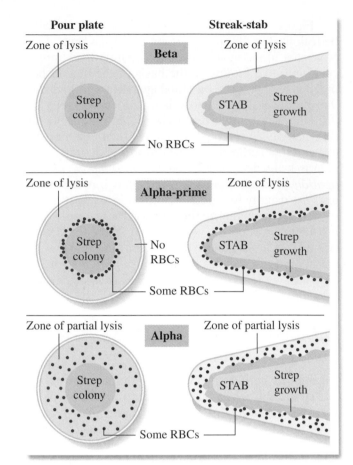

Figure 71.4 Comparison of hemolysis types as seen on pour plates and streak-stab plates.

Since the chances of isolating beta-hemolytic streptococci from the pharynx are usually quite slim, notify your instructor if you think you have isolated one.

6. Incubate the tubes at 37°C for 24 hours.
7. **Important:** At some time prior to the next laboratory session, review the material in Appendix E that pertains to this exercise.

Third Period

(Inoculations for Physiological Tests)

Presumptive identification of the various groups of streptococci is based on seven or eight physiological tests. Table 71.1 on page 461 reveals how they perform on these tests. Note that Groups A, B, and C are all beta-hemolytic; a few enterococci are also beta-hemolytic. The remainder are all alpha-hemolytic or nonhemolytic.

Since each of the physiological tests is specific for differentiating only two or three groups, it is not desirable to do all the tests on all unknowns. For economy and preciseness, only four tests that are mentioned for

Table 71.1 Physiological Tests for Streptococcal Differentiation

	Bergey's Group	Lancefield Group	Hemolysis	Bacitracin Susceptibility	CAMP Reaction or Hippurate Hydrolysis	SXT Sensitivity	Bile Esculin Hydrolysis	Tolerance to 6.5% NaCl	Optochin Susceptibility	Bile Solubility
S. pyogenes	Pyogenic	A	β	+	−	R	−	−	−	−	
S. agalactiae		B	β	−	+	R	−	±	−	−	
S. pneumoniae		none	α	−	−		−	−	+	+	
S. equi		C	β	−	−	S	−	−	−	−	
S. dysgalactiae[1]		C	β	−	−	S	−	−	−	−	
S. zooepidemicus[1]		C	β	−	−	S	−	−	−	−	
E. faecalis	Enterococci	D	β	−	−	R	+	+	−	−	
E. faecium		D	α	−	−	R	+	+	−	−	
S. bovis	Other	D	α[2]	−	−	R/S	+	−	−	−	
S. mitis	Oral (Viridans)	none	α[2]	−	−	S	−	−	−	−	
S. salivarius		none	α[2]	−	−	S	−	−	−	−	
S. mutans		none	none	±	−	S	−	−	−	−	

Note: R = resistant; S = sensitive; blank = not significant.
[1]Sub species of S. equi.
[2]Weakly alpha.

the third period in figure 71.2 should be performed on an isolate or unknown.

Before any inoculations are made, however, it is desirable to do a purity check on each TSB culture from the previous period. To accomplish this, it will be necessary to make a Gram-stained slide of each of the cultures.

Gram-Stained Slides (Purity Check)

Materials

- TSB cultures from previous period
- Gram-staining kit

1. Make a Gram-stained slide from your isolates and examine them under oil immersion lens. Do they appear to be pure cultures?
2. Draw the organisms in the appropriate circles in Laboratory Report 71.

Beta-Type Inoculations

Use the following procedure to perform tests on each isolate that has beta-type hemolysis:

Materials

for each isolate:
- 1 blood agar plate
- 1 tube of sodium hippurate broth
- 1 bacitracin differential disk
- 1 SXT sensitivity disk
- 1 broth culture of S. aureus
- dispenser or forceps for transferring disks

1. Label a blood agar plate and a tube of sodium hippurate broth with proper identification information of each isolate and unknown to be tested.
2. Follow the procedure outlined in figure 71.5 to inoculate each blood agar plate with the isolate (or unknown) and S. aureus.
   Note that a streak of the unknown is brought down perpendicular to the S. aureus streak, keeping the two organisms about 1 cm apart.
3. With forceps or dispenser, place one bacitracin differential disk and one SXT disk on the heavily streaked area at points shown in figure 71.5. Press down on each disk slightly.

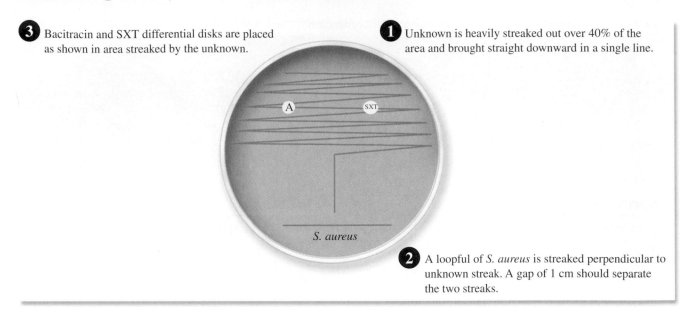

3 Bacitracin and SXT differential disks are placed as shown in area streaked by the unknown.

1 Unknown is heavily streaked out over 40% of the area and brought straight downward in a single line.

S. aureus

2 A loopful of *S. aureus* is streaked perpendicular to unknown streak. A gap of 1 cm should separate the two streaks.

Figure 71.5 **Blood agar inoculation technique for the CAMP, bacitracin, and SXT tests.**

4. Inoculate one tube of sodium hippurate broth for each isolate or unknown.

5. Incubate the blood agar plates at 37°C, aerobically, for 24 hours, and the hippurate broth tubes at 35°C, aerobically, for 24 hours. If the hippurate broths prove to be negative or weakly positive at 24 hours, they should be given more time to see if they change.

Alpha-Type Inoculations

As shown in figure 71.2, four inoculations will be made for each isolate or unknown that is alpha-hemolytic.

Materials

- 1 blood agar plate (for up to 4 unknowns)
- 1 6.5% sodium chloride broth
- 1 trypticase soy broth (TSB)
- 1 bile esculin (BE) slant
- 1 optochin (Taxo P) disk
- candle jar setup or CO_2 incubator

1. Mark the bottom of a blood agar plate to divide it into halves, thirds, or quarters, depending on the number of alpha-hemolytic organisms to be tested. Label each space with the code number of each test organism.

2. Completely streak over each area of the blood agar plate with the appropriate test organism, and place one optochin (Taxo P) disk in the center of each area. Press down slightly on each disk to secure it to the medium.

3. Inoculate one tube each of TSB, BE, and 6.5% NaCl broth with each test organism.

4. Incubate all media at 35°–37°C as follows:
   Blood agar plates: 24 hours in a candle jar
   6.5% NaCl broths: 24, 48, and 72 hours
   Bile esculin slants: 48 hours
   Trypticase soy broths: 24 hours
   Note: While the blood agar plates should be incubated in a candle jar or CO_2 incubator, the remaining cultures can be incubated aerobically.

Fourth Period

(Evaluation of Physiological Tests)

Once all of the inoculated media have been incubated for 24 hours, you are ready to examine the plates and tubes and add test reagents to some of the cultures. Some of the tests will also have to be checked at 48 and 72 hours.

After you have assembled all the plates and tubes from the last period, examine the blood agar plates first that were double-streaked with the unknowns and *S. aureus*. Note that the second, third, and fourth tests listed in table 71.1 can be read from these plates. Proceed as follows:

CAMP Reaction

If you have an unknown that produces an enlarged arrowhead-shaped hemolytic zone at the juncture where the unknown meets the *S. aureus* streak, as seen in figure 71.6, the organism is *S. agalactiae*. This phenomenon is due to what is called the *CAMP factor.* The only problem that can arise from this test is that if the plate is incubated anaerobically, a positive CAMP reaction can occur on *S. pyogenes* inoculated plates.

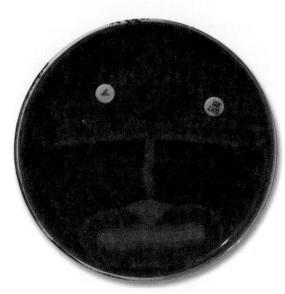

Figure 71.6 Note positive SXT disk on right, negative bacitracin disk on left, and positive CAMP reaction (arrowhead). Organism: *S. agalactiae.*

Record the CAMP reactions for each of your isolates or unknowns in Laboratory Report 71.

Bacitracin Susceptibility

Any size zone of inhibition seen around the bacitracin disks should be considered to be a positive test result. Note in table 71.1 that *S. pyogenes* is positive for this characteristic.

This test has two limitations: (1) the disks must be of the *differential type,* not sensitivity type, and (2) the test should not be applied to alpha-hemolytic streptococci. Reasons: Sensitivity disks have too high a concentration of the antibiotic, and many alpha-hemolytic streptococci are sensitive to these disks.

Record the results of this test in the table under number 4 of Laboratory Report 71.

SXT Sensitivity Test

The disks used in this test contain 1.25 mg of trimethoprim and 27.75 mg of sulfamethoxazole (SXT). The purpose of this test is to distinguish groups A and B from other beta-hemolytic streptococci. Note in table 71.1 that both groups A and B are uniformly resistant to SXT.

If a beta-hemolytic streptococcus proves to be bacitracin resistant and SXT susceptible, it is classified as being a **non-group A or B beta-hemolytic streptococcus.** This means that the organism is probably a species within group C. *Keep in mind that an occasional group A streptococcal strain is susceptible to both bacitracin and SXT disks.* One must always remember that exceptions to most tests do occur; that is why this identification procedure leads us only to *presumptive* conclusions.

Record any zone of inhibition (resistance) as positive for this test.

Hippurate Hydrolysis

Note in table 71.1 that hippurate hydrolysis and the CAMP test are grouped together as positive tests for *S. agalactiae.* If an organism is positive for both tests, or either one, one can assume with almost 100% certainty that the organism is *S. agalactiae.*

Proceed as follows to determine which of your isolates are able to hydrolyze sodium hippurate:

Materials

- serological test tubes
- serological pipettes (1 ml size)
- ferric chloride reagent
- centrifuge

1. Centrifuge the culture for 3 to 5 minutes.
2. Pipette 0.2 ml of the supernatant and 0.8 ml of ferric chloride reagent into an empty serological test tube. Mix well.
3. Look for a **heavy precipitate** to form. If the precipitate forms and persists for 10 minutes or longer, the test is positive. If the culture proves to be weakly positive, incubate the culture for another 24 hours and repeat the test.
4. Record your results in Laboratory Report 71.

Bile Esculin (BE) Hydrolysis

This is the best physiological test that we have for the identification of group D streptococci. Both enterococcal and nonenterococcal species of group D are able to hydrolyze esculin in the agar slant, causing the slant to blacken.

A positive BE test tells us that we have a group D streptococcus; differentiation of the two types of group D streptococci depends on the salt-tolerance test.

Examine the BE agar slants, looking for **blackening of the slant,** as illustrated in figure 71.7. If less than half of the slant is blackened, or if no blackening occurs within 24 to 48 hours, the test is negative.

Salt Tolerance (6.5% NaCl)

All enterococci of group D produce heavy growth in 6.5% NaCl broth. As indicated in table 71.1, none of the nonenterococci, group D, grow in this medium. This test, then, provides us with a good method for differentiating the two types of group D enterococci.

A positive result shows up as turbidity within 72 hours. A color change of **purple to yellow** may also be present. If the tube is negative at 24 hours, incubate it and check it again at 48 and 72 hours. *If the organism is salt tolerant and BE positive, it is considered*

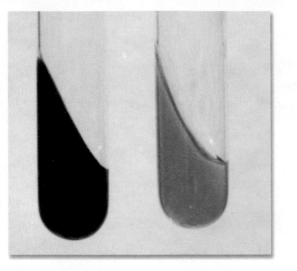

Figure 71.7 Positive bile esculin hydrolysis on left; negative on right.

to be an enterococcus. Parenthetically, it should be added here that approximately 80% of group B streptococci will grow in this medium.

Optochin Susceptibility

Optochin susceptibility is used for differentiation of the alpha-hemolytic viridans streptococci from the pneumococci. The pneumococci are sensitive to these disks; the viridans organisms are resistant.

Materials

- blood agar plates with optochin disks
- plastic metric ruler

1. Measure the diameters of zones of inhibition that surround each disk, evaluating whether the zones are large enough to be considered positive. The standards are as follows:
   - For 6 mm diameter disks, the zone must be at least 14 mm diameter to be considered positive.
   - For 10 mm diameter disks, the zone must be at least 16 mm diameter to be considered positive.
2. Record your results in Laboratory Report 71.

Bile Solubility

If an alpha-hemolytic streptococcal organism is soluble in bile and positive on the optochin test, presumptive evidence indicates that the isolate is *S. pneumoniae.* Perform the bile solubility test on each of your alpha-hemolytic isolates as follows:

Materials

- 2 empty serological tubes (per test)
- dropping bottle of phenol red indicator
- dropping bottle of 0.05N NaOH
- TSB culture of unknown
- 2% bile solution (sodium desoxycholate)
- bottle of normal saline solution
- 2 serological pipettes (1 ml size)
- water bath (37°C)

1. Mark one empty serological tube BILE and the other SALINE. Into their respective tubes, pipette 0.5 ml of 2% bile and 0.5 ml of saline.
2. Shake the TSB unknown culture to suspend the organisms and pipette 0.5 ml of the culture into each tube.
3. Add 1 or 2 drops of phenol red indicator to each tube and adjust the pH to 7.0 by adding drops of 0.05N NaOH.
4. Place both tubes in a 37°C water bath and examine periodically for 2 hours. If the turbidity clears in the bile tube, it indicates that the cells have disintegrated and the organism is *S. pneumoniae.* Compare the tubes side by side.
5. Record your results in Laboratory Report 71.

Final Confirmation

All the laboratory procedures performed so far lead us to presumptive identification. To confirm these conclusions, it is necessary to perform serological tests on each of the unknowns. If commercial kits are available for such tests, they should be used to complete the identification procedures.

Laboratory Report

Complete Laboratory Report 71.

Laboratory Report

Student: _____

Date: _____ Section: _____

71 The Streptococci: Isolation and Identification

A. Results

1. Tabulation of Pharynx Isolates

The instructor will construct a chart similar to this one on the chalkboard. After examining the blood agar plates that were inoculated with pharynx organisms, record the types and size range of colonies that are present on your plates. Record these data first on this table, then on the chalkboard. After all students have recorded their results on the board, complete the tabulation of their results here, also. The names of the organisms will not be recorded until all tests are completed.

STUDENT INITIALS	TYPE OF HEMOLYSIS (ALPHA, ALPHA-PRIME, BETA)	SIZE RANGE OF COLONIES (MM)	NAMES OF ORGANISMS

2. **Microscopy**
   Provide drawings here of the various pharyngeal isolates as seen under oil immersion (Gram staining).

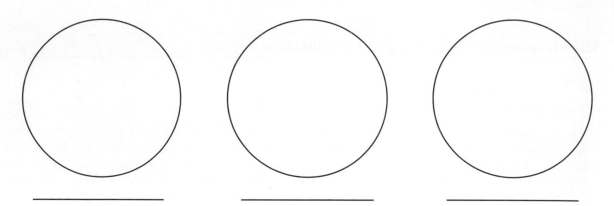

_____   _____   _____

3. **Percentages**
   From the data in the table on the previous page, calculate the percentages for each type of streptococci that were isolated from classmates.

   S. pyogenes: _____   **Group C streptococci:** _____

   S. agalactiae: _____   **Group D enterococci:** _____

   S. pneumoniae: _____   **Group D nonenterococci:** _____

   Oval (viridans) streptococci: _____

4. **Record of Test Results**
   Record here all information pertaining to the identification of pharyngeal isolates and unknowns.

SOURCE OF UNKNOWN	Hemolysis	Bacitracin Susceptibility	CAMP Reaction	Hippurate Hydrolysis	SXT Sensitivity	Bile Esculin Hydrolysis	Tolerance to 6.5% NaCl	Optochin Susceptibility	Bile Solubility

5. **Final Determination**
   Record here the identities of your various isolates and unknowns:

   Pharyngeal isolates: _____

   _____

   Unknowns: _____

   _____

   _____

B. Short Answer Questions

1. When bacteria from a throat swab are streaked on blood agar, why is the agar stabbed several times with the loop?

   _____

   _____

2. Differentiate between alpha- and beta-hemolysis.

   _____

   _____

3. What was Rebecca Lancefield's contribution to the study of streptococci?

   _____

   _____

4. A throat swab is typically performed on an individual suspected to have streptococcal pharyngitis, or strep throat.

   a. On blood agar, what type of hemolysis would be expected in an individual with streptococcal pharyngitis?

   _____

   b. What species causes streptococcal pharyngitis?

   _____

   c. Why is it important to first confirm the presence of this species before administering any therapy?

   _____

5. Bacteria are isolated from the blood of a patient with septicemia. You would like to determine whether they are *Streptococcus pyogenes* or *Staphylococcus aureus*.

   a. What characteristics do the two species share?

   _____

   b. What tests would be useful for the differentiation of the two species?

   _____

6. Name three tests that are useful for the differentiation of *S. pyogenes* and *S. agalactiae*.

   _____

   _____

7. Name two tests that are useful for the differentiation of pneumococci and oral viridans streptococci?

   _____

   _____

8. What test can be performed to differentiate the enterococci from other group D streptococci?

   _____

   _____

9. What test can be performed to differentiate between group A and group C streptococci?

   _____

   _____

10. Which streptococcal species includes cells that are arranged predominantly in pairs rather than chains?

    _____

    _____

11. Vaginal swabs are taken from pregnant women in their third trimester. Which streptococcal species is the focus of the investigation?

    _____

    _____

12. Which streptococci are implicated in the development of dental caries? What is the mechanism of their formation?

    _____

    _____

Gram-Negative Intestinal Pathogens

The enteric pathogens of primary medical importance are the **salmonella** and **shigella** as well as pathogenic strains of *Escherichia coli.* They cause gastroenteritis, bacillary dysentery, foodborne illnesses, and diarrhea in humans.

At present, the taxonomic differentiation of *salmonella* is undergoing revision. In the first edition of *Bergey's Manual of Systematic Bacteriology, Salmonella* is divided into two species, *S. choleraesuis* and *S. bongori. S. choleraesuis* is further divided into six subspecies based on different taxonomic and nucleic acid analyses. However, based on antigenic differences in the makeup of the bacterial cell, the *Salmonella* can be subdivided into 2500 different **serovars** or **serotypes,** most of which belong to *S. choleraesuis.* Serovars are strains that have similar biochemical characteristics but differ in their antigenic composition; hence, they are differentiated by serological typing. Identifying specific serovars is important in determining the identity of the pathogen causing outbreaks of *Salmonella* infections. For *Salmonella,* one of the most medically important strains has been *S. typhi,* the causative agent of typhoid fever. This organism is placed in subgenus I of *S. choleraesuis* and Typhi is its serovar.

The *Shigella* are divided into four species, *Shigella dysenteriae, S. boydii, S. flexeri,* and *S. sonnei.* Like the *Salmonella,* the four species are divided into a number of different serotypes. Various lines of evidence, including genetic analysis, suggests that the *Shigella* may be pathogenic variants or pathovars of *Escherichia coli.* However, the *Shigella* infect only primate hosts, and they therefore have a narrower host range than *E. coli.*

E. coli is a normal inhabitant of the human intestinal tract (coliform) and is used as an indicator organism for fecal contamination. However, pathogenic strains of *E. coli* can cause serious outbreaks of diarrhea in humans. Some are enterotoxigenic strains that produce toxins that are similar to the cholera toxin, which is responsible for the severe diarrhea seen in cholera. One strain, *E. coli* O 157:H 7, has been associated with various kinds of contaminated food, especially hamburger, and it has caused several deaths. *E. coli* is also responsible for urinary tract infections. The source of this organism in these infections is usually the intestinal tract of the patient.

Routine testing for the presence of these pathogens is a function of public health laboratories at various governmental levels. The isolation of these pathogenic enterics from feces is complicated by the fact that the colon contains a diverse population of bacteria. Species of such genera as *Escherichia, Proteus, Enterobacter, Pseudomonas,* and *Clostridium* exist in large numbers: hence it is necessary to use media that are differential and selective to favor the growth of the pathogens.

Figure 72.1 is a separation outline that is the basis for the series of tests that are used to demonstrate the presence of *Salmonella* or *Shigella* in a patient's blood, urine, or feces. Note that lactose fermentation separates these organisms from most of the other Enterobacteriaceae. Final differentiation of the two enteric pathogens from *Proteus* relies on motility, hydrogen sulfide production, and urea hydrolysis. The differentiation information of the positive lactose fermenters on the left side of the separation outline is provided here mainly for comparative references that can be used for the identification of other unknown enterics.

The procedural diagram in figure 72.2 reveals how we will apply these facts in the identification of an unknown *Salmonella* or *Shigella.* The entire process will involve four laboratory periods.

In this experiment, you will be given a mixed culture containing a coliform, *Proteus,* and *Salmonella* or *Shigella.* The pathogens will be of the less dangerous types, but their presence will, naturally, demand utmost caution in handling. Your problem will be to isolate the pathogen from the mixed culture and make a genus identification. There are five steps that are used to prove the presence of these pathogens in a stool sample: (1) enrichment, (2) isolation, (3) fermentation tests, (4) final physiological tests, and (5) serotyping.

Enrichment

There are two enrichment media that are most frequently used to inhibit the nonpathogens and favor the growth of pathogenic enterics. They are selenite F and gram-negative (GN) broths. While most *Salmonella* grow unrestricted in these two media, some of the *Shigella* are inhibited to some extent in selenite F broth; thus, for *Shigella* isolation, GN broth is preferred. In many cases, stool samples are plated directly on isolation media.

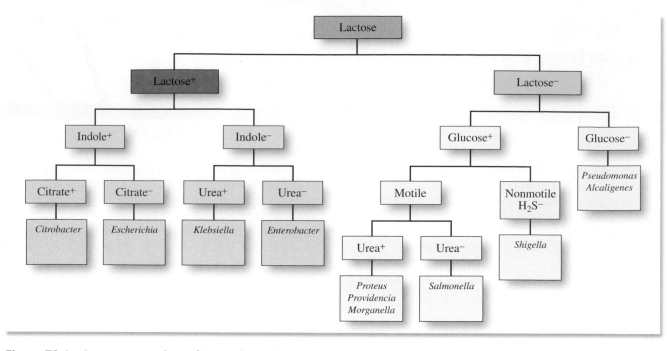

Figure 72.1 Separation outline of Enterobacteriaceae.

In actual practice, 1 to 5 grams of feces are placed in 10 ml of enrichment broth. In addition, plates of various kinds of selective media are inoculated directly. The broths are usually incubated for 4 to 6 hours.

Since we are not using stool samples in this exercise, the enrichment procedure is omitted. Instead, you will streak the isolation media directly from the unknown broth.

First Period

(Isolation)

There are several excellent selective differential media that have been developed for the isolation of these pathogens. Various inhibiting agents such as brilliant green, bismuth sulfite, sodium desoxycholate, and sodium citrate are included in them. For *Salmonella choleraesuis (typhi)*, bismuth sulfite agar appears to be the best medium. Colonies of *S. choleraesuis (typhi)* on this medium appear black due to the reduction of sulfite to sulfide.

Other widely used media are MacConkey agar, Hektoen Enteric agar (HE), and Xylose Lysine Desoxycholate (XLD) agar. These media may contain bile salts and/or sodium desoxycholate to inhibit gram-positive bacteria. To inhibit coliforms and other non-enterics, they may contain citrate. All of them contain lactose and a dye so that if an organism is a lactose fermenter, its colony will take on a color characteristic of the dye present.

Since the enrichment procedure is being omitted here, you will be issued an unknown broth culture

with a pathogenic enteric. Your instructor will indicate which selective media will be used. Proceed as follows to inoculate the selective media with your unknown mixture:

Materials

- unknown culture (mixture of a coliform, *Proteus*, and a *Salmonella* or *Shigella*)
- 1 or more petri plates of different selective media: MacConkey, Hektoen Enteric (HE), or Xylose Lysine Desoxycholate (XLD) agar

1. Label each plate with your name and unknown number.
2. With a loop, streak each plate with your unknown in a manner that will produce good isolation.
3. Incubate the plates at 37° C for 24 to 48 hours.

Second Period

(Fermentation Tests)

As stated above, the fermentation characteristic that separates the *Salmonella* and *Shigella* pathogens from the coliforms is their *inability to ferment lactose*. Once we have isolated colonies on differential media that look like *Salmonella* or *Shigella*, the next step is to determine whether the isolates can ferment lactose. All media for this purpose contain at least two sugars, glucose and lactose. Some contain a third sugar, sucrose. They also contain phenol red to indicate when fermentation occurs. Russell Double Sugar (RDS)

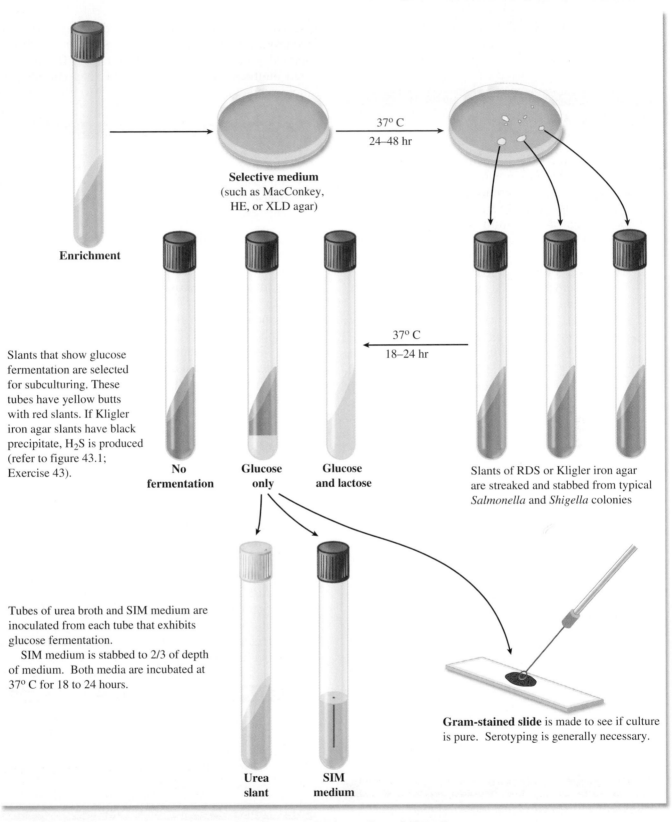

Enrichment

Slants that show glucose fermentation are selected for subculturing. These tubes have yellow butts with red slants. If Kligler iron agar slants have black precipitate, H₂S is produced (refer to figure 43.1; Exercise 43).

Selective medium
(such as MacConkey, HE, or XLD agar)

37° C
24–48 hr

37° C
18–24 hr

No fermentation

Glucose only

Glucose and lactose

Slants of RDS or Kligler iron agar are streaked and stabbed from typical *Salmonella* and *Shigella* colonies

Tubes of urea broth and SIM medium are inoculated from each tube that exhibits glucose fermentation.

SIM medium is stabbed to 2/3 of depth of medium. Both media are incubated at 37° C for 18 to 24 hours.

Gram-stained slide is made to see if culture is pure. Serotyping is generally necessary.

Urea slant

SIM medium

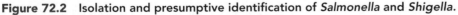

Figure 72.2 Isolation and presumptive identification of *Salmonella* and *Shigella*.

agar is one of the simpler media that works well. Kligler iron agar may also be used. It is similar to RDS with the addition of iron salts for detection of H₂S. Your instructor will indicate which one will be used.

Proceed as follows to inoculate three slants from colonies on the selective media that look like either *Salmonella* or *Shigella*. The reason for using three slants is that you may have difficulty distinguishing

Proteus from the SS pathogens. By inoculating three tubes from different colonies, you will be increasing your chances of success.

Materials

- 3 agar slants (RDS or Kligler iron)
- streak plates from first period

1. Label the three slants with your name and the number of your unknown.
2. Look for isolated colonies that look like *Salmonella* or *Shigella* organisms. The characteristics to look for on each medium are as follows:
   - **MacConkey agar**—*Salmonella, Shigella,* and other non-lactose-fermenting species produce smooth, colorless colonies. Coliforms that ferment lactose produce reddish, mucoid, or dark-centered colonies.
   - **Hektoen Enteric (HE) agar**—*Salmonella* and *Shigella* colonies are greenish-blue. Some species of *Salmonella* will have greenish-blue colonies with black centers due to H_2S production. Coliform colonies are salmon to orange and may have a bile precipitate.
   - **Xylose Lysine Desoxycholate (XLD) agar**—although most *Salmonella* produce red colonies with black centers, a few may produce red colonies that lack black centers. *Shigella* colonies are red. Coliform colonies are yellow. Some *Pseudomonas* produce false-positive red colonies.
3. With a straight wire, inoculate the three agar slants from separate SS-appearing colonies. Use the streak-stab technique. When streaking the surface of the slant before stabbing, move the wire over the entire surface for good coverage.
4. Incubate the slants at 37° C for 18 to 24 hours, longer incubation time may cause alkaline reversion. Even refrigeration beyond this time may cause reversion.

   Alkaline reversion is a condition in which the medium turns yellow during the first part of the incubation period and then changes to red later due to increased alkalinity.

Third Period

(Slant Evaluations and Final Inoculations)

During this period, you will inoculate tubes of SIM medium and urea broth with organisms from the slants of the previous period. Examination of the separation outline in figure 72.1 reveals that the final step in the differentiation of the SS pathogens is to determine whether a non-lactose fermenter can do three things:

(1) exhibit motility, (2) produce hydrogen sulfide, and (3) produce urease. You will also be making a Gram-stained slide to perform a purity check. If miniaturized multitest media are available, they can also be inoculated at this time.

Materials

- RDS or Kligler's iron agar slants from previous period
- 1 tube of SIM medium for each positive slant
- 1 urea slant for each positive slant
- miniaturized multitest media such as API 20E or Enterotube II (optional)

1. Examine the slants from the previous period and **select those tubes that have a yellow butt with a red slant.** These tubes contain organisms that ferment only glucose (non-lactose fermenters). If you used Kligler's iron agar, a black precipitate in the medium will indicate that the organism is a producer of H_2S (refer to figure 43.1).

   Note that slants in figure 72.1 that are completely yellow are able to ferment lactose as well as glucose. Tubes that are completely red are either nonfermenters or examples of alkaline reversion. Ignore those tubes.
2. With a loop, inoculate one urea slant from each slant that has a yellow butt and red slant (non-lactose fermenter).
3. With a straight wire, stab one tube of SIM medium from each of the same agar slants. Stab in the center to two-thirds of depth of medium.
4. Incubate these tubes at 37° C for 18 to 24 hours.
5. Make gram-stained slides from the same slants and confirm the presence of gram-negative rods.
6. If miniaturized multitest media are available, such as API 20E or Enterotube II, inoculate and incubate for evaluation in the next period. Consult Exercises 45 and 46 for instructions.
7. Refrigerate the positive RDS and Kligler iron slants for future use, if needed.

Fourth Period

(Final Evaluation)

During this last period, the tubes of SIM medium, urea broth, and any miniaturized multitest media from the last period will be evaluated. Serotyping can also be performed, if desired.

Materials

- tubes of urea broth and SIM medium from previous period
- Kovacs' reagent and chloroform
- 5 ml pipettes
- miniaturized multitest media from previous period
- serological testing materials (optional)

1. Examine the tubes of SIM medium, checking for motility and H$_2$S production. If you see cloudiness spreading from the point of inoculation, the organism is motile. A black precipitate will be evidence of H$_2$S production (refer to figure 43.1).

2. Test for indole production by pipetting 2 ml of chloroform into each SIM tube and then adding 2 ml of Kovacs' reagent. A **pink to deep red color** will form in the chloroform layer if indole is produced (figure 43.2).

   Salmonella are negative. Some *Shigella* may be positive. *Citrobacter* and *Escherichia* are positive.

3. Examine the urea slant tubes. If the medium has changed from yellow to **red** or **cerise color,** the organism is urease positive (figure 42.4).

4. If a miniaturized multitest media was inoculated in the last period, complete it now.

5. If time and materials are available, confirm the identification of your unknown with serological typing. Refer to Exercise 73.

Laboratory Report

Record the identity of your unknown in Laboratory Report 72 and answer all the questions.

Descriptive Chart

STUDENT: _____

LAB SECTION: _____

Habitat: _____ Culture No.: _____
Source: _____
Organism: _____

MORPHOLOGICAL CHARACTERISTICS

Cell Shape:

Arrangement:

Size:

Spores:

Gram's Stain:

Motility:

Capsules:

Special Stains:

CULTURAL CHARACTERISTICS

Colonies:

   Nutrient Agar:

   Blood Agar:

Agar Slant:

Nutrient Broth:

Gelatin Stab:

Oxygen Requirements:

Optimum Temp.:

PHYSIOLOGICAL CHARACTERISTICS

	TESTS	RESULTS
Fermentation	Glucose	
	Lactose	
	Sucrose	
	Mannitol	
Hydrolysis	Gelatin Liquefaction	
	Starch	
	Casein	
	Fat	
IMViC	Indole	
	Methyl Red	
	V-P (acetylmethylcarbinol)	
	Citrate Utilization	
	Nitrate Reduction	
	H_2S Production	
	Urease	
	Catalase	
	Oxidase	
	DNase	
	Phenylalanine Deaminase	

	REACTION	TIME
Litmus Milk	Acid	_____
	Alkaline	_____
	Coagulation	_____
	Reduction	_____
	Peptonization	_____
	No Change	_____

Laboratory Report

Student: _____

Date: _____ Section: _____

72 Gram-Negative Intestinal Pathogens

A. Results

1. What was the genus of your unknown?

   _____   _____

   Genus   No.

2. What problems, if any, did you encounter?

   _____

   _____

   _____

   _____

   _____

   _____

3. Now that you know the genus of your unknown, what steps would you follow to determine the species?

   _____

   _____

   _____

B. Short Answer Questions

1. Name three enteric pathogens of primary medical importance.

   _____

   _____

2. What precautions can be taken to prevent infection and illness caused by enteric pathogens?

   _____

   _____

3. What selective agents are added to media to preferentially grow enterobacteria for study? What type of growth is inhibited?

   _____

4. What characteristic separates *Salmonella* and *Shigella* from most of the other enterobacteria? What media can be used for this differentiation?

   _____

   _____

5. What two characteristics separate *Salmonella* from *Shigella?* What media can be used for this differentiation?

_____

_____

6. Which coliform bacteria are the most difficult to distinguish from the SS pathogens? What is the primary characteristic used to differentiate them?

_____

_____

7. How can acid production by glucose and lactose fermentation be differentiated in the same tube?

_____

_____

8. What is alkaline reversion? Why is it important?

_____

_____

Slide Agglutination Test:
Serological Typing

Organisms of different species differ not only in their morphology and physiology but also in the various components that make up their molecular structure. The proteins, polysaccharides, nucleic acids, and lipids define the molecular structure of an organism. Some of the macromolecules of a bacterial cell such as proteins, polysaccharides, lipoproteins, and nucleoproteins can act as **antigens** because when these substances are introduced into an animal, they elicit the formation of antibodies. In order to produce antibodies, antigens are usually foreign to the animal into which they are introduced. The antigenic structure of each species of bacteria is unique to that species, and much like the fingerprint of a human, can be used to identify the bacterium. The antigens that comprise lipopolysaccharide (O-antigens), capsules (K-antigens), and flagellar proteins (H-antigens) of the *Enterobacteriaceae* can be used to differentiate these bacteria into **serotypes** that can be identical physiologically but differ in their antigenic makeup. For example, *Escherichia coli* O157: H7, a serotype of *E. coli,* is a serious pathogen that has contaminated our food supply.

Specific cells of the immune system, called B-cells, are responsible for producing various kinds of antibodies. When an animal is challenged by the antigens on a bacterial cell, antibodies called immunoglobulins are produced that are specific for the bacterial antigens. The immunoglobulins occur in abundance in circulating blood in immunized animals. If blood cells from such animals are separated from the liquid portion of blood, the clear liquid that remains contains the immunoglobulins and is called **antiserum.**

The antigens of a microorganism can be determined by a procedure called **serological typing** (serotyping). It consists of adding a suspension of microorganism to antiserum that contains antibodies specific for antigens associated with the microorganism. If antigens are present, the antibodies in the antiserum will react with the antigens on the bacterial cell and cause the cells to **agglutinate,** or from visible clumps. Serotyping is particularly useful in the identification of *Salmonella* and *Shigella,* that cause infections in humans such as typhoid fever and bacillary dysentery.

For example, *Salmonella* can be differentiated into more than 2500 different serotypes based on antigenic differences associated with the cell. Serotyping of *Salmonella* or *Shigella* is useful in tracing epidemics caused by a particular strain or serotype of the respective organism. In the identification of these two genera, biochemical tests are first used to identify the organism as either *Salmonella* or *Shigella* (Exercise 72), followed by serotyping to identify specific strains.

In this exercise, you will be issued two unknown organisms, one of which is a *Salmonella.* By following the procedure shown in figure 73.1, you will determine which one of the unknowns is *Salmonella.* Note that you will use two test controls. A **negative test control** will be set up in depression A on the slide to see what the absence of agglutination looks like. The negative control is a mixture of antigen and saline (antibody is lacking). A **positive test control** will be performed in depression C with standardized

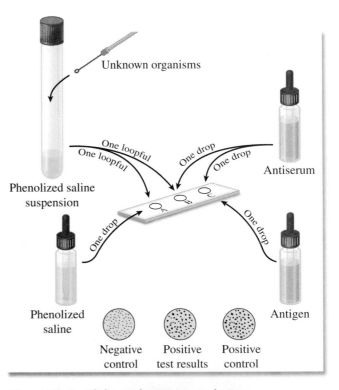

Figure 73.1 Slide agglutination technique.

antigen and antiserum to give you a typical reaction of agglutination.

Materials

- 2 numbered unknowns per student (slant cultures of a *salmonella* and a coliform)
- *salmonella* O antigen, group B (Difco #2840-56)
- *salmonella* O antiserum, poly A-I (Difco #2264-47)
- depression slides or spot plates
- dropping bottle of phenolized saline solution (0.85% sodium chloride, 0.5% phenol)
- 2 serological tubes per student
- 1 ml pipettes

> **CAUTION:** Keep in mind that *Salmonella typhimurium* is a pathogen and can cause gastroenteritis. Be careful!

1. Label three depressions on a spot plate or depression slide **A, B,** and **C,** as shown in figure 73.1.
2. Make a phenolized saline suspension of each unknown in separate serological tubes by suspending one or more loopfuls of organisms in 1 ml of phenolized saline. Mix the organisms sufficiently to ensure complete dispersion of clumps of bacteria. The mixture should be very turbid.

3. Transfer 1 loopful (0.05 ml) from the phenolized saline suspension of one tube to depressions A and B.
4. To depressions B and C, add 1 drop of *salmonella* O polyvalent antiserum. To depression A, add 1 drop of phenolized saline, and to depression C, add 1 drop of *salmonella* O antigen, group B.
5. Mix the organisms in each depression with a clean wire loop. Do not go from one depression to the other without washing the loop first.
6. Compare the three mixtures. Agglutination should occur in depression C (positive control), but not in depression A (negative control). If agglutination occurs in depression B, the organism is *Salmonella.*
7. Repeat this process on another slide for the other organism.

> **CAUTION:** Deposit all slides and serological tubes in a container of disinfectant provided by the instructor.

Laboratory Report

Record your results on the first portion of Laboratory Report 73–74.

Slide Agglutination (Latex) Test:
For *S. aureus* Identification

Many manufacturers of reagents for slide agglutination tests utilize polystyrene latex particles as carriers for the antibody molecules. By adsorbing reactive antibodies to these particles, an agglutination reaction results that occurs rapidly and is much easier to see than ordinary precipitin-type reactions that are used to demonstrate the presence of a soluble antigen.

In this exercise, we will use reagents manufactured by Difco Laboratories to determine if a suspected staphylococcus organism produces coagulase and/or protein A. The test reagent (*Difco Staph Reagent*) is a suspension of yellow latex particles sensitized with antibodies for coagulase and protein A. Reagents are also included to provide positive and negative controls in the test. Instead of using depression slides or spot plates, Difco provides disposable cards with eight black circles printed on them for performing the test. As indicated in figure 74.1, only three circles are used when performing the test on one unknown. The additional circles are provided for testing five additional unknowns at the same time. The black background of the cards facilitates rapid interpretation by providing good contrast for the agglutination reaction that occurs.

There are two versions of this test: direct and indirect. The procedure for the direct method is illustrated in figure 74.1. The indirect method differs in that saline is used to suspend the organism being tested.

It should be pointed out that the reliability correlation between this test for coagulase and the tube test (page 446) is very high. Studies reveal that a reliability correlation of over 97% exists. Proceed as follows to perform this test.

Materials

- plate culture of staphylococcus-like organism (trypticase soy agar plus blood)
- Difco Staph Latex Test kit #3850-32-7, which consists of:
  - bottle of Bacto Staph Latex Reagent
  - bottle of Bacto Staph Positive Control
  - bottle of Bacto Staph Negative Control
  - bottle of Bacto Normal Saline Reagent
  - disposable test slides (black circle cards)
  - mixing sticks (minimum of 3)
- slide rotator

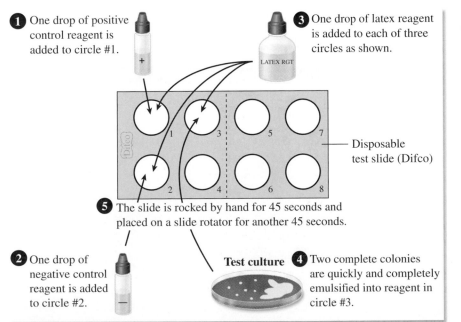

1 One drop of positive control reagent is added to circle #1.

3 One drop of latex reagent is added to each of three circles as shown.

LATEX RGT

Disposable test slide (Difco)

5 The slide is rocked by hand for 45 seconds and placed on a slide rotator for another 45 seconds.

2 One drop of negative control reagent is added to circle #2.

Test culture

4 Two complete colonies are quickly and completely emulsified into reagent in circle #3.

Figure 74.1 Slide agglutination test (direct method) for the presence of coagulase and/or protein A.

Direct Method

If the direct method is to be used, as illustrated in figure 74.1, follow this procedure:

1. Place 1 drop of Bacto Staph Positive Control reagent onto circle #1.
2. Place 1 drop of Bacto Staph Negative Control reagent on circle #2.
3. Place 1 drop of Bacto Staph Latex Reagent onto circles #1, #2, and #3.
4. Using a sterile inoculating needle or loop, quickly and completely emulsify *two isolated colonies* from the culture to be tested into the drop of Staph Latex Reagent in circle #3.

    Also, emulsify the Staph Latex Reagent in the positive and negative controls in circles #1 and #2 using separate mixing sticks supplied in the kit.

    All mixing in these three circles should be done quickly to minimize drying of the latex on the slide and to avoid extended reaction times for the first cultures emulsified.
5. Rock the slide by hand for 45 seconds.
6. Place the slide on a slide rotator capable of providing 110 to 120 rpm and rotate it for another 45 seconds.
7. Read the results immediately, according to the descriptions provided in the table at right. If agglutination occurs before 45 seconds, the results may be read at that time. *The slide should be read at normal reading distance under ambient light.*

Indirect Method

The only differences between the direct and indirect methods pertain to the amount of inoculum and the use of saline to emulsify the unknown being tested. Proceed as follows:

1. Place 1 drop of Bacto Staph Positive Control reagent onto test circle #1.
2. Place 1 drop of Bacto Staph Negative Control onto circle #2.
3. Place 1 drop of Bacto Normal Saline Reagent onto circle #3.

4. Using a sterile inoculating needle or loop, completely emulsify *four isolated colonies* from the culture to be tested into the circle containing the drop of saline (circle #3).
5. Add 1 drop of Bacto Staph Latex Reagent to each of the three circles.
6. Quickly mix the contents of each circle, using individual mixing sticks.
7. Rock the slide by hand for 45 seconds.
8. Place the slide on a slide rotator capable of providing 110 to 120 rpm and rotate it for another 45 seconds.
9. Read the results immediately according to the descriptions provided in the table below. If agglutination occurs before 45 seconds, the results may be read at that time. *The slide should be read at normal reading distance under ambient light.*

Positive Reactions	
4 +	Large to small clumps of aggregated yellow latex beads; clear background
3 +	Large to small clumps of aggregated yellow latex beads; slightly cloudy background
2 +	Medium to small but clearly visible clumps of aggregated yellow latex beads; moderately cloudy background
1 +	Fine clumps of aggregated yellow latex beads; cloudy background
Negative Reactions	
+	Smooth cloudy suspension; particulate grainy appearance that cannot be identified as agglutination
–	Smooth, cloudy suspension; free of agglutination or particles

Laboratory Report

Record your results on the last portion of Laboratory Report 73–74.

Laboratory Report

Student: _____

Date: _____ Section: _____

73   Slide Agglutination Test: Serological Typing

A. Results

1. Describe the appearance of the mixtures in each of the wells.

   _____

   _____

2. Which unknown number proved to be *Salmonella*?

   _____

   _____

B. Short Answer Questions

1. What types of compounds in bacterial cells can serve as antigens?

   _____

   _____

2. What are immunoglobulins?

   _____

   _____

3. What is a serotype of an organism?

   _____

   _____

4. What is agglutination?

   _____

   _____

5. What types of controls are used for the slide agglutination?

   _____

   _____

6. What is the importance of doing controls?

   _____

   _____

74   Slide Agglutination (Latex) Test: For *S. aureus* Identification

A. Results

1. Describe the appearance of the mixtures in each of the wells.

   _____

   _____

2. Was there a positive result for *S. aureus*? If so, what was the degree of the positive reaction?

   _____

   _____

B. Short Answer Questions

1. What two *S. aureus* antigens are being detected with the use of this test kit?

   _____

   _____

2. What definitive test for *S. aureus* is highly correlated with this agglutination test?

   _____

   _____

3. What advantages does the agglutination test have over the definitive *S. aureus* test?

   _____

   _____

4. What is the advantage of the latex test as compared to the ordinary precipitin-like reaction?

   _____

   _____

Infectious mononucleosis (IM) is a benign disease, occurring principally in individuals in the 13- to 25-year age group. It is caused by the Epstein-Barr virus (EBV), a herpesvirus, that is one of the most ubiquitous viruses in humans. Studies have shown that the virus can be isolated from saliva of patients with IM, as well as from some healthy, asymptomatic individuals. Between 80% and 90% of all adults possess antibodies for EBV.

The disease is characterized by a sudden onset of fever, sore throat, and pronounced enlargement of the cervical lymph nodes. There is also moderate leukocytosis with a marked increase in the number of lymphocytes (50% to 90%).

The serological test for IM takes advantage of an unusual property: the antibodies produced against the EBV coincidentally agglutinate sheep red blood cells. This is an example of a **heterophile antigen—** a substance isolated from an organism that stimulates the production of antibodies capable of reacting with tissues of other organisms. The antibodies are referred to as **heterophile antibodies.**

This test is performed by adding a suspension of sheep red blood cells to dilutions of inactivated patient's serum and incubating the tubes overnight in the refrigerator. Figure 75.1 illustrates the overall procedure. Agglutination titers of 320 or higher are considered significant. Titers of 40,960 have been obtained.

Proceed as follows to perform this test on a sample of test serum:

First Period

Materials

- test-tube rack (Wasserman type) with 10 clean serological tubes
- bottle of saline solution (0.85% NaCl), clear or filtered
- 1 ml pipettes
- 5 ml pipettes
- 2% suspension of sheep red blood cells
- patient's serum (known to be positive)

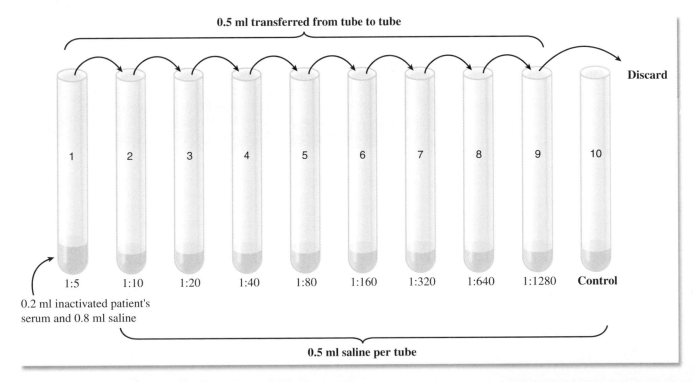

0.5 ml transferred from tube to tube

Discard

| 1 | 2 | 3 | 4 | 5 | 6 | 7 | 8 | 9 | 10 |

1:5   1:10   1:20   1:40   1:80   1:160   1:320   1:640   1:1280   **Control**

0.2 ml inactivated patient's serum and 0.8 ml saline

0.5 ml saline per tube

Figure 75.1   Procedure for setting up heterophile antibody test.

1. Place the test serum in a 56°C water bath for 30 minutes to inactivate the complement.
2. Set up a row of 10 serological tubes in the front row of a test-tube rack and number them from 1 to 10 (left to right) with a marking pencil.
3. Into tube 1, pipette 0.8 ml of physiological saline.
4. Dispense 0.5 ml of physiological saline to tubes 2 through 10. Use a 5 ml pipette.
5. With a 1 ml pipette add 0.2 ml of the inactivated serum to tube 1. Mix the contents of this tube by drawing into the pipette and expelling about five times.
6. Transfer 0.5 ml from tube 1 to tube 2, mix five times, and transfer 0.5 ml from tube 2 to tube 3, ard so on, through the ninth tube. *Discard 0.5 ml from the ninth tube after mixing.* Tube 10 is the **control.**
7. Add 0.2 ml of 2% sheep red blood cells to all tubes (1 through 10) and shake the tubes. Final dilutions of the serum are shown in figure 75.1.

8. Allow the rack of tubes to stand at room temperature for 1 hour, then transfer the tubes to a small wire basket, and place in a refrigerator to remain overnight.

Second Period

Set up the tubes in a tube rack in order of dilution and compare each tube with the control by holding the tubes overhead and looking up at the bottoms of the tubes. Nonagglutinated cells will tumble to the bottom of the tube and form a small button (as in control tube). Agglutinated cells will form a more-amorphous "blanket" (figure 75.2).

The **titer** should be recorded as the reciprocal of the last tube in the series that shows positive agglutination.

Laboratory Report

Complete Laboratory Report 75.

Figure 75.2 Agglutination is more readily seen when the tube is examined against a black surface.

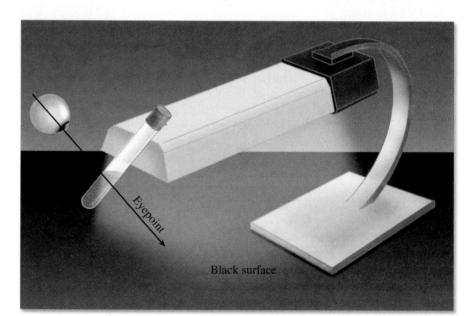

Laboratory Report

Student: _____

Date: _____ Section: _____

75 Tube Agglutination Test: The Heterophile Antibody Test

A. Results

1. What was the titer of the serum tested?

   _____

   _____

2. Is the result considered significant? Explain.

   _____

   _____

   _____

B. Short Answer Questions

1. For what disease is this diagnostic test used?

   _____

   _____

2. What is the name of the virus that causes this disease?

   _____

   _____

3. What percentage of all adults possesses antibodies for the virus?

   _____

   _____

4. What are heterophile antibodies? Why are they useful for this test?

   _____

   _____

5. What additional blood examination can be performed to confirm an infection?

   _____

   _____

   _____

White Blood Cell Study:
The Differential WBC Count

In 1883, at the Pasteur Institute in Paris, Metchnikoff published a paper proposing the **phagocytic theory of immunity.** On the basis of his studies performed on transparent starfish larvae, he postulated that amoeboid cells in the tissue fluid and blood of all animals are the major guardians of health against bacterial infection. He designated the large phagocytic cells of the blood as *macrophages* and the smaller ones as *microphages.* Today, Metchnikoff's macrophages are known as monocytes and his microphages as neutrophils or polymorphonuclear leukocytes.

Figure 76.1 illustrates the five types of leukocytes that are normally seen in the blood. Blood platelets and erythrocytes also are shown to present a complete picture of all formed elements in the blood. When observed as living cells under the microscope, they appear as refractile, colorless structures. As shown here, however, they reflect the dyes that are imparted by Wright's stain.

In this exercise, we will do a study of the white blood cells in human blood. This study may be made from a prepared stained microscope slide or from a slide made from your own blood. By scanning an entire slide and counting the various types, you will have an opportunity to encounter most, if not all, types. The erythrocytes and blood platelets will be ignored.

Figures 76.1 and 76.2 will be used to identify the various types of cells. Figure 76.3 illustrates the procedure for preparing a slide stained with Wright's stain. The relative percentages of each type will be determined after a total of 100 white blood cells have been identified. This method of white blood cell enumeration is called a **differential WBC count.**

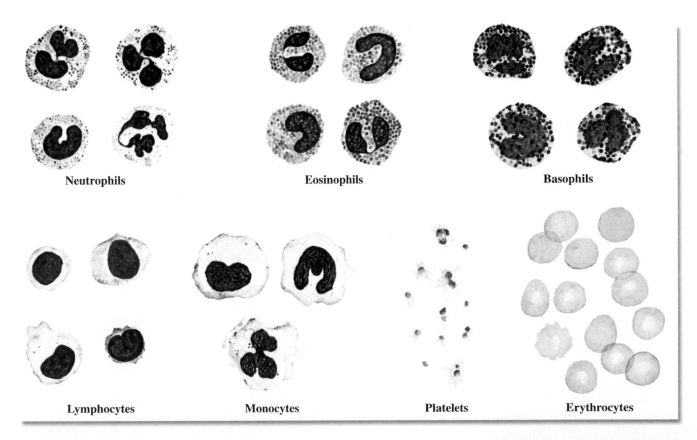

Neutrophils   Eosinophils   Basophils

Lymphocytes   Monocytes   Platelets   Erythrocytes

Figure 76.1   Formed elements of blood.

Granulocytes

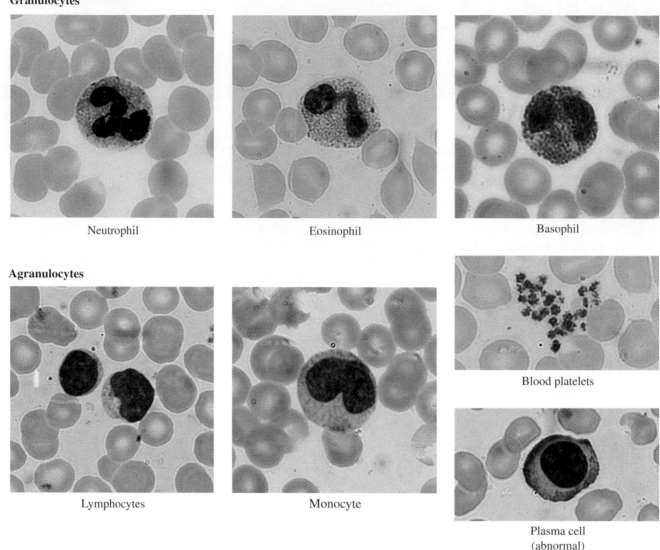

Neutrophil   Eosinophil   Basophil

Agranulocytes

Blood platelets

Lymphocytes   Monocyte

Plasma cell
(abnormal)

Figure 76.2 Photomicrographs of formed elements in blood.

As you proceed with this count, it will become obvious that the neutrophils are most abundant (50%–70%). The next most prominent cells are the lymphocytes (20%–30%). Monocytes comprise about 2%–6%; eosinophils, 1%–5%; and basophils, less than 1%.

A normal white blood cell count is between 5,000 and 10,000 white cells per cubic millimeter. Elevated white blood cell counts are referred to as *leukocytosis;* counts of 30,000 or 40,000 represent marked leukocytosis. When counts fall considerably below 5,000, *leukopenia* is said to exist. Both conditions can have grave implications.

The value of a differential count is immeasurable in the diagnosis of infectious diseases. High neutrophil counts, or *neutrophilia,* often signal localized infections, such as appendicitis or abscesses in some

other part of the body. *Neutropenia,* a condition in which there is a marked decrease in the numbers of neutrophils, occurs in typhoid fever, undulant fever, and influenza. *Eosinophilia* may indicate allergic conditions or invasions by parasitic roundworms such as *Trichinella spiralis,* the "pork worm." Counts of eosinophils may rise to as high as 50% in cases of trichinosis. High lymphocyte counts, or *lymphocytosis,* are present in whooping cough and some viral infections. Increased numbers of monocytes, or *monocytosis,* may indicate the presence of the Epstein-Barr virus, which causes infectious mononucleosis.

Note in the materials list that items needed for making a slide (option B) are listed separately. If a prepared slide (option A) is to be used, ignore the instructions under the heading "Preparation of Slide,"

(1) A small drop of blood is placed about 3/4 inch away from one end of slide. The drop should not exceed 1/8" diameter.

(2) The spreader slide is moved in direction of arrow, allowing drop of blood to spread along slide's back edge.

(3) The spreader slide is pushed along the slide, dragging the blood over the surface of the slide.

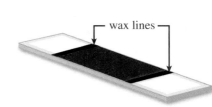

(4) A china marking pencil is used to mark off both ends of the smear to retain the staining solution on the slide.

Figure 76.3   Smear preparation technique for making a stained blood slide.

and proceed to the heading "Performing the Cell Count." Your instructor will indicate which option will be used. Proceed as follows:

PRECAUTIONS

When working with blood observe the following precautions:
1. Always disinfect the finger with alcohol prior to piercing it.
2. Use sterile disposable lancets only one time.
3. Dispose of used lancets by placing them into a beaker of disinfectant.
4. Avoid skin contact with blood of other students. Wear disposable latex gloves.
5. Disinfect finger with alcohol after blood has been taken.

Materials

prepared blood slide (option A):
- stained with Wright's or Giemsa's stains

for staining a blood smear (option B):
- 2 or 3 clean microscope slides (should have polished edges)
- sterile disposable lancets
- disposable latex gloves
- sterile absorbent cotton, 70% alcohol
- Wright's stain, wax pencil, bibulous paper
- distilled water in dropping bottle

Preparation of Slide

Figure 76.3 illustrates the procedure that will be used to make a stained slide of a blood smear. The most difficult step in making such a slide is getting a good spread of the blood, which is thick at one end and thin at the other end. If done properly, the smear will have a gradient of cellular density that will make it possible to choose an area that is ideal for study. The angle at which the spreading slide is held in making the smear will determine the thickness of the smear. It may be necessary for you to make more than one slide to get an ideal one.

1. Clean three or four slides with soap and water. Handle them with care to avoid getting their flat surfaces soiled by your fingers. Although only two slides may be used, it is often necessary to repeat the spreading process, thus the extra slides.
2. Scrub the middle finger with 70% alcohol and stick it with a lancet. Put a drop of blood on the slide $\frac{3}{4}''$ from one end and spread with another slide in the manner illustrated in figure 76.3.

   Note that the blood is dragged over the slide, not pushed. Do not pull the slide over the smear a second time. If you don't get an even smear the first time, repeat the process on a fresh clean slide. To get a smear that will be the proper thickness, hold the spreading slide at an angle somewhat greater than 45°.

3. Draw a line on each side of the smear with a wax pencil to confine the stain that is to be added. (Note: This step is helpful for beginners, and usually omitted by professionals.)

4. Cover the film with Wright's stain, *counting the drops* as you add them. Stain for **4 minutes** and then add the same number of drops of distilled water to the stain and let stand for another **10 minutes.** Blow gently on the mixture every few minutes to keep the solutions mixed.

5. Gently wash off the slide under running water for 30 seconds and shake off the excess. Blot dry with bibulous paper.

Performing the Cell Count

Whether you are using a prepared slide or one that you have just stained, the procedure is essentially the same. Although the high-dry objective can be used for the count, the oil immersion lens is much better. Differentiation of some cells is difficult with high-dry optics. Proceed as follows:

1. Scan the slide with the low-power objective to find an area where cell distribution is best. A good area is one in which the cells are not jammed together or scattered too far apart.

2. Systematically scan the slide, following the pathway indicated in figure 76.4. As each leukocyte is encountered, identify it, using figures 76.1 and 76.2 for reference.

3. Tabulate your count on the Laboratory Report sheet according to the instructions there. It is best to remove the lab report sheet from the back of the manual for this identification and tabulation procedure.

Laboratory Report

Place your results in the table in Part A of Laboratory Report 76.

Figure 76.4 Path to follow when seeking cells.

Laboratory Report

Student: _____

Date: _____ Section: _____

76 White Blood Cell Study: The Differential WBC Count

A. Results

As you move the slide in the pattern indicated in figure 76.4, record all the different types of cells in the following table. Refer to figures 76.1 and 76.2 for cell identification. Use this method of tabulation: ⊞ ⊞ |. Identify and tabulate 100 leukocytes. Divide the total of each kind of cell by 100 to determine percentages.

NEUTROPHILS	LYMPHOCYTES	MONOCYTES	EOSINOPHILS	BASOPHILS
Total				
Percent				

B. Short Answer Questions

1. Were your percentages for each type within the normal ranges? _____

2. What errors might one be likely to make when doing this count for the first time? _____
   _____
   _____
   _____

3. Differentiate between the following:

   Cellular immunity: _____

   Humoral immunity: _____

4. Do cellular and humoral immunity work independently? _____ Explain: _____
   _____
   _____
   _____
   _____
   _____
   _____

5. Define the following terminology. Also describe how these conditions are produced.
   a. neutrophilia _____
   _____
   b. neutropenia _____
   _____

c. eosinophilia _____

_____

d. lymphocytosis _____

_____

e. monocytosis _____

_____

Blood Grouping

Exercises 73 through 75 illustrate three uses of agglutination tests as related to (1) the identification of serological types, (2) species identification (*S. aureus*), and (3) disease identification (infectious mononucleosis and typhoid fever). The typing of blood is another example of a medical procedure that relies on this useful phenomenon.

The procedure for blood typing was developed by Karl Landsteiner around 1900. He is credited with having discovered that human blood types can be separated into four groups on the basis of two antigens that are present on the surface of red blood cells. These antigens are designated as A and B. The four groups (types) are A, B, AB, and O. The last group type O, which is characterized by the absence of A or B antigens, is the most common type in the United States (45% of the population). Type A is next in frequency, found in 39% of the population. The incidences of types B and AB are 12% and 4%, respectively.

Blood typing is performed with antisera containing high titers of anti-A and anti-B antibodies. The test may be performed by either slide or tube methods. In both instances, a drop of each kind of antiserum is added to separate samples of saline suspension of red blood cells. Figure 77.1 illustrates the slide technique. If agglutination occurs only in the suspension to which the anti-A serum was added, the blood is type A. If agglutination occurs only in the anti-B mixture, the blood is type B. Agglutination in both samples indicates that the blood is type AB. The absence of agglutination indicates that the blood is type O.

Between 1900 and 1940, a great deal of research was done to uncover the presence of other antigens in human red blood cells. Finally, in 1940, Landsteiner and Wiener reported that rabbit sera containing antibodies against the red blood cells of the rhesus monkey would agglutinate the red blood cells of 5% of white humans. This antigen in humans, which was first designated as the **Rh factor** (in due respect to the rhesus monkey), was later found to exist as six antigens: C, c, D, d, E, and e. Of these six antigens, the D factor is responsible for the Rh-positive condition and is found in 85% of Whites, 94% of Blacks, and 99% of Orientals.

Typing blood for the Rh factor can also be performed by both tube and slide methods, but there are certain differences in the two techniques. First of all, the antibodies in the typing sera are of the incomplete albumin variety, which *will not agglutinate human red cells when they are diluted with saline.* Therefore, it is necessary to use whole blood or dilute the cells with plasma. Another difference is that the test *must be performed at higher temperatures:* 37° C for tube test; 45° C for the slide test.

In this exercise, two separate slide methods are presented for typing blood. If only the Landsteiner ABO groups are to be determined, the first method may be preferable. If Rh typing is to be included, the second method, which utilizes a slide warmer, will be followed. The availability of materials will determine which method is to be used.

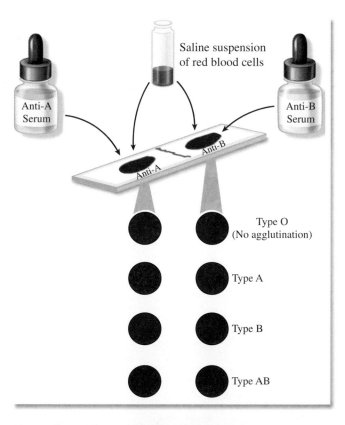

Figure 77.1   Typing of ABO blood groups.

PRECAUTIONS

Review the precautionary comments highlighted on page 485.

ABO Blood Typing

Materials

- small vial (10 mm dia × 50 mm long)
- disposable lancets (B–D Microlance, Serasharp, etc.)
- 70% alcohol and cotton
- china marking pencil
- microscope slides
- typing sera (anti-A and anti-B)
- applicators or toothpicks
- saline solution (0.85% NaCl)
- 1 ml pipettes
- disposable latex gloves

1. Mark a slide down the middle with a marking pencil, dividing the slide into two halves as shown in figure 77.1. Write "anti-A" on the left side and "anti-B" on the right side.
2. Pipette 1 ml of saline solution into a small vial or test tube.
3. Scrub the middle finger with a piece of cotton saturated with 70% alcohol and pierce it with a sterile disposable lancet. Allow 2 or 3 drops of blood to mix with the saline by holding the finger over the end of the vial and washing it with the saline by inverting the tube several times.
4. Place a drop of this red cell suspension on each side of the slide.

5. Add a drop of anti-A serum to the left side of the slide and a drop of anti-B serum to the right side.
   Do not contaminate the tips of the serum pipettes with the material on the slide.
6. After mixing each side of the slide with separate applicators or toothpicks, look for agglutination. The slide should be held about 6″ above an illuminated white background and rocked gently for 2 or 3 minutes. Record your results in Laboratory Report 77 as of 3 minutes.

Combined ABO and Rh Typing

As stated, Rh typing must be performed with heat on blood that has not been diluted with saline. A warming box such as the one in figure 77.2 is essential in this procedure. In performing this test, two factors are of considerable importance: first, only a small amount of blood must be used (a drop of about 3 mm diameter on the slide) and, second, proper agitation must be executed. The agglutination that occurs in this antibody-antigen reaction results in finer clumps; therefore, closer examination is essential. If the agitation is not properly performed, agglutination may not be as apparent as it should be.

In this combined method, we will use whole blood for the ABO typing as well as for the Rh typing. Although this method works satisfactorily as a classroom demonstration for the ABO groups, it is *not as reliable* as the previous method in which saline and room temperature are used. *This method is not recommended for clinical situations.*

Figure 77.2 Blood typing with warming box.

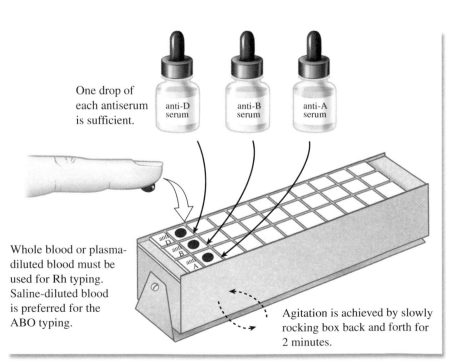

One drop of each antiserum is sufficient.

anti-D serum   anti-B serum   anti-A serum

Whole blood or plasma-diluted blood must be used for Rh typing. Saline-diluted blood is preferred for the ABO typing.

Agitation is achieved by slowly rocking box back and forth for 2 minutes.

Materials

- slide warming box with a special marked slide
- anti-A, anti-B, and anti-D typing sera
- applicators or toothpicks
- 70% alcohol and cotton
- disposable sterile lancets

1. Scrub the middle finger with a piece of cotton saturated with 70% alcohol and pierce it with a sterile disposable lancet. Place a small drop of blood in each of three squares on the marked slides on the warming box.

   To get the proper proportion of serum to blood, do not use a drop larger than 3 mm diameter on the slide.

2. Add a drop of anti-D serum to the blood in the anti-D square, mix with a toothpick, and note the time. **Only 2 minutes should be allowed for agglutination.**
3. Add a drop of anti-B serum to the anti-B square and a drop of anti-A serum to the anti-A square. Mix the sera and blood in both squares with *separate* fresh toothpicks.
4. Agitate the mixtures on the slide by slowly rocking the box back and forth on its pivot. At the end of 2 minutes, examine the anti-D square carefully for agglutination. If no agglutination is apparent, consider the blood to be Rh-negative. By this time the ABO type can also be determined.
5. Record your results in Laboratory Report 77.

Laboratory Report

Student: _____

Date: _____ Section: _____

77 Blood Grouping

A. Results

1. Describe each of the mixtures on the blood-typing slide.

    _____

    _____

2. Based upon these results, what is your blood type?

    _____

    _____

3. What antigens are present on your blood cells?

    _____

    _____

4. If you needed a transfusion, what blood types can you accept?

    _____

    _____

5. To which recipient blood types can you donate blood?

    _____

    _____

B. Short Answer Questions

1. What happens when an individual receives an incompatible blood type in the course of a transfusion?

    _____

    _____

2. Which blood type is known as the universal recipient? Explain.

    _____

    _____

3. Which blood type is known as the universal donor? Explain.

    _____

    _____

4. Why might a pregnant woman want to know her Rh type? Why is this information increasingly important for subsequent pregnancies?

    _____

    _____

A Synthetic Epidemic

A disease caused by microorganisms that enter the body and multiply in the tissues at the expense of the host is said to be an **infectious disease.** Infectious diseases that are transmissible to other persons are considered to be **communicable.** The transfer of communicable infectious agents between individuals can be accomplished by direct contact, such as in handshaking, kissing, and sexual intercourse, or they can be spread indirectly through food, water, objects, animals, and so on.

Epidemiology is the study of how, when, where, what, and who are involved in the spread and distribution of diseases in human populations. An epidemiologist is, in a sense, a medical detective who searches out the sources of infection so that the transmission cycle can be broken.

Whether an epidemic actually exists is determined by the epidemiologist by comparing the number of new cases with previous records. If the number of newly reported cases in a given period of time in a specific area is excessive, an **epidemic** is considered to be in progress. If the disease spreads to one or more continents, a **pandemic** is occurring.

In this experiment, we will have an opportunity to approximate, in several ways, the work of the epidemiologist. Each member of the class will take part in the spread of a "synthetic infection." The mode of transmission will be handshaking. For obvious safety reasons, the agent of transmission will not be a pathogen.

Two different approaches to this experiment are given: procedures A and B. In procedure A, a white powder is used. In procedure B, two nonpathogens (*Micrococcus luteus* and a yeast *Rhodotorula mucilaginosa*) will be used. The advantage of procedure A is that it can be completed in one laboratory session. Procedure B, on the other hand, is more realistic in that viable organisms are used; however, it involves two periods. Your instructor will indicate which procedure is to be followed.

Procedure A

In this experiment, each student will be given a numbered container of white powder. Only one member in the class will be given a powder that is to be considered the infectious agent. The other members will be

issued a transmissible agent that is considered noninfectious. After each student has spread the powder on his or her hands, all members of the class will engage in two rounds of handshaking, directed by the instructor. A record of the handshaking contacts will be recorded on a chart similar to the one in the Laboratory Report. After each round of handshaking, the hands will be rubbed on blotting paper so that a chemical test can be applied to it to determine the presence or absence of the infectious agent.

Once all the data are compiled, an attempt will be made to determine two things: (1) the original source of the infection, and (2) who the carriers are. The type of data analysis used in this experiment is similar to the procedure that an epidemiologist would employ. Proceed as follows:

Materials

- 1 numbered container of starch or flour
- 1 piece of white blotting paper
- spray bottles of iodine

Preliminaries

1. After assembling your materials, write your name and unknown number at the top of your sheet of blotting paper. In addition, draw a line down the middle, top to bottom, and label the left side ROUND 1 and the right side ROUND 2.
2. Wash and dry your hands thoroughly.
3. Moisten the right hand with water and prepare it with the agent by thoroughly coating it with the white powder, especially on the palm surface. This step is similar to the contamination that would occur to one's hand if it were sneezed into during a cold.

   IMPORTANT: Once the hand has been prepared, do not rest it on the tabletop or allow it to touch any other object.

Round 1

1. On the cue of the instructor, you will begin the first round of handshaking. Your instructor will inform you when it is your turn to shake hands with someone. You may shake with anyone, but it is best not to shake your neighbor's hand. *Be sure*

to use only your treated hand, and avoid extracurricular glad-handing.

2. In each round of handshaking, you will be selected by the instructor *only once* for handshaking; however, due to the randomness of selection by the handshakers, it is possible that you may be selected as the "shakee" several times.

3. After every member of the class has shaken someone's hand, you need to assess just who might have picked up the "microbe." To accomplish this, wipe your fingers and palm of the contaminated hand on the left side of your blotting paper. Press fairly hard, but don't tear the surface.

   IMPORTANT: Don't allow your hand to touch any other object. A second round of handshaking follows.

Round 2

1. On the cue of your instructor, shake hands with another person. Avoid contact with any other objects.

2. Once the second handshaking episode is finished, rub the fingers and palm of the contaminated hand on the right side of the blotting paper.

   CAUTION: Keep your contaminated hand off the left side of the blotting paper.

Chemical Identification

1. To determine who has been "infected," we will now spray the iodine solution on the handprints of both rounds. One at a time, each student, with the help of the instructor, will spray his or her blotting paper with iodine solution.

2. Color interpretation is as follows:
   • Blue— positive for infectious agent
   • Brown or yellow— negative

Tabulation of Results

1. Tabulate the results on the chalkboard, using a table similar to the one in the Laboratory Report.

2. Once all results have been recorded, proceed to determine the originator of the epidemic. The easiest way to determine this is to put together a flowchart of shaking.

3. Identify those persons that test positive. You will be working backward with the kind of information an epidemiologist has to work with (contacts and infections). Eventually, a pattern will emerge that shows which person started the epidemic.

4. Complete Laboratory Report 78.

Procedure B

In this experiment, each student will be given a piece of hard candy that has had a drop of *Micrococcus luteus* or *Rhodotorula mucilaginosa* applied to it.

Only one person in the class will receive candy with *R. mucilaginosa,* the presumed pathogen. All others will receive *M. luteus.*

After each student has handled the piece of candy with a glove-covered right hand, he or she will shake hands (glove to glove) with another student as directed by the instructor. A record will be kept of who takes part in each contact. Two rounds of handshaking will take place. After each round, a plate of trypticase soy agar will be streaked.

After incubating the plates, a tabulation will be made for the presence or absence of *R. mucilaginosa* on the plates. From the data collected, an attempt will be made to determine two things: (1) the original source of the infection and (2) who the carriers are. The type of data analysis used in this experiment is similar to the procedure that an epidemiologist would employ. Proceed as follows:

Materials

- sterile rubber surgical gloves (1 per student)
- hard candy contaminated with *M. luteus*
- hard candy contaminated with *R. mucilaginosa*
- sterile swabs (2 per student)
- TSA plates (1 per student)

Preliminaries

1. Draw a line down the middle of the bottom of a TSA plate, dividing it into two halves. Label one half ROUND 1 and the other ROUND 2.

2. Put a sterile rubber glove on your right hand. Avoid contaminating the palm surface.

3. Grasp the piece of candy in your gloved hand, rolling it around the surface of your palm. Discard the candy into a beaker of disinfectant set aside for disposal. You are now ready to do the first-round handshake.

Round 1

1. *On the cue of your instructor,* select someone to shake hands with. You may shake with anyone, but it is best not to shake hands with your neighbor.

2. In each round of handshaking, you will be selected by the instructor *only once* for handshaking; however, due to the randomness of selection by the handshakers, it is possible that you may be selected as the "shakee" several times. The instructor or a recorder will record the initials of the shaker and shakee each time.

3. After you have shaken someone's hand, swab the surface of your palm and transfer the organisms to the side of your plate designated as ROUND 1. Discard this swab into the appropriate container for disposal.

Round 2

1. Again, on the cue of your instructor, select some-one at random to shake hands with. Be sure not to contaminate your gloved hand by touching something else.
2. With a fresh swab, swab the palm of your hand and transfer the organisms to the side of your plate designated as ROUND 2. Make sure that your initials and the initials of the shakee are re-corded by the instructor or recorder.
3. Incubate the TSA plate at room temperature for 48 hours.

Tabulation and Analysis

1. After 48 hours incubation, look for typical red *R. mucilaginosa* colonies on your petri plate. If such colonies are present, record them as positive on your Laboratory Report chart and on the chart on the chalkboard.
2. Fill out the chart on Laboratory Report 78 with all the information from the chart on the chalkboard.
3. Identify those persons that test positive. You will be working backward with the kind of information an epidemiologist has to work with (contacts and infections). Eventually a pattern will emerge that shows which person started the epidemic.

Laboratory Report

Complete Laboratory Report 78.

Laboratory Report

Student: _____

Date: _____ Section: _____

78   A Synthetic Epidemic

A. Results

Record in the table below the information that has been tabulated on the chalkboard. The SHAKER is the person designated by the instructor to shake hands with another class member. The SHAKEE is the individual chosen by the shaker. For **Procedure A,** a blue color is positive; yellow or brown is negative. For **Procedure B,** red colonies (*R. mucilaginosa*) is positive; no red colonies is negative.

SHAKER Round 1	RESULT + or –	SHAKEE Round 1	RESULT + or –	SHAKER Round 2	RESULT + or –	SHAKEE Round 2	RESULT + or –
1.				1.			
2.				2.			
3.				3.			
4.				4.			
5.				5.			
6.				6.			
7.				7.			
8.				8.			
9.				9.			
10.				10.			
11.				11.			
12.				12.			
13.				13.			
14.				14.			
15.				15.			
16.				16.			
17.				17.			
18.				18.			
19.				19.			
20.				20.			
21.				21.			
22.				22.			
23.				23.			
24.				24.			

B. Short Answer Questions

1. Who in the group was "patient zero," the starter of the epidemic? _____

2. How many carriers resulted after

   Round 1? _____   Round 2? _____

3. If this were a real infectious agent, such as a cold virus or influenza, list some other factors in transmission besides the ones we tested:

   _____

   _____

4. How would it have been possible to stop this infection cycle? _____

   _____

   _____

5. For each of the following diseases, describe strategies to stop the transmission of the disease.

   a. gonorrhea _____

   b. infectious mononucleosis _____

   c. salmonellosis _____

   d. tuberculosis _____

   e. West Nile encephalitis _____

Tables

TABLE I International Atomic Weights

Element	Symbol	Atomic Number	Atomic Weight
Aluminum	Al	13	26.97
Antimony	Sb	51	121.76
Arsenic	As	33	74.91
Barium	Ba	56	137.36
Beryllium	Be	4	9.013
Bismuth	Bi	83	209.00
Boron	B	5	10.82
Bromine	Br	35	79.916
Cadmium	Cd	48	112.41
Calcium	Ca	20	40.08
Carbon	C	6	12.010
Chlorine	Cl	17	35.457
Chromium	Cr	24	52.01
Cobalt	Co	27	58.94
Copper	Cu	29	63.54
Fluorine	F	9	19.00
Gold	Au	79	197.2
Hydrogen	H	1	1.0080
Iodine	I	53	126.92
Iron	Fe	26	55.85
Lead	Pb	82	207.21
Magnesium	Mg	12	24.32
Manganese	Mn	25	54.93
Mercury	Hg	80	200.61
Nickel	Ni	28	58.69
Nitrogen	N	7	14.008
Oxygen	O	8	16.0000
Palladium	Pd	46	106.7
Phosphorus	P	15	30.98
Platinum	Pt	78	195.23
Potassium	K	19	39.096
Radium	Ra	88	226.05
Selenium	Se	34	78.96
Silicon	Si	14	28.06
Silver	Ag	47	107.880
Sodium	Na	11	22.997
Strontium	Sr	38	87.63
Sulfur	S	16	32.066
Tin	Sn	50	118.70
Titanium	Ti	22	47.90
Tungsten	W	74	183.92
Uranium	U	92	238.07
Vanadium	V	23	50.95
Zinc	Zn	30	65.38
Zirconium	Zr	40	91.22

TABLE II Autoclave Steam Pressures and Corresponding Temperatures

Steam Pressure lb/sq in	Temperature °C	Temperature °F	Steam Pressure lb/sq in	Temperature °C	Temperature °F	Steam Pressure lb/sq in	Temperature °C	Temperature °F
0	100.0	212.0						
1	101.9	215.4	11	116.4	241.5	21	126.9	260.4
2	103.6	218.5	12	117.6	243.7	22	127.8	262.0
3	105.3	221.5	13	118.8	245.8	23	128.7	263.7
4	106.9	224.4	14	119.9	247.8	24	129.6	265.3
5	108.4	227.1	15	121.0	249.8	25	130.4	266.7
6	109.8	229.6	16	122.0	251.6	26	131.3	268.3
7	111.3	232.3	17	123.0	253.4	27	132.1	269.8
8	112.6	234.7	18	124.1	255.4	28	132.9	271.2
9	113.9	237.0	19	125.0	257.0	29	133.7	272.7
10	115.2	239.4	20	126.0	258.8	30	134.5	274.1

Figures are for steam pressure only and the presence of any air in the autoclave invalidates temperature readings from the above table.

TABLE III Autoclave Temperatures as Related to the Presence of Air

Gauge Pressure, lb	Pure steam, complete air discharge °C	°F	Two-thirds air discharge, 20-in. vacuum °C	°F	One-half air discharge, 15-in. vacuum °C	°F	One-third air discharge, 10-in. vacuum °C	°F	No air discharge °C	°F
5	109	228	100	212	94	202	90	193	72	162
10	115	240	109	228	105	220	100	212	90	193
15	121	250	115	240	112	234	109	228	100	212
20	126	259	121	250	118	245	115	240	109	228
25	130	267	126	259	124	254	121	250	115	240
30	135	275	130	267	128	263	126	259	121	250

TABLE IV Antibiotic Susceptibility Test Discs

Antibiotic Agent	Concentration	Individual / 10 Pack
Amikacin	30 µg	231596 / 231597
Amoxicillin / Clavulanic Acid	30 µg	231628 / 23629
Ampicillin	10 µg	230705 / 231264
Ampicillin / Subactam	10/10 µg	231659 / 231660
Azlocillin	75 µg	231624 / 231625
Bacitracin	2 units	230717 / 231267
Carbenicillin	100 µg	231235 / 231555
Cefaclor	30 µg	231652 / 231653
Cefazolin	30 µg	231592 / 231593
Cefixime	5 µg	231663 / NA
Cefoperazone	75 µg	231612 / 231613
Cefotaxime	30 µg	231606 / 231607
Cefotetan	30 µg	231655 / 231656
Cefoxitin	30 µg	231590 / 231591
Ceftazidime	30 µg	231632 / 231633
Ceftriaxone	30 µg	231634 / 231635
Cefuroxime	30 µg	231620 / 231621
Cephalothin	30 µg	230725 / 231271
Chloramphenicol	30 µg	230733 / 231274
Clindamycin	2 µg	231213 / 231275
Doxycycline	30 µg	230777 / 231286
Erythromycin	15 µg	230793 / 231290
Gentamicin	10 µg	231227 / 231299
Imipenem	10 µg	231644 / 231645
Kanamycin	30 µg	230825 / 230829
Mezlocillin	75 µg	231614 / 231615
Minocycline	30 µg	231250 / 231251

Courtesy and © Becton, Dickinson, and Company

(continued)

TABLE IV (Antibiotic Susceptibility Test Discs continued)

Antibiotic Agent	Concentration	Individual / 10 Pack
Moxalactam	30 µg	231610 / 231611
Nafcillin	1 µg	230866 / 231309
Nalidixic Acid	30 µg	230870 / 230874
Netilimicin	30 µg	231602 / 231603
Nitrofurantoin	100 µg	230801 / 231292
Penicillin	2 units	230914 / 231320
Piperacillin	100 µg	231608 / 231609
Rifampin	5 µg	231541 / 231544
Streptomycin	10 µg	230942 / 231328
Sulfisoxazole	0.25 mg	230813 / 231296
Tetracycline	5 µg	230994 / 231343
Ticarcillin	75 µg	231618 / 231619
Tobramycin	10 µg	231568 / 213569
Trimethoprim	5 µg	231600 / 231601
Vancomycin	30 µg	231034 / 231353

TABLE V Indicators of Hydrogen Ion Concentration

Many of the following indicators are used in the media of certain exercises in this manual. This table indicates the pH range of each indicator and the color changes that occur. To determine the exact pH within a particular range one should use a set of standard colorimetric tubes that are available from the prep room. Consult your lab instructor.

Indicator	Full Acid Color	Full Alkaline Color	pH Range
Cresol Red	red	yellow	0.2 – 1.8
Metacresol Purple (acid range)	red	yellow	1.2 – 2.8
Thymol Blue	red	yellow	1.2 – 2.8
Bromphenol Blue	yellow	blue	3.0 – 4.6
Bromcresol Green	yellow	blue	3.8 – 5.4
Chlorcresol Green	yellow	blue	4.0 – 5.6
Methyl Red	red	yellow	4.4 – 6.4
Chlorphenol Red	yellow	red	4.8 – 6.4
Bromcresol Purple	yellow	purple	5.2 – 6.8
Bromthymol Blue	yellow	blue	6.0 – 7.6
Neutral Red	red	amber	6.8 – 8.0
Phenol Red	yellow	red	6.8 – 8.4
Cresol Red	yellow	red	7.2 – 8.8
Metacresol Purple (alkaline range)	yellow	purple	7.4 – 9.0
Thymol Blue (alkaline range)	yellow	blue	8.0 – 9.6
Cresolphthalein	colorless	red	8.2 – 9.8
Phenolphthalein	colorless	red	8.3 – 10.0

B

Indicators, Stains, Reagents

Indicators

All the indicators used in this manual can be made by (1) dissolving a measured amount of the indicator in 95% ethanol, (2) adding a measured amount of water, and (3) filtering with filter paper. The following chart provides the correct amounts of indicator, alcohol, and water for various indicator solutions.

Indicator Solution	Indicator (gm)	95% Ethanol (ml)	Distilled H$_2$0 (ml)
Bromcresol green	0.4	500	500
Bromcresol purple	0.4	500	500
Bromthymol blue	0.4	500	500
Cresol red	0.2	500	500
Methyl red	0.2	500	500
Phenolphthalein	1.0	50	50
Phenol red	0.2	500	500
Thymol blue	0.4	500	500

Stains and Reagents

Acid-alcohol (for Kinyoun stain)
Ethanol (95%) . . . . .97 ml
Concentrated HCl . . . . .3 ml

Alcohol, 70% (from 95%)
Alcohol, 95%. . . . . 368.0 ml
Distilled water . . . . 132.0 ml

Barritt's Reagent (Voges-Proskauer test)
Solution A: 6 g alpha-naphthol in 100 ml 95% ethyl alcohol.
Solution B: 16 g potassium hydroxide in 100 ml water.

Note that no creatine is used in these reagents as is used in O'Meara's reagent for the V-P test.

Carbolfuchsin Stain (Kinyoun stain)
Basic fuchsin . . . . 4 gm
Phenol . . . . .8 ml
Alcohol (95%) . . . . .20 ml
Distilled/deionized water. . . . . .100 ml

Dissolve the basic fuchsin in the alcohol, and add the water while slowly shaking. Melt the phenol in a 56° C water bath and carefully add 8 mls to the stain.

Note: To facilitate staining of acid-fast bacteria, 1 drop of Tergitol No. 7 (Sigma Chemical Co.) can be added to 30 to 40 mls of the Kinyoun carbolfuchsin stain.

Crystal Violet Stain (Hucker modification)
Solution A: Dissolve 2.0 g of crystal violet (85% dye content) in 20 ml of 95% ethyl alcohol.
Solution B: Dissolve 0.8 g ammonium oxalate in 80.0 ml distilled water.
Mix solutions A and B.

Diphenylamine Reagent (nitrate test)
Dissolve 0.7 g diphenylamine in a mixture of 60 ml of concentrated sulfuric acid and 28.8 ml of distilled water.
Cool and add slowly 11.3 ml of concentrated hydrochloric acid. After the solution has stood for 12 hours, some of the base separates, showing that the reagent is saturated.

Ferric Chloride Reagent (Ex. 78)

$FeCl_3 \cdot 6H_2O$ . . . . . 12 g

2% Aqueous HCl . .100 ml

Make up the 2% aq. HCl by adding 5.4 ml of concentrated HCl (37%) to 94.6 ml H_2O. Inoculate with two or three colonies of beta-hemolytic streptococci, incubate at 35° C for 20 or more hours. Centrifuge the medium to pack the cells, and pipette 0.8 ml of the clear supernate into a Kahn tube. Add 0.2 ml of the ferric chloride reagent to the Kahn tube and mix well. If a heavy precipitate remains longer than 10 minutes, the test is positive.

Gram's Iodine (Lugol's)

Dissolve 2.0 g of potassium iodide in 300 ml of distilled water and then add 1.0 g iodine crystals.

Iodine, 5% Aqueous Solution (Ex. 37)

Dissolve 4 g of potassium iodide in 300 ml of distilled water and then add 2.0 g iodine crystals.

Kovacs' Reagent (indole test)

n-amyl alcohol. 75.0 ml

Hydrochloric acid (conc.) . . . . 25.0 ml

ρ-dimethylamine-benzaldehyde . . . . 5.0 g

Lactophenol Cotton Blue Stain

Phenol crystals . . . . .20 g

Lactic acid . . . . .20 ml

Glycerol . . . . .40 ml

Cotton blue . . . . 0.05 g

Dissolve the phenol crystals in the other ingredients by heating the mixture gently under a hot water tap.

Malachite Green Solution (spore stain)

Dissolve 5.0 g malachite green oxalate in 100 ml distilled water.

McFarland Nephelometer Barium Sulfate Standards (Ex. 48)

Prepare 1% aqueous barium chloride and 1% aqueous sulfuric acid solutions.

Add the amounts indicated in table 1 to clean, dry ampoules. Ampoules should have the same diameter as the test tube to be used in subsequent density determinations.

Seal the ampoules and label them.

Methylene Blue (Loeffler's)

Solution A: Dissolve 0.3 g of methylene blue (90% dye content) in 30.0 ml ethyl alcohol (95%).

Table 1 Amounts for Standards

Tube	Barium Chloride 1% (ml)	Sulfuric Acid 1% (ml)	Corresponding Approx. Density of Bacteria (million/ml)
1	0.1	9.9	300
2	0.2	9.8	600
3	0.3	9.7	900
4	0.4	9.6	1200
5	0.5	9.5	1500
6	0.6	9.4	1800
7	0.7	9.3	2100
8	0.8	9.2	2400
9	0.9	9.1	2700
10	1.0	9.0	3000

Solution B: Dissolve 0.01 g potassium hydroxide in 100.0 ml distilled water. Mix solutions A and B.

Naphthol, alpha

5% alpha-naphthol in 95% ethyl alcohol

Caution: Avoid all contact with human tissues. Alpha-naphthol is considered to be carcinogenic.

Nessler's Reagent (ammonia test)

Dissolve about 50 g of potassium iodide in 35 ml of cold ammonia-free distilled water. Add a saturated solution of mercuric chloride until a slight precipitate persists. Add 400 ml of a 50% solution of potassium hydroxide. Dilute to 1 liter, allow to settle, and decant the supernatant for use.

Nigrosin Solution (Dorner's)

Nigrosin, water soluble . . . . .10 g

Distilled water . . . . .100 ml

Boil for 30 minutes. Add as a preservative 0.5 ml formaldehyde (40%). Filter twice through double filter paper and store under aseptic conditions.

Nitrate Test Reagent
(see Diphenylamine)

Nitrite Test Reagents

Solution A: Dissolve 8 g sulfanilic acid in 1000 ml 5N acetic acid (1 part glacial acetic acid to 2.5 parts water).

Solution B: Dissolve 5 g dimethyl-α naphthylamine in 1000 ml 5N acetic acid. Do not mix solutions.

Caution: Although at this time it is not known for sure, there is a possibility that dimethyl-α-naphthylamine in solution B may be carcinogenic. For reasons of safety, avoid all contact with tissues.

Oxidase Test Reagent

Mix 1.0 g of dimethyl-ρ-phenylenediamine hydrochloride in 100 ml of distilled water.

Preferably, the reagent should be made up fresh, daily. It should not be stored longer than one week in the refrigerator. Tetramethyl-ρ-phenylenediamine dihydrochloride (1%) is even more sensitive, but is considerably more expensive and more difficult to obtain.

Phenolized Saline

Dissolve 8.5 g sodium chloride and 5.0 g phenol in 1 liter distilled water.

Physiological Saline

Dissolve 8.5 g sodium chloride in 1 liter distilled water.

Potassium permanganate
(for fluorochrome staining)

$KMnO_4$ . . . . 2.5 g

Distilled water . . . . 500.0 ml

Safranin (for gram staining)

Safranin O (2.5% sol'n in 95% ethyl alcohol) . . . . 10.0 ml

Distilled water . . . . 100.0 ml

Trommsdorf's Reagent (nitrite test)

Add slowly, with constant stirring, 100 ml of a 20% aqueous zinc chloride solution to a mixture of 4.0 g of starch in water. Continue heating until the starch is dissolved as much as possible, and the solution is nearly clear. Dilute with water and add 2 g of potassium iodide. Dilute to 1 liter, filter, and store in amber bottle.

Vaspar

Melt together 1 pound of Vaseline and 1 pound of paraffin. Store in small bottles for student use.

Voges-Proskauer Test Reagent
(see Barritt's)

White Blood Cell (WBC) Diluting Fluid

Hydrochloric acid   5 ml

Distilled water   495 ml

Add 2 small crystals of thymol as a preservative.

Media

Conventional Media The following media are used in the experiments of this manual. All of these media are available in dehydrated form from either Difco Laboratories, Detroit, Michigan, or Baltimore Biological Laboratory (BBL), a division of Becton, Dickinson & Co., Cockeysville, Maryland. Compositions, methods of preparation, and usage will be found in their manuals, which are supplied upon request at no cost. The source of each medium is designated as (B) for BBL and (D) for Difco.

Bile esculin (D)
Brewer's anaerobic agar (D)
Desoxycholate citrate agar (B,D)
Desoxycholate lactose agar (B,D)
DNase test agar (B,D)
Endo agar (B,D)
Eugonagar (B,D)
Fluid thioglycollate medium (B,D)
Heart infusion agar (D)
Hektoen Enteric Agar (B,D)
Kligler iron agar (B,D)
Lead acetate agar (D)
Levine EMB agar (B,D)
Lipase reagent (D)
Litmus milk (B,D)
Lowenstein-Jensen medium (B,D)
MacConkey Agar (B,D)
Mannitol salt agar (B,D)
MR-VP medium (D)
Mueller-Hinton medium (B,D)
Nitrate broth (D)
Nutrient agar (B,D)
Nutrient broth (B,D)

Nutrient gelatin (B,D)
Phenol red sucrose broth (B,D)
Phenylalanine agar (D)
Phenylethyl alcohol medium (B)
Russell double sugar agar (B,D)
Sabouraud's glucose (dextrose) agar (D)
Semisolid medium (B)
Simmons citrate agar (B,D)
Snyder test agar (D)
Sodium hippurate (D)
Spirit blue agar (D)
SS agar (B,D)
m-Staphylococcus broth (D)
Staphylococcus medium 110 (D)
Starch agar (D)
Trypticase soy agar (B)
Trypticase soy broth (B)
Tryptone glucose extract agar (B,D)
Urea (urease test) broth (B,D)
Veal infusion agar (B,D)
Xylose Lysine Desoxycholate Agar (B,D)

Special Media The following media are not included in the manuals that are supplied by Difco and BBL; therefore, methods of preparation are presented here.

Bile Esculin Slants (Ex. 71)

Heart infusion agar . . . . 40.0 g
Esculin. . . . . 1.0 g
Ferric chloride . . . . 0.5 g
Distilled water . . . . 1000.0 ml
Dispense into sterile 15 × 125 mm screw-capped tubes, sterilize in autoclave at 121 ° C for 15 minutes, and slant during cooling.

Blood Agar

Trypticase soy agar power. . . . . .40 g
Distilled water .1000 ml
    Final pH of 7.3
Defibrinated sheep or rabbit blood . . . . . .50 ml
Liquefy and sterilize 1000 ml of trypticase soy agar in a large Erlenmeyer flask. While the TSA is being sterilized, warm up 50 ml of defibrinated blood to

50° C. After cooling the TSA to 50° C, aseptically transfer the blood to the flask and mix by gently rotating the flask (cold blood may cause lumpiness).

Pour 10–12 ml of the mixture into sterile petri plates. If bubbles form on the surface of the medium, flame the surface gently with a Bunsen burner before the medium solidifies. It is best to have an assistant to lift off the petri plate lids while pouring the medium into the plates. A full flask of blood agar is somewhat cumbersome to handle with one hand.

Bromthymol Blue Carbohydrate Broths

Make up stock indicator solution:

Bromthymol blue. .8 g
95% ethyl alcohol . . . . .250 ml
Distilled water .250 ml
Indicator is dissolved first in alcohol and then water is added.

Make up broth:

Sugar base (lactose, sucrose, glucose, etc.). .5 g
Tryptone . . . . .10 g
Yeast extract . . . . .5 g
Indicator solution. . . . . .2 ml
Distilled water .1000 ml
      Final pH 7.0

Emmons' Culture Medium for Fungi

C. W. Emmons developed the following recipe as an improvement over Sabouraud's glucose agar for the cultivation of fungi. Its principal advantage is that a neutral pH does not inhibit certain molds that have difficulty growing on Sabouraud's agar (pH 5.6). Instead of relying on a low pH to inhibit bacteria, it contains chloramphenicol, which does not adversely affect the fungi.

Glucose . . . . .20 g
Neopeptone . . . . .10 g
Agar. . . . . .20 g
Chloramphenicol . . . . 40 mg
Distilled water .1000 ml

After the glucose, peptone, and agar are dissolved, heat to boiling, add the chloramphenicol which has been suspended in 10 ml of 95% alcohol and remove quickly from the heat. Autoclave for only 10 minutes.

Freshwater Enrichment Medium (*Desulfovibrio*; Ex. 56)

Sodium lactate. . . . . .5 g
$CaSO_4$ . . . . .1 g
$MgSO_4 \cdot H_2O$. . . . . .2 g

NH_4Cl . . . . .1 g
K_2HPO_4 . . . . 0.5 g
Water . . . . .1000 ml
Autoclave at standard conditions. Just prior to inoculation, boil and cool the medium.

Glucose Peptone Acid Agar

Glucose . . . . .10 g
Peptone . . . . .5 g
Monopotassium phosphate . . . . .1 g
Magnesium sulfate ($MgSO_4 \cdot 7H_2O$). . . . . 0.5 g
Agar. . . . . .15 g
Water . . . . .1000 ml
While still liquid after sterilization, add sufficient sulfuric acid to bring the pH down to 4.0.

Glycerol Yeast Extract Agar

Glycerol. . . . . .5 ml
Yeast extract . . . . .2 g
Dipotassium phosphate . . . . .1 g
Agar. . . . . .15 g
Water . . . . .1000 ml

m Endo MF Broth (Ex. 62)

This medium is extremely hygroscopic in the dehydrated form and oxidizes quickly to cause deterioration of the medium after the bottle has been opened. Once a bottle has been opened it should be dated and discarded after one year. If the medium becomes hardened within that time it should be discarded. Storage of the bottle inside a larger bottle that contains silica gel will extend shelf life.

Failure of Exercise 62 can often be attributed to faulty preparation of the medium. It is best to make up the medium the day it is to be used. It should not be stored over 96 hours prior to use. The Millipore Corporation recommends the following method for preparing this medium. (These steps are not exactly as stated in the Millipore Application Manual AM302.)

1. Into a 250-ml screw-cap Erlenmeyer flask place the following:
   Distilled water . . . . .50 ml
   95% ethyl alcohol . . . . .2 ml
   Dehydrated medium (*m* Endo MF broth) . . 4.8 g
   Shake the above mixture by swirling the flask until the medium is dissolved and then add another 50 ml of distilled water.
2. Cap the flask loosely and immerse it into a pan of boiling water. As soon as the medium begins to simmer, remove the flask from the water bath. Do not boil the medium any further.

3. Cool the medium to 45° C, and adjust the pH to between 7.1 and 7.3.

4. If the medium must be stored for a few days, place it in the refrigerator at 2°–10° C, with screw-cap tightened securely.

LB Broth (Ex. 67)

Tryptone . . . . 10 g
Yeast Extract . . . . .5 g
NaCl. . . . . .5 g
Distilled water .1000 ml

Add 1 *M* NaOH to adjust the pH to 7. Autoclave at standard conditions.

Milk Salt Agar (15% NaCl)

Prepare three separate beakers of the following ingredients:

1. Beaker containing 200 grams of sodium chloride.
2. Large beaker (2000 ml size) containing 50 grams of skim milk powder in 500 ml of distilled water.
3. Glycerol-peptone agar medium:

$MgSO_4 \cdot 7H_2O$ . . . . 5.0 g
$MgNO_3 \cdot 6H_2O$ . . . . 1.0 g
$FeCl_3 \cdot 7H_2O$ . . . . 0.025 g
Difco proteose-peptone #3 . . . . 5.0 g
Glycerol. . . . . 10.0 g
Agar. . . . . 30.0 g
Distilled water . . . . 500.0 ml

Sterilize the above three beakers separately. The milk solution should be sterilized at 113°–115° C (8 lb pressure) in autoclave for 20 minutes. The salt and glycerol-peptone agar can be sterilized at conventional pressure and temperature. After the milk solution has cooled to 55° C, add the sterile salt, which should also be cooled down to a moderate temperature. If the salt is too hot, coagulation may occur. Combine the milk-salt and glycerol-peptone agar solutions by gently swirling with a glass rod. Dispense aseptically into petri plates.

Nitrate Broth

Beef extract . . . . .3 g
Peptone . . . . .5 g
Potassium nitrate .1 g
Distilled water .1000 ml
    Final pH 7.0 at 25° C

Nitrate Agar

Beef extract . . . . .3 g
Peptone . . . . .5 g
Potassium nitrate .1 g
Agar. . . . . 12 g
Distilled water .1000 ml
    Final pH 6.8 at 25° C

Nitrate Succinate–Mineral Salts Broth (Ex. 53)

This medium is used as an enrichment medium for isolating denitrifying bacteria from soil. Note that two stock solutions (A and B) should be made up before attempting to put together the complete medium.

Solution A (Trace Mineral Salts)

$FeSO_4 \cdot 7H_2O$. 300 mg
$MnCl_2 \cdot 4H_2O$ . . . . 180 mg
$Co(NO_3)_2 \cdot 7H_2O$. . . . . 130 mg
$ZnSO_4 \cdot 7H_2O$ . . . . 40 mg
H_2MoO_4 . . . . 20 mg
$CuSO_4 \cdot 5H_2O$ . . . . 1 mg
$CaCl_2$ . . . . 1000 mg

This solution should be stored at 4° C until used.

Solution B

NH_4Cl . . . . .1 g
Na_2HPO_4 . . . . 2.14 g
KH_2PO_4 . . . . 1.09 g
$MgSO_4 \cdot 7H_2O$ . . . . 0.2 g
Trace mineral salts (Sol A) . . . . .10 ml
Water to . . . . .1000 ml

Complete Medium

Solution B . . . . .1000 ml
Sodium succinate. .2 g
Potassium nitrate .3 g

Adjust the pH to 6.8, dispense into bottles, and autoclave at standard conditions.

Nitrate Succinate–Mineral Salts Agar (Ex. 53)

Add 15 g agar to 1000 ml of the above complete medium. Dispense into petri plates and sterilize in the autoclave.

Nitrogen-Free Base for *Azotobacter*

K_2HPO_4 . . . . .1 g
KH_2PO_4 . . . . 0.2 g
$MgSO_4 \cdot 7H_2O$ . . . . 0.2 g
$FeSO_4 \cdot 7H_2O$ . . . . 0.05 g
$CaCl_2 \cdot 2H_2O$ . . . . 0.1 g
$Na_2MoO_4 \cdot 2H_2O$ . . . . 0.05 g
Water . . . . 1000 mls
*Carbohydrate . . . . .10 gm

Adjust pH to 7.2. **Note:** If the medium is to be used immediately, sterilization is not necessary. However, if it is to be stored it should be autoclaved.

Phage Growth Medium (Ex. 27)

KH_2PO_4 . . . . 1.5 g
Na_2HPO_4 . . . . 3.0 g
NH_4Cl . . . . 1.0 g
$MgSO_4 \cdot 7H_2O$ . . . . 0.2 g
Glycerol . . . . 10.0 g
Acid-hydrolyzed casein . . . . 5.0 g
dl-Tryptophan . . . . 0.01 g
Gelatin . . . . 0.02 g
Tween-80 . . . . 0.2 g
Distilled water . . . . 1000.0 ml
Sterilize in autoclave at 121° C for 20 minutes.

Phage Lysing Medium (Ex. 27)

Add sufficient sodium cyanide (NaCN) to the above growth medium to bring the concentration up to 0.02 M. For 1 liter of lysing medium this will amount to about 1 gram (actually 0.98 g) of NaCN. When an equal amount of this lysing medium is added to the growth medium during the last 6 hours of incubation, the concentration of NaCN in the combined medium is 0.01 M.

Purple Nonsulfur Medium (Ex. 55)

This culture medium is used for the enrichment and culture of anaerobic phototrophic bacteria. To make up this medium you need to first prepare three stock solutions (A, B, and C) before putting together the entire batch.

---

*Carbohydrate: Either glucose or mannitol can be used to as a carbon source to isolate Azotobacter. Depending on the carbohydrate, different species of Azotobacter can be obtained. The glucose or mannitol should be sterilized separate from the salts. Dissolve the suger in 100 ml of water and sterilize at 1218 C for 15 minutes. Sterilize the salts in the same manner. After sterilization, the two solutions are mixed aseptically and dispensed into sterile 8-oz, flat-sided bottles (50 ml per bottle).

A. Iron Citrate Solution

Ammonium ferrous sulfate . . . . 748 mg
Sodium citrate . . . . 1180 mg
Water to . . . . .500 ml
Store this stock solution at 4° C until needed.

B. Vitamin B_{12} Solution

Certain strains require this vitamin. To make up 100 ml of this solution, add 1 mg to 100 ml of water. Store at 4° C until needed.

C. Trace Metals Solution

H_3BO_4 . . . . 2.86 g
$MnCl_2 \cdot H_2O$ . . . . 1.81 g
$ZnSO_4 \cdot 7H_2O$ . . . . 0.222 g
$Na_2MoO_4 \cdot 2H_2O$ . . . . 0.390 g
$CuSO_4 \cdot 5H_2O$ . . . . 0.079 g
$Co(NO_3)_2 \cdot 6H_2O$ . . . . 0.0494 g
Water to . . . . .1000 ml
Store at 4° C until needed.

Complete Enrichment Medium

The final batch of this medium has the following ingredients. The succinate provides the organic carbon and the yeast extract provides essential vitamins for certain strains.

KH_2PO_4 . . . . 0.5 g
$MgSO_4 \cdot 7H_2O$ . . . . 0.2 g
NaCl . . . . 0.4 g
NH_4Cl . . . . 0.4 g
$CaCl_2 \cdot 2H_2O$ . . . . 0.05 g
Sodium succinate . . . . 1.0 g
Yeast extract . . . . 0.2 g
Iron citrate solution (A) . . . . .5 ml
Vitamin B_{12} solution (B) . . . . 0.1 ml
Trace elements solution (C) . . . . .1 ml
Water to . . . . .1000 ml

Adjust the pH to 6.8, dispense into bottles and autoclave at standard conditions.

Russell Double Sugar Agar (Ex. 80)

Beef extract . . . . .1 g
Proteose Peptone No. 3 (Difco) . . . . .12 g
Lactose . . . . .10 g
Dextrose . . . . .1 g
Sodium chloride . . . . .5 g
Agar . . . . .15 g
Phenol red (Difco) . . . . 0.025 g
Distilled water . . . . .1000 ml

Final pH 7.5 at 25° C

Dissolve ingredients in water, and bring to boiling. Cool to 50°–60° C, and dispense about 8 ml per tube (16-mm dia tubes). Slant tubes to cool. butt depth should be about $\frac{1}{2}''$.

Skim Milk Agar

Skim milk powder . . . . 100 g
Agar . . . . 15 g
Distilled water . . . . 1000 ml

Dissolve the 15 g of agar into 700 ml of distilled water by boiling. Pour into a large flask and sterilize at 121° C, 15 lb pressure.

In a separate container, dissolve the 100 g of skim milk powder into 300 ml of water heated to 50° C. Sterilize this milk solution at 113°–115° C (8 lb pressure) for 20 minutes.

After the two solutions have been sterilized, cool to 55° C and combine in one flask, swirling gently to avoid bubbles. Dispense into sterile petri plates.

SOC Medium (Ex. 67)

Tryptone . . . . 20 g
Yeast extract . . . . 5 g
NaCl . . . . 0.6 g
KCl . . . . 0.18 g
$MgCl_2$ . . . . 2 g
$MgSO_4$ . . . . 2.5 g
Glucose . . . . 3.6 g

Add 950 ml of distilled water to dissolve the solutes. Adjust the pH to 7 with 1 M NaOH and bring the volume to 1000 ml. Autoclave at standard conditions.

Sodium Chloride (6.5%) Tolerance Broth (Ex. 71)

Heart infusion broth . . . . 25 g
NaCl . . . . 60 g
Indicator (1.6 g bromcresol purple in 100 ml
   95% ethanol) . . . . 1 ml
Dextrose . . . . 1 g
Distilled water . . . . 1000 ml

Add all reagents together up to 1000 ml (final volume). Dispense in 15 × 125 mm screw-capped tubes and sterilize in an autoclave 15 minutes at 121° C.

A positive reaction is recorded when the indicator changes from purple to yellow or when growth is obvious even though the indicator does not change.

Sodium Hippurate Broth (Ex. 71)

Heart infusion broth . . . . 25 g
Sodium hippurate . . . . 10 g
Distilled water . . . . 1000 ml

Sterilize in autoclave at 121° C for 15 minutes after dispensing in 15 × 125 mm screw-capped tubes. Tighten caps to prevent evaporation.

Soft Nutrient Agar (for bacteriophage)

Dehydrated nutrient broth . . . . 8 g
Agar . . . . 7 g
Distilled water . . . . 1000 ml

Sterilize in autoclave at 121° C for 20 minutes.

Spirit Blue Agar (Ex. 42)

This medium is used to detect lipase production by bacteria. Lipolytic bacteria cause the medium to change from pale lavender to deep blue.

Spirit blue agar (Difco) . . . . 35 g
Lipase reagent (Difco) . . . . 35 ml
Distilled water . . . . 1000 ml

Dissolve the spirit blue agar in 1000 ml of water by boiling. Sterilize in autoclave for 15 minutes at 15 psi (121° C). Cool to 55° C and slowly add the 35 ml of lipase reagent, agitating to obtain even distribution. Dispense into sterile petri plates.

Tryptone Agar

Tryptone . . . . 10 g
Agar . . . . 15 g
Distilled water . . . . 1000 ml

Tryptone Broth

Tryptone . . . . 10 g
Distilled water . . . . 1000 ml

Tryptone Yeast Extract Agar

Tryptone . . . . 10 g
Yeast extract . . . . 5 g
Dipotassium phosphate . . . . 3 g
Sucrose . . . . 50 g
Agar . . . . 15 g
Water . . . . 1000 ml
   pH 7.4

Identification Charts

appendix

CHART I   Interpretation of Test Results of API 20E System

PROCEDURE

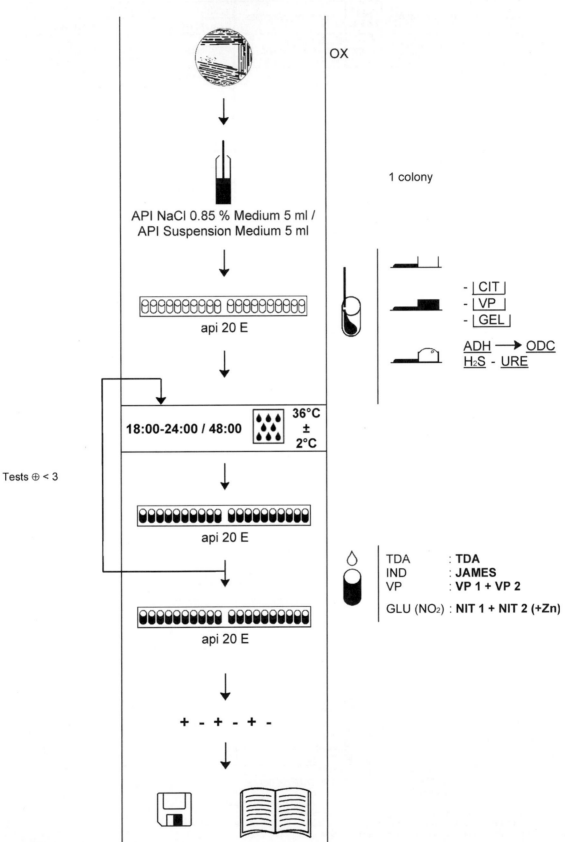

CHART II Symbol Interpretation of API 20E System

READING TABLE

TESTS	ACTIVE INGREDIENTS	QTY (mg/cup.)	REACTIONS/ENZYMES	RESULTS	
				NEGATIVE	POSITIVE
ONPG	2-nitrophenyl-ßD-galactopyranoside	0.223	ß-galactosidase (Ortho NitroPhenyl-ßD-Galactopyranosidase)	colorless	yellow (1)
ADH	L-arginine	1.9	Arginine DiHydrolase	yellow	red / orange (2)
LDC	L-lysine	1.9	Lysine DeCarboxylase	yellow	red / orange (2)
ODC	L-ornithine	1.9	Ornithine DeCarboxylase	yellow	red / orange (2)
CIT	trisodium citrate	0.756	CITrate utilization	pale green / yellow	blue-green / blue (3)
H2S	sodium thiosulfate	0.075	H2S production	colorless / greyish	black deposit / thin line
URE	urea	0.76	UREase	yellow	red / orange (2)
TDA	L-tryptophane	0.38	Tryptophane DeAminase	TDA / immediate yellow	reddish brown
IND	L-tryptophane	0.19	INDole production	JAMES / immediate colorless pale green / yellow	pink
VP	sodium pyruvate	1.9	acetoin production (Voges Proskauer)	VP 1 + VP 2 / 10 min colorless	pink / red (5)
GEL	Gelatin (bovine origin)	0.6	GELatinase	no diffusion	diffusion of black pigment
GLU	D-glucose	1.9	fermentation / oxidation (GLUcose) (4)	blue / blue-green	yellow / greyish yellow
MAN	D-mannitol	1.9	fermentation / oxidation (MANnitol) (4)	blue / blue-green	yellow
INO	inositol	1.9	fermentation / oxidation (INOsitol) (4)	blue / blue-green	yellow
SOR	D-sorbitol	1.9	fermentation / oxidation (SORbitol) (4)	blue / blue-green	yellow
RHA	L-rhamnose	1.9	fermentation / oxidation (RHAmnose) (4)	blue / blue-green	yellow
SAC	D-sucrose	1.9	fermentation / oxidation (SACcharose) (4)	blue / blue-green	yellow
MEL	D-melibiose	1.9	fermentation / oxidation (MELibiose) (4)	blue / blue-green	yellow
AMY	amygdalin	0.57	fermentation / oxidation (AMYgdalin) (4)	blue / blue-green	yellow
ARA	L-arabinose	1.9	fermentation / oxidation (ARAbinose) (4)	blue / blue-green	yellow
OX	(see oxidase test package insert)		cytochrome-OXidase	(see oxidase test package insert)	

SUPPLEMENTARY TESTS

TESTS	ACTIVE INGREDIENTS	QTY (mg/cup.)	REACTIONS/ENZYMES	RESULTS	
				NEGATIVE	POSITIVE
Nitrate reduction GLU tube	potassium nitrate	0.076	NO2 production	NIT 1 + NIT 2 / 2-5 min yellow	red
			reduction to N2 gas	Zn / 5 min orange-red	yellow
MOB	API M Medium or microscope		motility	non-motile	motile
McC	MacConkey medium		growth	absence	presence
OF-F	glucose (API OF Medium)		fermentation : under mineral oil	green	yellow
OF-O			oxidation : exposed to the air	green	yellow

(1) A very pale yellow should also be considered positive.
(2) An orange color after 36-48 hours incubation must be considered negative.
(3) Reading made in the cupule (aerobic).
(4) Fermentation begins in the lower portion of the tubes, oxidation begins in the cupule.
(5) A slightly pink color after 10 minutes should be considered negative.
• The quantities indicated may be adjusted depending on the titer of the raw materials used.
• Certain cupules contain products of animal origin, notably peptones.

Courtesy of bioMérieux, Inc.

CHART III   Characterization of Gram-Negative Rods—The API 20E System

IDENTIFICATION TABLE

% of positive reactions after 18-24/48 hrs. at 36°C ± 2°C

API 20 E V4.0	ONPG	ADH	LDC	ODC	CIT	H2S	URE	TDA	IND	VP	GEL	GLU	MAN	INO	SOR	RHA	SAC	MEL	AMY	ARA	OX	NO2	N2	MOB	McC	OF/O	OFF
Butiauxella agrestis	100	0	0	85	25	0	0	0	0	0	0	100	100	0	99	99	0	92	99	100	0	100	0	100	100	100	100
Cedecea davisae	99	89	0	99	75	0	0	0	0	89	0	100	100	10	0	0	100	0	100	1	0	99	0	87	100	100	100
Cedecea lapagei	99	99	0	0	75	0	0	0	0	90	0	99	99	0	0	0	1	0	100	1	0	100	0	87	100	100	100
Citrobacter braakii	50	45	0	99	75	81	1	0	4	0	0	100	99	1	100	99	1	91	99	99	0	100	0	95	100	100	100
Citrobacter freundii	90	24	0	0	75	75	1	0	1	0	0	100	100	25	99	99	82	82	40	99	0	98	0	95	100	100	100
Citrobacter koseri/amalonaticus	99	75	0	100	97	0	1	0	99	0	0	100	100	25	99	99	0	99	98	99	0	100	0	95	100	100	100
Citrobacter koseri/farmeri	100	2	0	100	25	0	1	0	1	0	0	100	100	0	95	99	80	99	0	100	0	85	0	95	100	100	100
Citrobacter youngae	100	50	0	1	80	80	1	0	0	0	0	100	100	0	95	100	1	80	25	100	0	100	0	100	100	100	100
Edwardsiella hoshinae	0	0	99	99	50	94	0	0	99	0	0	100	100	0	0	0	100	0	0	1	0	100	0	98	100	100	100
Edwardsiella tarda	0	0	100	99	1	75	0	0	99	0	0	100	0	0	0	0	0	0	0	0	0	100	0	97	100	100	100
Enterobacter aerogenes	99	0	98	98	82	0	1	0	0	85	0	99	99	99	99	99	99	99	99	99	0	100	0	92	100	100	100
Enterobacter amnigenus 1	99	25	0	99	82	0	1	0	0	75	0	100	100	1	1	100	99	99	99	99	0	100	0	100	100	100	100
Enterobacter amnigenus 2	99	80	0	99	80	0	1	0	0	75	0	100	100	25	99	100	0	100	100	99	0	100	0	95	100	100	100
Enterobacter asburiae	100	25	0	99	80	0	0	0	0	10	0	99	100	25	100	99	99	100	100	100	0	100	0	0	100	100	100
Enterobacter cancerogenus	100	75	0	99	99	0	0	0	0	89	0	100	100	0	0	100	1	1	100	99	0	100	0	99	100	100	100
Enterobacter cloacae	98	82	0	92	90	0	0	0	0	85	0	99	99	12	90	85	96	90	99	99	0	100	0	95	100	100	100
Enterobacter gergoviae	99	32	99	100	75	0	99	0	0	90	0	99	100	23	1	100	90	100	99	99	0	100	0	90	100	100	100
Enterobacter intermedius	99	0	0	75	75	0	0	0	0	2	0	100	97	0	88	99	40	100	99	99	0	100	0	92	100	100	100
Enterobacter sakazakii	100	96	0	91	94	0	1	0	25	91	10	100	98	75	8	99	99	99	99	99	0	100	0	96	100	100	100
Escherichia coli 1	90	1	74	70	0	0	3	0	89	0	0	99	98	1	91	82	36	75	3	70	0	98	0	5	100	100	100
Escherichia coli 2	26	1	45	20	0	1	1	0	50	0	0	99	90	1	42	30	3	3	1	70	0	100	0	93	100	100	100
Escherichia fergusonii	96	1	99	100	1	0	0	0	99	0	0	100	99	1	0	87	1	1	99	99	0	100	0	99	100	100	100
Escherichia hermannii	100	30	0	100	1	0	0	0	99	0	0	100	100	0	0	99	25	95	99	99	0	100	0	100	100	100	100
Escherichia vulneris	100	75	50	0	0	0	0	0	0	0	0	100	100	0	1	95	7	95	10	99	0	100	0	60	100	100	100
Ewingella americana	98	0	0	0	75	0	0	0	0	95	1	99	99	0	1	0	0	1	50	1	0	100	0	85	100	100	100
Hafnia alvei 1	75	0	99	98	50	0	0	0	0	50	0	99	99	0	0	99	0	0	25	99	0	100	0	94	100	100	100
Hafnia alvei 2	50	0	99	99	1	0	10	0	0	10	0	99	98	1	1	93	0	0	1	99	1	95	0	100	100	100	100
Klebsiella ornithinolytica	100	0	99	99	99	0	85	0	100	65	0	100	100	99	100	100	100	100	100	100	0	100	0	0	100	100	100
Klebsiella oxytoca	99	0	99	0	89	0	78	0	100	80	0	100	100	99	99	99	99	99	100	100	0	90	0	0	100	100	100
Klebsiella pneumoniae ssp ozaenae	94	18	25	1	18	0	1	0	0	0	0	99	96	57	66	58	20	80	97	85	0	92	0	0	100	100	100
Klebsiella pneumoniae ssp pneumoniae	99	0	73	0	86	0	75	0	1	90	0	100	100	90	99	75	75	99	99	10	0	100	0	0	100	100	100
Klebsiella pneumoniae ssp rhinoscleromatis	1	0	0	0	0	0	0	0	0	0	0	100	100	90	7	1	1	100	99	99	0	100	0	0	100	100	100
Klebsiella terrigena	100	0	99	6	52	0	0	0	0	75	0	99	99	90	99	93	89	99	100	99	0	100	0	94	100	100	100
Kluyvera spp	95	0	25	99	60	0	0	0	80	0	0	99	99	0	25	99	66	99	99	100	0	95	0	100	100	100	100
Leclercia adecarboxylata	99	0	0	0	1	0	1	0	99	0	0	99	98	0	2	100	66	99	1	100	0	99	0	85	100	100	100
Moellerella wisconsensis	97	0	0	98	40	0	0	0	15	1	0	100	1	1	1	1	100	1	0	100	0	90	0	0	100	100	100
Morganella morganii	1	0	0	98	0	1	99	93	99	0	0	99	0	0	0	0	0	0	0	0	0	88	0	95	100	100	100
Pantoea spp 1	85	1	0	0	13	0	1	0	0	9	1	100	99	1	26	0	98	26	59	61	0	85	0	85	100	100	100
Pantoea spp 2	99	0	0	0	99	0	1	0	53	62	4	99	99	36	82	90	98	81	99	99	0	85	0	85	100	100	100
Pantoea spp 3	99	1	0	0	21	0	1	0	0	86	15	100	99	34	1	97	93	23	65	97	0	85	0	85	100	100	100
Pantoea spp 4	86	1	0	0	29	0	0	0	59	0	0	99	99	10	32	99	72	89	99	99	0	93	0	95	100	100	100
Proteus mirabilis	1	0	0	99	50	75	99	98	1	0	82	98	0	0	0	0	100	0	0	0	0	99	0	85	100	100	100
Proteus penneri	1	0	0	0	1	20	99	99	0	0	50	99	0	0	0	0	89	0	0	0	0	99	0	94	100	100	100
Proteus vulgaris	1	0	0	0	12	83	99	99	92	0	74	99	1	1	0	0	100	1	66	0	0	100	0	96	100	100	100
Providencia alcalifaciens/rustigianii	0	0	0	0	80	0	99	100	99	0	0	99	0	1	0	0	0	0	0	0	0	100	0	94	100	100	100
Providencia rettgeri	1	0	0	0	74	0	99	100	90	0	0	82	1	78	0	50	0	0	40	1	0	98	0	85	100	100	100
Providencia stuartii	1	0	0	0	85	0	30	98	95	0	0	98	3	80	0	0	15	0	0	0	0	100	0	85	100	100	100
Rahnella aquatilis	100	0	0	50	50	0	0	0	0	99	0	100	100	0	98	99	97	100	98	100	0	100	0	6	100	100	100
Salmonella arizonae	98	75	97	98	75	99	0	0	1	0	0	100	99	0	98	99	78	97	99	98	0	100	0	99	100	100	100
Salmonella choleraesuis	0	15	99	99	6	64	0	0	0	0	0	100	99	0	99	99	20	0	99	99	0	100	0	95	100	100	100
Salmonella gallinarum	0	1	100	1	0	25	0	0	0	0	0	100	100	0	100	1	0	100	0	100	0	100	0	0	100	100	100

Courtesy of bioMérieux, Inc.

CHART III Characterization of Gram-Negative Rods—The API 20E System cont.

API 20 E V4.0	ONPG	ADH	LDC	ODC	CIT	H2S	URE	TDA	IND	VP	GEL	GLU	MAN	INO	SOR	RHA	SAC	MEL	AMY	ARA	OX	NO2	N2	MOB	McC	OF/O	OF/F
Salmonella paratyphi A	0	5	0	99	0	1	0	0	0	0	0	100	99	0	99	98	0	96	0	99	0	100	0	95	100	100	100
Salmonella pullorum	0	1	75	100	0	85	0	0	0	0	0	100	100	0	99	100	0	99	0	75	0	100	0	0	100	100	100
Salmonella typhi	0	1	99	0	0	8	0	0	0	0	0	100	99	40	99	0	0	99	0	0	0	100	0	97	100	100	100
Salmonella ssp	1	56	82	93	65	83	0	0	1	0	1	99	100	50	99	86	1	90	1	99	0	100	0	95	100	100	100
Serratia ficaria	99	0	0	0	100	0	0	0	0	40	90	100	100	50	99	74	99	99	100	99	0	92	0	100	100	100	100
Serratia fonticola	99	0	73	99	75	0	0	0	0	0	0	100	100	97	100	99	30	99	99	99	0	100	0	91	100	100	100
Serratia liquefaciens	95	0	78	98	80	0	2	0	0	59	65	100	99	80	98	2	99	72	97	97	0	95	0	95	100	100	100
Serratia marcescens	94	0	95	95	96	0	0	0	1	70	87	100	99	85	98	1	99	68	97	25	0	100	0	97	100	100	100
Serratia odorifera 1	95	0	95	99	95	0	0	0	99	50	99	100	99	99	99	1	99	99	99	99	0	95	0	95	100	100	100
Serratia odorifera 2	95	0	96	1	95	0	0	0	99	50	99	100	99	99	99	1	1	99	98	95	0	99	0	97	100	100	100
Serratia plymuthica	99	0	0	0	65	0	1	0	0	65	50	100	90	70	70	1	99	85	98	98	0	99	0	50	100	100	100
Serratia rubidaea	99	0	30	0	92	0	0	0	0	71	82	99	99	75	7	3	99	95	98	99	0	100	0	85	100	100	100
Shigella spp	1	0	0	1	0	0	0	0	29	0	0	99	63	0	7	7	1	0	0	50	0	100	0	0	100	100	100
Shigella sonnei	96	0	0	93	0	0	0	0	0	0	0	99	99	0	1	75	0	20	0	99	0	100	0	0	100	100	100
Yersinia enterocolitica	80	0	0	90	0	0	98	0	50	5	0	99	99	25	98	1	97	4	75	75	0	98	0	2	100	97	99
Yersinia frederiksenii/intermedia	99	0	0	75	1	0	99	0	99	80	0	100	99	25	99	99	99	75	99	99	0	98	0	5	100	99	99
Yersinia kristensenii	80	0	0	80	0	0	99	0	97	0	0	100	99	10	99	1	0	0	99	99	0	98	0	5	100	99	99
Yersinia pestis	68	0	0	0	0	0	1	0	0	1	0	99	99	75	70	1	0	0	30	99	0	47	0	0			100
Yersinia pseudotuberculosis	98	0	0	0	0	0	99	0	0	0	0	99	97	0	0	75	0	50	25	30	0	95	0	95	99	97	99
Aeromonas hydrophila gr. 1	98	90	25	1	25	0	0	0	85	25	90	99	99	1	3	5	97	1	75	75	100	97	0	95	99	99	99
Aeromonas hydrophila gr. 2	99	97	80	1	80	0	0	0	85	80	97	97	99	9	9	1	80	1	75	5	100	97	0	95	96	99	99
Aeromonas salmonicida ssp salmonicida	1	60	1	0	1	0	0	0	0	10	75	97	54	0	0	0	0	0	0	0	100	98	0	1			99
Photobacterium damsela	1	99	75	100	1	0	98	0	0	10	1	50	0	0	0	0	0	0	0	0	100	99	0	25			33
Plesiomonas shigelloides	95	99	100	100	0	0	0	0	100	0	0	99	1	99	0	0	0	0	0	0	100	99	0	95	100	98	100
Vibrio alginolyticus	0	1	94	97	60	0	1	0	99	58	92	98	98	0	0	0	94	5	10	1	100	47	0	100	96	94	94
Vibrio cholerae	98	0	99	97	75	0	0	0	80	58	75	75	80	0	1	0	75	36	0	0	100	96	0	100	91	90	10
Vibrio fluvialis	95	99	0	0	1	0	0	0	0	1	80	10	99	0	1	0	0	0	0	75	100	100	0	100	57		6
Vibrio hollisae	1	0	0	0	0	0	0	0	94	1	1	75	0	0	0	0	0	0	0	75	100	100	0	100	48	93	49
Vibrio mimicus	99	0	99	99	50	0	0	0	99	1	99	99	99	0	0	0	0	0	0	0	99	100	0	100	99	99	49
Vibrio parahaemolyticus	0	0	100	99	50	0	1	0	100	1	75	100	99	0	0	0	1	0	12	50	100	63	0	100	98	90	6
Vibrio vulnificus	99	0	91	80	25	0	0	0	99	1	99	99	75	0	0	1	0	0	90	0	99	54	0	100	100	99	49
Pasteurella aerogenes	99	0	0	80	0	0	99	0	0	0	0	99	0	0	0	0	99	10	0	75	75	100	0	0	2	100	100
Pasteurella multocida 1	4	0	0	25	0	1	0	0	99	0	0	29	1	0	1	0	75	0	0	0	99	90	0	0	2	23	23
Pasteurella multocida 2	7	0	0	45	0	0	0	0	99	0	0	44	99	0	99	0	99	0	0	0	99	90	0	0	9	33	33
Pasteurella pneumotropica/haemolytica	60	0	1	10	51	0	25	0	15	5	3	35	12	12	12	1	35	1	2	1	80	3	0	0	9	98	98
Acinetobacter baumannii/calcoaceticus	0	0	0	0	51	0	14	0	0	0	5	0	0	0	0	0	0	0	0	99	0	62	1	0	75	97	0
Bordetella/Alcaligenes/Moraxella spp *	0	0	0	0	52	0	0	0	0	0	1	0	1	0	0	0	0	0	0	1	95	40	75	75	88	99	0
Burkholderia cepacia	50	0	25	0	78	0	0	0	0	0	43	60	1	99	13	0	99	7	0	20	90	75	0	99	91	94	99
Chromobacterium violaceum	0	99	0	0	75	0	0	0	14	0	80	99	0	0	0	0	10	0	0	0	99	75	94	99	88	97	0
Chryseomonas luteola	86	75	0	0	94	0	0	0	0	25	13	84	100	0	0	0	1	1	85	99	99	100	0	100	91	99	99
Chryseobacterium indologenes	5	0	0	0	12	0	0	0	75	0	80	0	0	0	0	0	0	0	0	0	99	30	0	0	57	94	10
Chryseobacterium meningosepticum	77	0	0	0	20	0	1	0	85	0	90	0	0	0	0	0	15	1	0	0	99	6	0	0	48	93	6
Eikenella corrodens	0	0	75	99	0	0	0	0	0	0	0	0	0	0	0	0	0	1	0	0	100	95	0	1	1	49	49
Flavimonas oryzihabitans	0	0	0	0	89	0	0	0	0	25	1	10	0	0	0	0	0	0	0	45	0	7	0	100	99	99	2
Myroides /Chryseobacterium indologenes	0	0	0	0	50	0	75	0	0	0	75	0	0	0	0	0	0	10	0	0	99	42	60	0	84	2	2
Ochrobactrum anthropi	15	0	0	30	1	0	25	0	1	15	1	1	1	0	1	1	1	0	1	10	90	12	56	99	99	47	0
Pseudomonas aeruginosa	0	89	0	0	92	0	25	0	0	1	75	50	0	0	0	1	0	10	1	25	97	26	48	97	96	98	0
Pseudomonas fluorescens/putida	0	75	0	0	37	0	0	0	10	10	27	25	0	0	0	0	0	25	1	20	99	26	56	100	85	93	0
Non-fermenter spp	1	0	1	0	0	0	0	0	0	15	9	9	0	0	0	0	1	1	1	2	93	48	35	99	85	49	0
Shewanella putrefaciens	0	0	75	80	75	75	1	0	0	0	75	9	0	0	0	0	1	1	1	0	93	96	35	100	96	9	0
Stenotrophomonas maltophilia	70	0	75	1	75	1	0	0	0	0	90	1	0	0	0	0	0	0	0	1	1	26	1	100	91	49	49

Courtesy of bioMérieux, Inc.

CHART IV Characterization of Enterobacteriaceae—The Enterotube II System

Groups		Reactions	Glucose	Gas Production	Lysine	Ornithine	H2S	Indole	Adonitol	Lactose	Arabinose	Sorbitol	Voges-Proskauer	Dulcitol	Phenylalanine Deaminase	Urea	Citrate
ESCHERICHIEAE		Escherichia	+ 100.0	+J 92.0	d 80.6	d 57.8	−K 4.0	+ 96.3	− 5.2	+J 91.6	+ 91.3	± 80.3	− 0.0	d 49.3	− 0.1	− 0.1	− 0.2
ESCHERICHIEAE		Shigella	+ 100.0	−A 2.1	− 0.0	∓B 20.0	− 0.0	∓ 37.8	− 0.0	−B 0.3	± 67.8	∓ 29.1	− 0.0	d 5.4	− 0.0	− 0.0	− 0.0
EDWARDSIELLEAE		Edwardsiella	+ 100.0	+ 99.4	+ 100.0	+ 99.0	+ 99.6	+ 99.0	− 0.0	− 0.0	∓ 10.7	− 0.2	− 0.0	− 0.0	− 0.0	− 0.0	− 0.0
SALMONELLEAE		Salmonella	+ 100.0	+C 91.9	+H 94.6	+I 92.7	+E 91.6	− 1.1	− 0.0	− 0.8	± 89.2	+ 94.1	− 0.0	dD 86.5	− 0.0	− 0.0	dF 80.1
SALMONELLEAE		Arizona	+ 100.0	+ 99.7	+ 99.4	+ 100.0	+ 98.7	− 2.0	− 0.0	d 69.8	+ 99.1	+ 97.1	− 0.0	− 0.0	− 0.0	− 0.0	+ 96.8
SALMONELLEAE	CITROBACTER	freundii	+ 100.0	+ 91.4	− 0.0	d 17.2	± 81.6	− 6.7	− 0.0	d 39.3	+ 100.0	+ 98.2	− 0.0	d 59.8	− 0.0	dw 89.4	+ 90.4
SALMONELLEAE	CITROBACTER	amalonaticus	+ 100.0	+ 97.0	− 0.0	+ 97.0	− 0.0	+ 99.0	− 0.0	± 70.0	+ 99.0	+ 97.0	− 0.0	∓ 11.0	− 0.0	± 81.0	+ 94.0
SALMONELLEAE	CITROBACTER	diversus	+ 100.0	+ 97.3	− 0.0	+ 99.8	− 0.0	+ 100.0	+ 100.0	d 40.3	+ 98.0	+ 98.2	− 0.0	± 52.2	− 0.0	dw 85.8	+ 99.7
PROTEEAE	PROTEUS	vulgaris	+ 100.0	±G 86.0	− 0.0	− 0.0	+ 95.0	+ 91.4	− 0.0	− 0.0	− 0.0	− 0.0	− 0.0	− 0.0	+ 100.0	+ 95.0	d 10.5
PROTEEAE	PROTEUS	mirabilis	+ 100.0	+G 96.0	− 0.0	+ 99.0	+ 94.5	− 3.2	− 0.0	− 2.0	− 0.0	− 0.0	∓ 16.0	− 0.0	+ 99.6	± 89.3	± 58.7
PROTEEAE	MORGANELLA	morganii	+ 100.0	±G 86.0	− 0.0	+ 97.0	− 0.0	+ 99.5	− 0.0	− 0.0	− 0.0	− 0.0	− 0.0	− 0.0	+ 95.0	+ 97.1	−L 0.0
PROTEEAE	PROVIDENCIA	alcalifaciens	+ 100.0	dG 85.2	− 0.0	− 1.2	− 0.0	+ 99.4	+ 94.3	− 0.3	− 0.7	− 0.6	− 0.0	− 0.0	+ 97.4	− 0.0	+ 97.9
PROTEEAE	PROVIDENCIA	stuartii	+ 100.0	− 0.0	− 0.0	− 0.0	− 0.0	+ 98.6	∓ 12.4	− 3.6	− 4.0	− 3.4	− 0.0	− 0.0	+ 94.5	∓ 20.0	+ 93.7
PROTEEAE	PROVIDENCIA	rettgeri	+ 100.0	∓G 12.2	− 0.0	− 0.0	− 0.0	+ 95.9	+ 99.0	d 10.0	− 0.0	− 1.0	− 0.0	− 0.0	+ 98.0	+ 100.0	+ 96.0
KLEBSIELLEAE	ENTEROBACTER	cloacae	+ 100.0	+ 99.3	− 0.0	+ 93.7	− 0.0	− 0.0	∓ 28.0	± 94.0	+ 99.4	+ 100.0	+ 100.0	d 15.2	− 0.0	± 74.6	+ 98.9
KLEBSIELLEAE	ENTEROBACTER	sakazakii	+ 100.0	+ 97.0	− 0.0	+ 97.0	− 0.0	∓ 16.0	− 0.0	+ 100.0	+ 100.0	− 0.0	+ 97.0	6.0	− 0.0	− 0.0	+ 94.0
KLEBSIELLEAE	ENTEROBACTER	gergoviae	+ 100.0	+ 93.0	± 64.0	+ 100.0	− 0.0	− 0.0	− 0.0	∓ 42.0	+ 100.0	− 0.0	+ 100.0	− 0.0	− 0.0	+ 100.0	+ 96.0
KLEBSIELLEAE	ENTEROBACTER	aerogenes	+ 100.0	+ 95.9	+ 97.5	+ 95.9	− 0.0	− 0.8	+ 97.5	+ 92.5	+ 100.0	+ 98.3	+ 100.0	− 4.1	− 0.0	− 0.0	+ 92.6
KLEBSIELLEAE	ENTEROBACTER	agglomerans	+ 100.0	∓ 24.1	− 0.0	− 0.0	− 0.0	∓ 19.7	− 7.5	d 52.9	+ 97.5	d 26.3	± 64.8	d 12.9	∓ 27.6	d 34.1	d 84.2
KLEBSIELLEAE	HAFNIA	alvei	+ 100.0	+ 98.9	+ 99.6	+ 98.6	− 0.0	− 0.0	− 0.0	d 2.8	+ 99.3	− 0.0	± 65.0	− 2.4	− 0.0	d 3.0	d 5.6
KLEBSIELLEAE	SERRATIA	marcescens	+ 100.0	±G 52.6	+ 99.6	+ 99.6	− 0.0	−w 0.1	∓ 56.0	− 1.3	− 0.0	+ 99.1	+ 98.7	− 0.0	− 0.0	dw 39.7	+ 97.6
KLEBSIELLEAE	SERRATIA	liquefaciens	+ 100.0	d 72.5	± 64.2	+ 100.0	− 0.0	−w 1.8	− 8.3	d 15.6	+ 97.3	+ 97.3	∓ 49.5	− 0.0	− 0.9	dw 3.7	+ 93.6
KLEBSIELLEAE	SERRATIA	rubidaea	+ 100.0	dG 35.0	± 61.0	− 0.0	− 0.0	−w 2.0	+ 88.0	+ 100.0	+ 100.0	− 8.0	+ 92.0	− 0.0	− 0.0	dw 4.0	± 88.0
KLEBSIELLEAE	KLEBSIELLA	pneumoniae	+ 100.0	+ 96.0	+ 97.2	− 0.0	− 0.0	− 0.0	± 89.0	+ 98.7	+ 99.9	+ 99.4	+ 93.7	∓ 33.0	− 0.0	+ 95.4	+ 96.8
KLEBSIELLEAE	KLEBSIELLA	oxytoca	+ 100.0	+ 96.0	+ 97.2	− 0.0	− 0.0	+ 100.0	± 89.0	∓ 98.7	+ 100.0	+ 98.0	+ 93.7	∓ 33.0	− 0.0	∓ 95.4	∓ 96.8
KLEBSIELLEAE	KLEBSIELLA	ozaenae	+ 100.0	d 55.0	∓ 35.8	− 1.0	− 0.0	− 0.0	+ 91.8	± 26.2	+ 100.0	± 78.0	− 0.0	− 0.0	− 0.0	d 14.8	d 28.1
KLEBSIELLEAE	KLEBSIELLA	rhinoscleromatis	+ 100.0	− 0.0	− 0.0	− 0.0	− 0.0	− 0.0	+ 98.0	d 6.0	+ 100.0	+ 98.0	− 0.0	− 0.0	− 0.0	− 0.0	− 0.0
YERSINIAE	YERSINIA	enterocolitica	+ 100.0	− 0.0	− 0.0	+ 90.7	− 0.0	∓ 26.7	− 0.0	− 0.0	+ 98.7	+ 98.7	− 0.1	− 0.0	− 0.0	+ 90.7	− 0.0
YERSINIAE	YERSINIA	pseudotuberculosis	+ 100.0	− 0.0	− 0.0	− 0.0	− 0.0	− 0.0	− 0.0	− 0.0	± 55.0	− 0.0	− 0.0	− 0.0	− 0.0	+ 100.0	− 0.0

E. *S. enteritidis* bioserotype Paratyphi A and some rare biotypes may be H2S negative.

F. *S. typhi, S. enteritidis* bioserotype Paratyphi A and some rare biotypes are citrate-negative and *S. cholerae-suis* is usually delayed positive.

G. The amount of gas produced by *Serratia, Proteus* and *Providencia alcalifaciens* is slight; therefore, gas production may not be evident in the ENTEROTUBE II.

H. *S. enteritidis* bioserotype Paratyphi A is negative for lysine decarboxylase.

I. *S. typhi* and *S. gallinarum* are ornithine decarboxylase-negative.

J. The Alkalescens-Dispar (A-D) group is included as a biotype of *E. coli.* Members of the A-D group are generally anaerogenic, non-motile and do not ferment lactose.

K. An occasional strain may produce hydrogen sulfide.

L. An occasional strain may appear to utilize citrate.

Courtesy and © Becton, Dickinson, and Company

CHART V Reaction Interpretations for API Staph

PROCEDURE

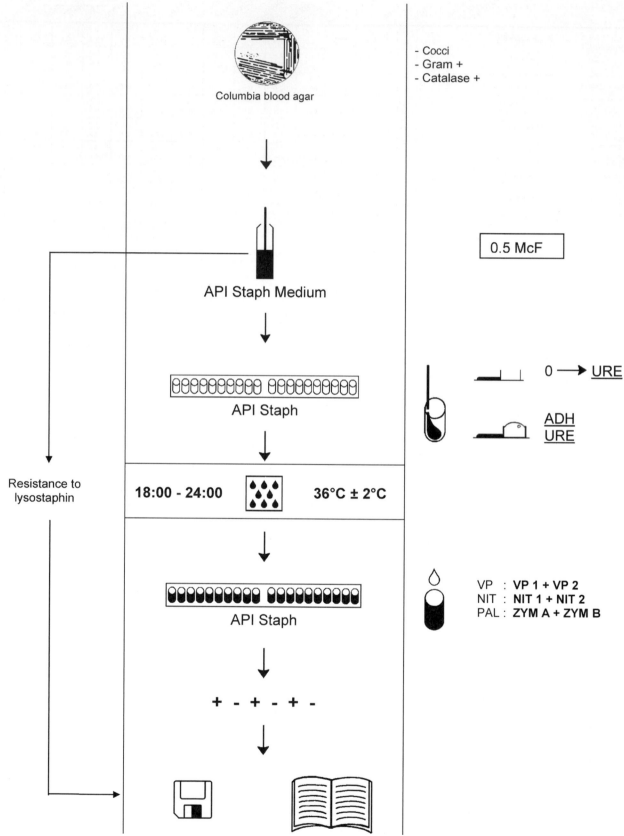

- Cocci
- Gram +
- Catalase +

Columbia blood agar

0.5 McF

API Staph Medium

API Staph

$0 \longrightarrow$ <u>URE</u>

<u>ADH</u>
<u>URE</u>

Resistance to
lysostaphin

18:00 - 24:00   **36°C ± 2°C**

API Staph

VP : **VP 1 + VP 2**
NIT : **NIT 1 + NIT 2**
PAL : **ZYM A + ZYM B**

+ - + - + -

Courtesy of bioMérieux, Inc.

CHART VI   Biochemistry of API Staph Tests

READING TABLE

TESTS	ACTIVE INGREDIENTS	QTY (mg/cup.)	REACTIONS / ENZYMES	RESULT	
				NEGATIVE	POSITIVE
0	No substrate		Negative control	red	—
GLU	D-glucose	1.56	(Positive control) (D-GLUcose)	red *	yellow
FRU	D-fructose	1.4	acidification (D-FRUctose)		
MNE	D-mannose	1.4	acidification (D-ManNosE)		
MAL	D-maltose	1.4	acidification (MALtose)		
LAC	D-lactose (bovine origin)	1.4	acidification (LACtose)		
TRE	D-trehalose	1.32	acidification (D-TREhalose)		
MAN	D-mannitol	1.36	acidification (D-MANnitol)		
XLT	xylitol	1.4	acidification (XyLiTol)		
MEL	D-melibiose	1.32	acidification (D-MELibiose)		
NIT	potassium nitrate	0.08	Reduction of NITrates to nitrites	NIT 1 + NIT 2 / 10 min	
				colorless-light pink	red
PAL	ß-naphthyl phosphate	0.0244	ALkaline Phosphatase	ZYM A + ZYM B / 10 min	
				yellow	violet
VP	sodium pyruvate	1.904	Acetyl-methyl-carbinol production (Voges Proskauer)	VP 1 + VP 2 / 10 min	
				colorless-light pink	violet-pink
RAF	D-raffinose	1.56	acidification (RAFfinose)	red	yellow
XYL	D-xylose	1.4	acidification (XYLose)		
SAC	D-saccharose (sucrose)	1.32	acidification (SACcharose)		
MDG	methyl-αD-glucopyranoside	1.28	acidification (Methyl-αD-Glucopyranoside)		
NAG	N-acetyl-glucosamine	1.28	acidification (N-Acetyl-Glucosamine)		
ADH	L-arginine	1.904	Arginine DiHydrolase	yellow	orange-red
URE	urea	0.76	UREase	yellow	red-violet

The acidification tests should be compared to the negative (0) and positive (GLU) controls.

* When MNE and XLT are preceded or followed by positive tests, then an orange test should be considered negative.

• The quantities indicated may be adjusted depending on the titer of the raw materials used.

• Certain cupules contain products of animal origin, notably peptones.

Lysostaphin resistance test

Determine resistance to lysostaphin on P agar, as indicated in the following procedure or according to the manufacturer's recommendations.

To perform the test, inoculate the surface of a P agar plate, by flooding it with a bacterial suspension (approximately 10^7 organisms/ml).

Leave to dry for 10-20 minutes at 36°C ± 2°C.

Place a drop of lysostaphin solution (200 μg/ml) on the surface of the agar.

Incubate for 18-24 hrs. at 35-37°C.

Total or partial lysis of the bacterial culture indicates susceptibility to the enzyme.

This test constitutes the 21st test of the strip. It is considered positive if resistance to lysostaphin is determined.

Courtesy of bioMérieux, Inc.

CHART VII   API Staph Profile Register

IDENTIFICATION TABLE
% of reactions positive after 18-24 hrs. at 36°C ± 2°C

API STAPH V4.0	0	GLU	FRU	MNE	MAL	LAC	TRE	MAN	XLT	MEL	NIT	PAL	VP	RAF	XYL	SAC	MDG	NAG	ADH	URE	LSTR
Staphylococcus aureus	0	100	100	95	96	88	91	80	0	0	83	97	78	1	0	97	2	90	80	80	0
Staphylococcus auricularis	0	100	99	36	72	10	90	9	0	0	81	0	1	0	0	40	0	15	90	1	0
Staphylococcus capitis	0	100	99	80	43	22	2	36	0	0	86	23	90	0	0	50	0	1	85	35	0
Staphylococcus caprae	0	100	99	70	10	75	74	10	0	0	99	95	99	0	0	0	0	1	99	60	0
Staphylococcus carnosus	0	100	100	99	0	99	99	99	0	0	99	83	83	0	0	0	0	100	100	0	0
Staphylococcus chromogenes	0	100	100	99	79	100	100	13	0	0	96	96	1	0	1	100	0	31	89	95	0
Staphylococcus cohnii ssp cohnii	0	100	99	66	99	2	97	88	33	0	21	66	94	0	0	2	0	9	2	1	0
Staph.cohnii ssp urealyticum	0	100	100	99	98	98	100	94	64	0	1	94	87	0	0	0	0	98	0	99	0
Staphylococcus epidermidis	0	100	99	70	99	81	2	0	0	1	80	84	68	1	0	97	4	18	73	88	0
Staphylococcus haemolyticus	0	99	75	5	99	80	91	60	0	1	78	3	57	0	0	98	13	83	85	1	0
Staphylococcus hominis	0	98	94	41	97	50	86	28	0	1	82	27	70	1	0	97	4	50	43	84	0
Staphylococcus hyicus	0	100	99	99	0	87	99	0	0	0	90	90	15	0	0	99	2	93	100	68	0
Staphylococcus lentus	0	100	100	100	100	100	100	100	7	99	92	21	57	100	100	100	28	100	0	1	0
Staphylococcus lugdunensis	0	100	89	88	99	66	99	0	0	0	99	16	99	0	0	100	0	90	1	50	0
Staphylococcus saprophyticus	0	100	99	2	97	90	99	88	22	0	35	14	79	1	0	96	1	70	30	65	0
Staphylococcus schleiferi	0	100	80	100	0	1	71	0	0	0	99	97	99	0	0	0	0	94	99	0	0
Staphylococcus sciuri	0	99	99	99	99	70	93	98	0	0	83	67	30	0	16	95	7	68	0	0	0
Staphylococcus simulans	0	100	100	57	11	95	92	73	4	0	83	27	38	0	4	97	2	90	97	84	0
Staphylococcus warneri	0	99	99	50	98	19	96	70	0	0	23	16	90	0	0	99	0	6	77	97	0
Staphylococcus xylosus	0	100	100	92	81	85	95	90	30	9	82	75	67	11	82	87	10	80	5	90	0
Kocuria kristinae	0	99	96	99	90	9	84	3	0	0	6	3	93	0	0	90	12	0	0	0	97
Kocuria varians/rosea	0	91	92	8	1	1	8	1	0	0	75	4	8	4	8	4	0	1	1	29	95
Micrococcus spp	0	2	4	0	1	0	1	0	0	0	8	15	1	0	0	1	0	1	11	11	91

Courtesy of bioMérieux, Inc.

The Streptococci: Classification, Habitat, Pathology, and Biochemical Characteristics

To fully understand the characteristics of the various species of medically important streptococci, this appendix has been included as an adjunct to Exercise 71.

The first system that was used for grouping the streptococci was based on the type of hemolysis and was proposed by J. H. Brown in 1919. In 1933, R. C. Lancefield proposed that these bacteria be separated into groups A, B, C, and so on, on the basis of precipitation-type serological testing. Both hemolysis and serological typing still play predominant roles today in our classification system. Note below that the Lancefield groups are categorized with respect to the type of hemolysis that is produced on blood agar.

Beta-Hemolytic Groups

Using a streak-stab technique, a blood agar plate is incubated aerobically at 37° C for 24 hours. Isolates that have colonies surrounded by clear zones completely free of red blood cells are characterized as being *beta-hemolytic*. Three serological groups of streptococci fall in this category: groups A, B, and C; a few species in group D are also beta-hemolytic.

Group A Streptococci

This group is represented by only one species: *Streptococcus pyogenes.* Approximately 25% of all upper respiratory infections (URIs) are caused by this species; another 10% of URIs are caused by other streptococci; most of the remainder (65%) are caused by viruses. Since no unique clinical symptoms can be used to differentiate viral from streptococcal URIs, and since successful treatment relies on proper identification, it becomes mandatory that throat cultures be taken in an attempt to prove the presence or absence of streptococci. It should be added that if streptococcal URIs are improperly treated, serious sequelae such as pneumonia, acute endocarditis, rheumatic fever, and glomerulonephritis can result.

S. pyogenes is the only beta-hemolytic streptococcus that is primarily of *human origin.* Although the pharynx is the most likely place to find this species, it may be isolated from the skin and rectum. Asymptomatic pharyngeal and anal carriers are not uncommon. Outbreaks of postoperative streptococcal infections have been traced to both pharyngeal and anal carriers among hospital personnel.

These bacteria (0.6–1.0 μm diameter) occur as cocci in pairs and as short to moderate-length chains in clinical specimens; in broth cultures, the chains are often longer.

When grown on blood agar, the colonies are small (0.5 mm dia.), transparent to opaque, and domed; they have a smooth or semimatte surface and an entire edge; complete hemolysis (beta-type) occurs around each colony, usually two to four times the diameter of the colony.

S. pyogenes produces two hemolysins: streptolysin S and streptolysin O. The beta-type hemolysis on blood agar is due to the complete destruction of red blood cells by the streptolysin S.

There is no group of physiological tests that can be used with *absolute* certainty to differentiate *S. pyogenes* from other streptococci; however, if an isolate is beta-hemolytic and sensitive to bacitracin, one can be 95% certain that the isolate is *S. pyogenes.*

Group B Streptococci

The only recognized species of this group is *S. agalactiae.* Although this organism is frequently found in milk and associated with *mastitis in cattle,* the list of human infections caused by it is as long as the one for *S. pyogenes*: abscesses, acute endocarditis, impetigo, meningitis, neonatal sepsis, and pneumonia are just a few. Like *S. pyogenes,* this pathogen may also be found in the pharynx, skin, and rectum; however, it is more likely to be found in the genital and intestinal tracts of healthy adults and infants. It is not unusual to find the organism in vaginal cultures of third-trimester pregnant women.

Cells are spherical to ovoid (0.6–1.2 μm dia) and occur in chains of seldom less than four cells; long chains are frequently present. Characteristically, the chains appear to be composed of paired cocci.

Colonies of *S. agalactiae* on blood agar often produce double zone hemolysis. After 24 hours incubation, colonies exhibit zones of beta-hemolysis. After cooling, a second ring of hemolysis forms, which is separated from the first by a ring of red blood cells.

Reference to table E.1 emphasizes the significant characteristics of *S. agalactiae.* Note that this organism gives a positive CAMP reaction, hydrolyzes hippurate, and is not (usually) sensitive to bacitracin. It is also resistant to SXT. Presumptive identification of this species relies heavily on a positive CAMP test or hippurate hydrolysis, even if beta-hemolysis is not clearly demonstrated.

Group C Streptococci

Three species fall in this group: *S. equisimilis, S. equi,* and *S. zooepidemicus.* Although all of these species may cause human infections, the diseases are not usually as grave as those caused by groups A and B. Some group C species have been isolated from impetiginous lesions, abscesses, sputum, and the pharynx. There is no evidence that they are associated with acute glomerulonephritis, rheumatic fever, or even pharyngitis.

Presumptive differentiation of this group from *S. pyogenes* and *S. agalactiae* is based primarily on (1) resistance to bacitracin, (2) inability to hydrolyze hippurate or bile esculin, and (3) a negative CAMP test. There are other groups that have some of these same characteristics, but they will not be studied here. Tables 12.16 and 12.17 on page 1049 of *Bergey's Manual,* vol. 2, provide information about these other groups.

Alpha-Hemolytic Groups

Streptococcal isolates that have colonies with zones of incomplete lysis around them are said to be **alpha-hemolytic.** These zones are often greenish; sometimes they are confused with beta-hemolysis. *The only way to be certain that such zones are not beta-hemolytic is to examine the zones under* 60 X *microscopic magnification.* Figure 71.4, page 420, illustrates the differences between alpha- and beta-hemolysis. If some red blood cells are seen in the zone, the isolate is classified as being alpha-hemolytic.

The grouping of streptococci on the basis of alpha-hemolysis is not as clear-cut as it is for beta-hemolytic groups. Note in table E1 that the bottom four groups that have alpha-hemolytic types may also have beta-hemolytic or nonhemolytic strains. Thus, we see that hemolysis in these four groups can be a misleading characteristic in identification.

Alpha-hemolytic isolates from the pharynx are usually *S. pneumoniae,* viridans streptococci, or group D. Our primary concern here in this experiment is to identify isolates of *S. pneumoniae.* To accomplish this goal, it will be necessary to differentiate any alpha-hemolytic isolate from group D and viridans streptococci.

Streptococcus pneumoniae
(Pneumococcus)

This organism is the most frequent cause of bacterial pneumonia, a disease that has a high mortality rate among the aged and debilitated. It is also frequently implicated in conjunctivitis, otitis media, pericarditis, subacute endocarditis, meningitis, septicemia, empyema, and peritonitis. Thirty to 70% of normal individuals carry this organism in the pharynx.

Spherical or ovoid, these cells (0.5–1.25 μm dia) occur typically as pairs, sometimes singly, often in short chains. Distal ends of the cells are pointed or lancet-shaped and are heavily encapsulated with polysaccharide on primary isolation.

Colonies on blood agar are small, mucoidal, opalescent, and flattened with entire edges surrounded by a zone of greenish discoloration (alpha-hemolysis). In contrast, the viridans streptococcal colonies are smaller, gray to whitish gray, and opaque with entire edges.

Presumptive identification of *S. pneumoniae* can be made with the optochin and bile solubility tests. On the optochin test, the pneumococci exhibit sensitivity to ethylhydrocupreine (optochin). With the bile solubility test, pneumococci are dissolved in bile (2% sodium desoxycholate). Except for bacitracin susceptibility (±), *S. pneumoniae* is negative on all other tests used for differentiation of streptococci.

Viridans Group

Streptococci that fall in this group are primarily alpha-hemolytic; some are nonhemolytic. Approximately 10 species are included in this group. All of them are highly adapted parasites of the upper respiratory tract. Although usually regarded as having low pathogenicity, they are opportunistic and sometimes cause serious infections. Two species (*S. mutans* and *S. sanguis*) are thought to be the primary cause of dental caries, since they have the ability to form dental plaque. Viridans streptococci are implicated more often than any other bacteria in subacute bacterial endocarditis.

When it comes to differentiation of bacteria of this group from the pneumococci and enterococci, we will use the optochin, bile solubility, and salt-tolerance tests.

Group D Enterococci

Members of this group are, currently, considered by most taxonomists to belong to the genus *Enterococcus.*

The enterococci of serological group D may be alpha-hemolytic, beta-hemolytic, or nonhemolytic. The principal species of this enterococcal group are *E. faecalis, E. faecium, E. durans,* and *E. avium.*

Subacute endocarditis, pyelonephritis, urinary tract infections, meningitis, and biliary infections are caused by these organisms. All five of these species have been isolated from the intestinal tract. Approximately 20% of subacute bacterial endocarditis and 10% of urinary tract infections are caused by members of this group. Differentiation of this group from other streptococci in systemic infections is mandatory because *E. faecalis, E. faecium,* and *E. durans* are resistant to penicillin and require combined antibiotic therapy.

Since *E. faecalis* can be isolated from many food products (not connected with fecal contamination), it can be a transient in the pharynx and show up as an isolate in throat cultures. Morphologically, the cells are ovoid (0.5–1.0 μm dia) occurring as pairs in short chains. Hemolytic reactions of *E. faecalis* on blood agar will vary with the type of blood used in the medium. Some strains produce beta-hemolysis on agar with horse, human, and rabbit blood; on sheep blood agar the colonies will always exhibit alpha-hemolysis. Other streptococci are consistently either beta, alpha or nonhemolytic.

Cells of *E. faecium* are morphologically similar to *E. faecalis* except that motile strains are often encountered. A strong alpha-type hemolysis is usually seen around colonies of *E. faecium* on blood agar.

Although presumptive differentiation of group D enterococci from groups A, B, and C is not too difficult with physiological tests, it is more laborious to differentiate the individual species within group D. As indicated in table E.1, the enterococci (1) hydrolyze bile esculin, (2) are CAMP negative, and (3) grow well in 6.5% NaCl broth.

Differentiation of the five species within this group involves nine or ten physiological tests.

Group D Streptococci (Nonenterococci)

The only medically significant nonenterococcal species of group D is *S. bovis.* This organism is found in the intestinal tract of humans as well as in cows, sheep, and other ruminants. It can cause meningitis, subacute endocarditis, and urinary tract infections. On blood agar, the organism is usually alpha-hemolytic; occasionally, it is nonhemolytic. The best way to differentiate it from the group D enterococci is to test its tolerance to 6.5% NaCl.

Reading References

General Information

Alcamo, I. Edward. *Fundamentals of Microbiology,* 7th ed. Reading, Mass.: Addison-Wesley, 2001.

Atlas, R. M., and Bartha, R. *Microbial Ecology: Fundamentals and Applications,* 4th ed. Menlo Park, Calif.: Benjamin/Cummings Publishing, 1997.

Baron, Samuel, ed. *Medical Microbiology,* 4th ed. Reading, Mass.: Addison-Wesley, 1996.

Brock, Thomas D. *Robert Koch: A Life in Medicine and Bacteriology.* Herndon, Va.: ASM Press, 2000.

Brogden, Kim A., et al. *Virulence Mechanisms of Bacterial Pathogens,* 3rd ed. Herndon, Va.: ASM Press, 2000.

Brun, Yves V., and Shimkets, L. J. *Prokaryotic Development.* Herndon, Va.: ASM Press, 2000.

Burlage, Robert S., Atlas, R., Stahl, D., Geesey, G., and Saylor, G. *Techniques in Microbial Ecology.* Cary, N.C.: Oxford University Press, 1998.

Chan, Pelczar, and Krieg. *Laboratory Exercises in Microbiology,* 6th ed. New York: McGraw-Hill, 1993.

Collier, Leslie H., et al. *Topley and Wilson's Microbiology and Microbial Infections.* Six Volumes. Herndon, Va.: ASM Press, 1998.

Colwell, Rita R. *Nonculturable Microorganisms in the Environment.* Herndon, Va.: ASM Press, 2000.

Doyle, Michael P., et al. *Food Microbiology: Fundamentals and Frontiers.* Herndon, Va.: ASM Press, 2001.

Dubos, Rene. *Pasteur and Modern Science.* Paperback. Herndon, Va.: ASM Press, 1998.

Flint, S. J., et al. *Principles of Virology: Molecular Biology, Pathogenesis, and Control of animal viruses,* 2nd-ed. Herndon, Va.: ASM Press, 2003.

Gerhardt, Philipp, et al. *Methods for General and Molecular Bacteriology.* Herndon, Va.: ASM Press, 1994.

Hurst, Christon J., et al. *Manual of Environmental Microbiology,* 2nd ed. Herndon, Va.: ASM Press, 2002.

Karam, Jim D., et al. *Molecular Biology of Bacteriophage T-4.* Herndon, Va.: ASM Press, 1994.

Lacey, Alan J. *Light Microscopy in Biology,* 2nd ed. Cary, N.C.: Oxford University Press, 1999.

Lederberg, Joshua, et al. *Encyclopedia of Microbiology.* New York: Academic Press, 1997.

Lovley, Derek R. *Environmental Microbe-Metal Interactions.* Herndon, Va.: ASM Press, 2000.

Madigan, Michael T., and Marrs, Barry L. *Extremophiles.* New York: Scientific American Vol. 276, Number 4: pp. 82–87, 1997.

Madigan, Michael T., Martinko, John M., and Parker, Jack. *Brock Biology of Microorganisms,* 11th ed. Englewood Cliffs, N.J.: Prentice-Hall, 2005.

Mobley, Harry L. T., and Warren, John W. *Urinary Tract Infections.* Herndon, Va.: ASM Press, 1995.

Needham, Cynthia, et al. *Intimate Strangers: Unseen Life on Earth.* Herndon, Va.: ASM Press, 2000.

Prescott, Lansing M., Harley, John P., and Klein, Donald A. *Microbiology,* 6th ed. New York: McGraw-Hill, 2005.

Rose, Noel R. *Manual of Clinical Laboratory Immunology,* 6th ed. Herndon, Va.: ASM Press, 2002.

Rosenburg, Eugene. *Microbial Ecology and Infectious Disease.* ASM Press, 1999.

Salyers, Abigail A., and Whitt, D. D. *Bacterial Pathogenesis,* 2nd ed. Herndon, Va.: ASM Press, 2001.

Smith, A. D., et al. *Oxford Dictionary of Biochemistry and Molecular Biology.* Cary, N.C.: Oxford University Press, Revised ed. 2000.

Snyder, Larry, and Champness, Wendy. *Molecular Genetics of Bacteria,* 2nd ed. Herndon, Va.: ASM Press, 2002.

Talaro, K., and Talaro, A. *Foundations in Microbiology,* 5th ed. Dubuque, IA; New York: McGraw-Hill, 2005.

Tortora, Gerard J., Funke, B. R., and Case, C. L. *Microbiology: An Introduction,* 8th ed. Menlo Park, Calif.: Benjamin/Cummings Publishing, 2003.

Walker, Graham C., and Kaiser, Dale. *Frontiers in Microbiology: A Collection of Minireviews from the Journal of Bacteriology.* Herndon, Va.: ASM Press, 1993.

White, David. *The Physiology and Biochemistry of Prokaryotes,* 2nd ed. Cary, N.C.: Oxford University Press, 1999.

Laboratory Procedures

American Type Culture Collections. *Catalog of Cultures,* 8th ed. Rockville, Md. www.atcc.org

Atlas, R. M., and Snyder, J. W. *Handbook of Media for Clinical Microbiology.* Boca Raton, Fla.: CRC Press, 1996.

Chart, Henrik. *Methods in Practical Laboratory Bacteriology.* Boca Raton, Fla.: CRC Press, 1994.

Difco Laboratory Staff. *Difco Manual of Dehydrated Culture Media and Reagents,* 11th ed. Detroit, Mich.: Difco Laboratories, 1998.

Flemming, D. O., Richardson, J. H., Tulis, J. J., and Vesley, D. *Laboratory Safety: Principles and Practices,* 2nd ed. Herndon, Va.: ASM Press, 1995.

Garcia, Lynne S., and Brukner, David A. *Diagnostic Medical Parasitology,* 4th ed. Herndon, Va.: ASM Press, 2001.

Isenberg, Henry D., et al. *Clinical Microbiology Procedures Handbook,* Vols. 1 and 2, 2nd ed. Herndon, Va.: ASM Press, 2004.

MacFaddin, Jean *Biochemical Tests for Identification of Medical Bacteria,* 3rd ed. Lippincott Williams & Wilkins NY.-1999.

Miller, Michael J. *A Guide to Specimen Management in Clinical Microbiology,* 2nd ed. Herndon, Va.: ASM Press, 1999.

Murray, Patrick R., et al. *Manual of Clinical Microbiology,* 8th ed. Herndon, Va.: ASM Press, 2003.

Murray, Patrick R., et al. *Manual of Clinical Microbiology,* 7th ed. Herndon, Va.: ASM Press, 1999.

Shapton, D. A., and Shapton, N. F. *Principles and Practices of Safe Processing of Food.* New York: Academic Press, 1994.

Identification of Microorganisms

Fischetti, Vincent A., et al. *Gram-Positive Pathogens.* Herndon, Va.: ASM Press, 2000.

Holt, John G., Kreig, N. R., et al. *Bergey's Manual of Systematic Bacteriology,* vol. 1, 1st ed. Baltimore, Md.: Williams & Wilkins, 1984.

Jahn, Theodore L., et al. *Protozoa,* 2nd ed. Dubuque, Ia.: WCB/McGraw-Hill, 1978.

Lapage, S. P., et al. *International Code of Nomenclature of Bacteria.* Herndon, Va.: ASM Press, 1992.

Larone, Davise. *Medically Important Fungi: A Guide to Identification,* 4th ed. Herndon, Va.: ASM Press, 2002.

Murray, Patrick, et al. *Manual of Clinical Microbiology,* 8th-ed. Bethesda, Md.: American Society for Microbiology, 2003.

Piggot, Patrick J., et al. *Regulation of Bacterial Differentiation.* Herndon, Va.: ASM Press, 1993.

Skerman, V. B. D., and Sneath, P. H. A. *Approved Lists of Bacterial Names.* Herndon, Va.: ASM Press, 1998.

Sneath, Peter H. A., et al. *Bergey's Manual of Systematic Bacteriology,* vol. 2, 1st ed. Baltimore, MD.: Williams & Wilkins, 1986.

Staley, James T., et al. *Bergey's Manual of Systematic Bacteriology,* vol. 3, 1st ed. Baltimore, MD.: Williams & Wilkins, 1989.

Sanitary and Medical Microbiology

Balows, Albert et al. *Manual of Clinical Microbiology,* 5th ed. Herndon, Va.: ASM Press, 1991.

Flemming, D. O., Richardson, J. H., Tulis, J. J., and Vesley, D. *Laboratory Safety: Principles and Practices,* 2nd ed. Herndon, Va.: ASM Press, 1995.

Greenberg, Arnold E., et. al. *Standard Methods for the Examination of Water and Wastewater,* 19th ed. Washington, D.C.: American Public Health Association, 1995.

Jay, James M. *Modern Food Microbiology,* 6th ed. New York: Chapman-Hall, 2000.

Kneip, Theodore, and Crable, John V. *Methods for Biological Monitoring: A Manual for Assessing Human Exposure to Hazardous Substances.* Washington, D.C.: American Public Health Association, 1988.

Marshall, Robert T. *Standard Methods for the Examination of Dairy Products,* 16th ed. Washington, D.C.: American Public Health Association, 1992.

Miller, Michael J. *A Guide to Specimen Management in Clinical Microbiology,* 2nd ed. Herndon, Va.: ASM Press, 1998.

Ray, Bibek. *Fundamental Food Microbiology,* 3rd ed. Boca Raton, Fla.: CRC Press, 2003.

Vanderzant, Carl, and Splittstoesser, Don. *Compendium of Methods for the Microbiological Examination of Foods,* 3rd ed. Washington, D.C.: American Public Health Association, 1992.

Index

INDEX

H

Hair, tying back, x
"Hair-lock"-like margins, 268
Halobacterium salinarium, 218
Halobacterium species, 298
Halophiles, 217
Halos, 22
Halotolerant microorganisms, 217. *See also* Salt tolerance
Hand scrubbing, 249–52
Handshaking exercise, 499–501
Hanging drop slides, 22
Hanging drop technique, 124, 125, 165
Heat. *See* Temperature
Heat filters, 26, 27
Heat-fixing slides, 96, 103
Heat resistance of endospores, 113, 209
Hektoen Enteric agar, 472
Helicobacter pylori, 145
Helminthosporium, 64
Hemolysis, 457, 459, 460
Heteronema, 43
Heterophile antibody test, 483–84
Heterophile antigens, 483
Heterotrophs, 61, 131
High-dry lenses
    basic features, 5
    calibrating, 35
    procedures for using, 7–8
    to view hanging drop slides, 125
Hilly elevations, 268
Hippurate hydrolysis, 458, 463
Horizontal gene transfer, 423
Houseflies, 191–94
Hydrochloric acid, 134
Hydrogen ion concentration. *See* pH
Hydrogen peroxide, 145, 277–78, 306
Hydrogen sulfide
    detecting in Kligler's iron agar, 288, 472
    detecting with Enterotube II, 314
    formation in canned foods, 409
    from Gram-negative intestinal pathogens, 472, 473
    production by sulfate reducers, 373
    in Winogradsky columns, 365, 366
Hydrolytic enzymes, 281–84, 285
Hyperthermophiles, 203
Hypertonic environments, 217
Hyphae
    of actinomyces, 337
    culturing, 170–71
    structure, 61, 62, 63
Hypotonic environments, 217
Hypotrichidium, 43

I

Identifying microorganisms. *See also* Applied microbiology;
        Medical microbiology exercises
    cultural characteristics, 265–68
    by hydrolytic and degradative reactions, 281–85
    from morphology, 259–62
    multiple physiological test media for, 287–90
    overview, 257
    oxidation and fermentation tests, 269–79
    rapid systems for Enterobacteriaceae, 303–6, 311–15
    rapid systems for O/F Gram-negative rods, 321–25
    rapid systems for staphylococci, 329–31
    rapid systems overview, 301–2
    using *Bergey's Manual*, 295–99
Imipenem, 238
Immersion oil, 6, 8, 14, 29
Immunoglobulins, 477
IMViC tests, 289, 395
Incubating plates, 75, 83, 146–47
India ink, 101
Indirect counting methods, 152
Indirect method of slide agglutination test, 480
Indole, 288, 289

Indole test
    conventional approach, 281, 284, 288
    on Gram-negative intestinal pathogens, 473
    use with Enterotube II, 313, 314
    use with Oxi/Ferm Tube II, 323, 324–25
Infections, preventing in laboratory, ix–xi. *See also* Aseptic technique
Infection threads, 349
Infectious diseases, defined, 499
Infectious mononucleosis, 483, 488
Influenza, 488
Infundibular liquefaction, 267
Ingrowing elevations, 268
Inhibitory substances, 381
Inoculating loops
    sterilizing and using, 71, 72, 73
    when to use, 75
Inoculation techniques
    agar cubes, 169–70
    culture tubes, 71–73
    Petri plates, 71, 85
    slant cultures, 74–77
Insects, isolating phages from, 191–94
Insulin, 418
Interference of light rays, 18
Intestinal pathogens, 469–73
Iodine
    Gram staining with, 109, 111
    starch hydrolysis detection with, 282–83
    use in handshaking exercise, 500
Iris controls, 29
Irregular culture configurations, 268
Irregular margins, 268
Isotonic environments, 217

K

Kanamycin, 238
Kinyoun method, 117, 118
Kirby-Bauer method, 235–40
Klebsiella pneumoniae, 104, 112
Kligler's iron agar, 287–88, 471, 472
Koch, Robert, 81, 132
Kovac's reagent
    conventional use, 284, 288, 473
    use with API 20E, 305
    use with Oxi/Ferm Tube II, 323, 324–25
Kurthia species, 297

L

Lab coats, x
Labeling plates, 75
Laboratory safety. *See* Safety rules
Lacrymaria, 43
Lactobacillus species, 297
Lactophenol cotton blue, 171
Lactose fermentation
    coliform counts based on, 391
    detecting with Enterotube II, 315
    detecting with Oxi/Ferm Tube II, 324
    Enterobacteriaceae identification via, 469, 470–72
    overview, 287, 289
Lancefield, Rebecca, 457
Lancets, 489, 494
Landsteiner, Karl, 493
Latex agglutination slide test, 479–80
Lectins, 349
Leghemoglobin, 349
Legumes, 349–50
Lens care, 3, 6–7
Lens systems, 4–7
Lepocinclis, 44
Leucosin, 46
Leukocytosis, 483, 488
Leukopenia, 488
Levels of biosafety, ix–x, xii
Levine EMB agar, 395